Probability and Mathematical Statistics (Continued)

PURI and SEN · Nonparametric Me
PURI and SEN · Nonparametric Me
PURI, VILAPLANA, and WERTZ
  and Applied Statistics
RANDLES and WOLFE · Introducti
  Statistics

RAO · Asymptotic Theory of Statistic
RAO · Linear Statistical Inference an ...ations, *Second Edition*
ROBERTSON, WRIGHT and DYKSTRA · Order Restricted Statistical Inference
ROGERS and WILLIAMS · Diffusions, Markov Processes, and Martingales, Volume II: Îto Calculus
ROHATGI · Statistical Inference
ROSS · Stochastic Processes
RUBINSTEIN · Simulation and the Monte Carlo Method
RUZSA and SZEKELY · Algebraic Probability Theory
SCHEFFE · The Analysis of Variance
SEBER · Linear Regression Analysis
SEBER · Multivariate Observations
SEBER and WILD · Nonlinear Regression
SEN · Sequential Nonparametrics: Invariance Principles and Statistical Inference
SERFLING · Approximation Theorems of Mathematical Statistics
SHORACK and WELLNER · Empirical Processes with Applications to Statistics
STOYANOV · Counterexamples in Probability
WHITTAKER · Graphical Models in Applied Multivariate Statistics

Applied Probability and Statistics

ABRAHAM and LEDOLTER · Statistical Methods for Forecasting
AGRESTI · Analysis of Ordinal Categorical Data
AGRESTI · Categorical Data Analysis
AICKIN · Linear Statistical Analysis of Discrete Data
ANDERSON and LOYNES · The Teaching of Practical Statistics
ANDERSON, AUQUIER, HAUCK, OAKES, VANDAELE, and WEISBERG · Statistical Methods for Comparative Studies
ARTHANARI and DODGE · Mathematical Programming in Statistics
ASMUSSEN · Applied Probability and Queues
*BAILEY · The Elements of Stochastic Processes with Applications to the Natural Sciences
BARNETT · Interpreting Multivariate Data
BARNETT and LEWIS · Outliers in Statistical Data, *Second Edition*
BARTHOLOMEW · Stochastic Models for Social Processes, *Third Edition*
BARTHOLOMEW and FORBES · Statistical Techniques for Manpower Planning
BATES and WATTS · Nonlinear Regression Analysis and Its Applications
BHATTACHARYA and WAYMIRE · Stochastic Processes with Applications
BECK and ARNOLD · Parameter Estimation in Engineering and Science
BELSLEY, KUH, and WELSCH · Regression Diagnostics: Identifying Influential Data and Sources of Collinearity
BHAT · Elements of Applied Stochastic Processes, *Second Edition*
BHATTACHARYA and WAYMIRE · Stochastic Processes with Applications
BLOOMFIELD · Fourier Analysis of Time Series: An Introduction
BOLLEN · Structural Equations with Latent Variables
BOX · R. A. Fisher, the Life of a Scientist
BOX and DRAPER · Empirical Model-Building and Response Surfaces
BOX and DRAPER · Evolutionary Operation: A Statistical Method for Process Improvement
BOX, HUNTER, and HUNTER · Statistics for Experimenters: An Introduction to Design, Data Analysis, and Model Building

*Now available in a lower priced paperback edition in the Wiley Classics Library.

Applied Probability and Statistics (Continued)

BROWN and HOLLANDER · Statistics: A Biomedical Introduction
BUCKLEW · Large Deviation Techniques in Decision, Simulation, and Estimation
BUNKE and BUNKE · Nonlinear Regression, Functional Relations and Robust Methods: Statistical Methods of Model Building
BUNKE and BUNKE · Statistical Inference in Linear Models, Volume I
CHAMBERS · Computational Methods for Data Analysis
CHATTERJEE and HADI · Sensitivity Analysis in Linear Regression
CHATTERJEE and PRICE · Regression Analysis by Example
CHOW · Econometric Analysis by Control Methods
CLARKE and DISNEY · Probability and Random Processes: A First Course with Applications, *Second Edition*
COCHRAN · Sampling Techniques, *Third Edition*
COCHRAN and COX · Experimental Designs, *Second Edition*
CONOVER · Practical Nonparametric Statistics, *Second Edition*
CONOVER and IMAN · Introduction to Modern Business Statistics
CORNELL · Experiments with Mixtures, Designs, Models, and the Analysis of Mixture Data, *Second Edition*
COX · A Handbook of Introductory Statistical Methods
COX · Planning of Experiments
DANIEL · Applications of Statistics to Industrial Experimentation
DANIEL · Biostatistics: A Foundation for Analysis in the Health Sciences, *Fourth Edition*
DANIEL and WOOD · Fitting Equations to Data: Computer Analysis of Multifactor Data, *Second Edition*
DAVID · Order Statistics, *Second Edition*
DAVISON · Multidimensional Scaling
DEGROOT, FIENBERG and KADANE · Statistics and the Law
*DEMING · Sample Design in Business Research
DILLON and GOLDSTEIN · Multivariate Analysis: Methods and Applications
DODGE · Analysis of Experiments with Missing Data
DODGE and ROMIG · Sampling Inspection Tables, *Second Edition*
DOWDY and WEARDEN · Statistics for Research
DRAPER and SMITH · Applied Regression Analysis, *Second Edition*
DUNN · Basic Statistics: A Primer for the Biomedical Sciences, *Second Edition*
DUNN and CLARK · Applied Statistics: Analysis of Variance and Regression, *Second Edition*
ELANDT-JOHNSON and JOHNSON · Survival Models and Data Analysis
FLEISS · The Design and Analysis of Clinical Experiments
FLEISS · Statistical Methods for Rates and Proportions, *Second Edition*
FLURY · Common Principal Components and Related Multivariate Models
FRANKEN, KÖNIG, ARNDT, and SCHMIDT · Queues and Point Processes
GALLANT · Nonlinear Statistical Models
GIBBONS, OLKIN, and SOBEL · Selecting and Ordering Populations: A New Statistical Methodology
GNANADESIKAN · Methods for Statistical Data Analysis of Multivariate Observations
GREENBERG and WEBSTER · Advanced Econometrics: A Bridge to the Literature
GROSS and HARRIS · Fundamentals of Queueing Theory, *Second Edition*
GROVES · Survey Errors and Survey Costs
GROVES, BIEMER, LYBERG, MASSEY, NICHOLLS, and WAKSBERG · Telephone Survey Methodology
GUPTA and PANCHAPAKESAN · Multiple Decision Procedures: Theory and Methodology of Selecting and Ranking Populations

Continued on back end papers

*Now available in a lower priced paperback edition in the Wiley Classics Library.

Robust Estimation and Testing

Robust Estimation and Testing

ROBERT G. STAUDTE

Department of Statistics
La Trobe University
Melbourne, Victoria, Australia

SIMON J. SHEATHER

Australian Graduate School of Management
University of New South Wales
Sydney, New South Wales, Australia

A Wiley-Interscience Publication
John Wiley & Sons, Inc.
New York / Chichester / Brisbane / Toronto / Singapore

Library of Congress Cataloging in Publication Data:

Staudte, Robert G.
Robust estimation and testing/Robert G. Staudte, Simon J. Sheather.
p. cm.—(Wiley series in probability and mathematical statistics. Probability and mathematical statistics)
"A Wiley-Interscience publication."
Includes bibliographical references.
ISBN 0-471-85547-2
1. Estimation theory. 2. Robust statistics. I. Sheather, Simon J. II. Title. III. Series.
QA276.8.S74 1990
519.5'44–dc20 89-36770
CIP

Printed in the United States of America

10 9 8 7 6 5 4 3 2 1

To Wendy, with affection.

—Bob

To Samantha, a constant source of inspiration.

—Simon

Preface

This textbook provides an introduction to the theory and methods of robust statistics at an intermediate level. It is intended specifically for students in senior undergraduate or first year graduate statistics courses. The primary goals of the text are first, to convince the student of the need for robust statistics; second, to provide the student with practical methods for carrying out robust procedures in a variety of statistical contexts; and third, to develop the techniques and concepts that are likely to be useful in the future analysis of new statistical models and procedures.

In particular, this textbook emphasizes the concepts of breakdown point and influence function of an estimator. It demonstrates the technique of expressing an estimator as a descriptive measure, from which one may easily derive its influence function. The latter is then used to explore the efficiency and robustness properties of the estimator. Mathematical techniques are complemented by Minitab macros for finding bootstrap and influence function estimates of standard errors of the estimators, robust confidence intervals, and robust regression estimates and their standard errors.

For the last three years this text has formed the basis for a one-semester course in robust statistical inference, which is part of the honors program in Statistics at La Trobe University and a Master of Science Degree Program. The Master of Science program is administered and taught by members of The Key Centre for Statistical Sciences, which is composed of the statistics departments of La Trobe University, Monash University, the Royal Melbourne Institute of Technology, and the University of Melbourne. The students are either in their fourth year honors program or enrolled in the Master of Science Degree by coursework program at one of these four institutions.

Chapter 1 places the course in the framework of scientific methodology, as well as giving the historical development of statistics during the last half-century.

Chapter 2 reviews some traditional finite sample methods of analyzing estimators of a scale parameter, with the goal of easing the transition from what the student is familiar with to alternative ways of analyzing estimators. The context of estimating the scale parameter of an exponential distribution, rather than the location parameter for a symmetric distribution, was chosen for three reasons. First, pure location problems are rare in practice; there is at least the need for simultaneous scale estimates to help in estimating the standard error of the location estimate. Second, robustness ideas are more easily explained and understood in a context where outliers only arise on one side of the data. Third, inference for the location of a symmetric distribution—no matter how important—is an unusual problem in statistics, because there is no question of *what* to estimate for the location of a symmetric distribution. This is a very special situation in statistics; it is much more common that *what* to estimate is a difficult problem, which is all too often "solved" by adopting a mathematically convenient parametric model. Returning to the scale parameter problem, what is the meaning of a "scale" in a distribution-free context? What is a measure of association in a distribution-free context? It is important to answer such questions when little is known about distributional shape. By starting with the scale parameter context, we will be facing some of these realities from the beginning.

In Chapter 3 the context of scale for an exponential distribution is enlarged to include other shapes. The finite sample results of Chapter 2 are complemented by asymptotic results, and efficiency and robustness concepts for estimators are introduced. Breakdown points and influence functions are derived and interpreted for several estimators. In the process of comparing robust estimators or in constructing confidence intervals based on them, the need to find estimates for their standard errors arises. Particular attention is given to both the influence function method and the bootstrap method. Certain starred sections in this chapter may be omitted without loss of continuity; they will require more mathematical background than the main text.

The important joint location–scale estimation problem is treated in Chapter 4. This chapter includes only a fraction of the available material on this subject, since the authors have chosen to concentrate on a few, absolutely essential robust estimators. The emphasis is on basic properties of L-estimators and M-estimators, plus the Hodges-Lehmann estimator for location.

Continuing the location-scale context of Chapter 4, Chapter 5 explains how the presence of outliers or local dependence will affect the level and power of standard tests of location. An indispensable tool, the asymptotic power function, is derived and interpreted for a number of tests.

In Chapter 6 the two-sample problem for comparing two populations is studied to see how certain tests cope with unequal variances as well as outliers. Robust alternatives to the classical t-tests and t-intervals are presented.

Chapter 7 treats the linear regression models in great detail. It contains a thorough treatment of important material on least squares diagnostics, which is otherwise scattered in the literature. There is a section on choosing designs in higher dimensions which avoid high leverage points. In addition there is a discussion of weighted least squares and its connection to robust M-estimates. The influence function for weighted least squares is derived, and proofs given for asymptotic normality of estimators. Computational methods for finding robust estimates of regression parameters and their standard errors are motivated and illustrated.

The appendix includes all the larger data sets and a number of Minitab macros which enable the student to easily calculate robust estimators and their standard errors, robust confidence intervals, and P-values for various robust tests. These data sets and macros, plus instructions, are provided in easy-to-use form on the floppy disk accompanying the book.

One obstacle in this endeavor has been the sheer volume of available material, most of it published only in the last 25 years. It has been necessary to severely restrict the number of "robust" procedures to the few that are considered essential for pedagogical reasons. Complements and references at the end of each chapter will guide the reader to further developments, but it seems impossible to give adequate credit to all the numerous contributors to the subject. Certain areas have been omitted: Bayesian robust statistics, adaptive procedures, and R-estimates and tests, with the notable exceptions of the Hodges-Lehmann estimator and the Wilcoxon tests. Estimates and tests based on ranks are thoroughly covered in the recent textbook by E. L. Lehmann (1975), *Statistical Methods Based on Ranks*, and by T. P. Hettmansperger, (1984a) *Statistical Inference Based on Ranks.* A distribution-free approach to estimation and testing in one- and two-sample problems and simple linear regression is given in *Distribution-Free Statistical Methods* by J. S. Maritz (1981); this book includes many parametric estimators and tests, but is motivated and unified by permutation methods.

There are two excellent books on robust procedures by pioneers in the field: *Robust Statistics* (1981) by P. J. Huber and *Robust Statistics, The Approach Based on Influence Functions* (1986) by F. R. Hampel, E. M. Ronchetti, P. J. Rousseeuw and W. A. Stahel. Both books are written for a mathematically mature audience and contain a large number of valuable statistical results and insights. In the text to follow the mathematical level is gradually increased, so that after completion of it, the student should be well prepared for reading these more advanced books. The approach to robust statistics developed by Hampel and Huber is based on parametric models, together with the realistic assumption that the data will not conform to the desired model. In a variety of contexts much progress has been made in finding alternative procedures to the classical maximum likelihood procedures which are *robust* in the sense that they are still reasonably efficient in the presence of outliers and other departures from the assumptions.

The reason so many classical procedures are nonrobust to outliers is that the parameters of the model are expressed in terms of moments, and their classical estimators are expressed in terms of sample moments, which are very sensitive to outliers. Another approach to robustness is to concentrate on the parameters of interest suggested by the scientific problem under study. It may well turn out that these parameters can be expressed as functions of the underlying distribution independently of a particular parametric model; that is, as descriptive measures. If these descriptive measures are judiciously chosen, their naturally induced estimators are robust to aberrations in the data. This approach was introduced by P. J. Bickel and E. L. Lehmann in a series of papers in the 1970s and is also developed here as an important alternative to the methods that emphasize parametric models.

This text grew out of a set of lecture notes, *Robust Estimation* (1980) (Queen's University Series in Pure and Applied Mathematics, no. 53). A first draft of those notes was written while one of the authors, R. G. Staudte, was visiting professor in the Mathematics and Statistics Department at Queen's University, Kingston, Ontario; many thanks to Colin R. Blyth and Queen's University for the opportunity to prepare the material for a graduate class in statistics, plus several faculty members. This textbook includes some of the material from those notes, and also much more material at an intermediate level. The impetus to expand the notes into a textbook came from John Wiley and Sons, after encouraging comments from Erich L. Lehmann.

More material developed as a result of a Seminar on Robust Testing held at La Trobe University in 1983, in which both graduate students and faculty (including the two authors) participated. A substantial number of problems ranging from simple exercises to lengthy projects were developed for the text as it was used in the classroom over the past ten years.

ROBERT G. STAUDTE
SIMON J. SHEATHER

Melbourne, Victoria, Australia
Sydney, New South Wales, Australia
October 1989

Acknowledgments

The authors are indebted to Michael B. Dollinger of Pacific Lutheran University, Tacoma, Washington, who visited the Mathematics Department at La Trobe University in 1987. His enthusiastic research for and criticism of the chapter on linear regression has provided many insights and improvements. Thanks also to Elvezio Ronchetti of the University of Geneva, who during a 1988 visit to the Statistics Department at La Trobe read much of the text and provided encouragement and useful suggestions. The authors also thank Geoff Watterson of Monash University, whose detailed reading of the manuscript caught many minor and some major errors. Of course the authors assume full responsibility for mistakes that remain.

There are other scholars who have provided stimulating discussions and inspiration; in particular Bruce M. Brown of the University of Tasmania, Thomas P. Hettmansperger of Pennsylvania State University, Robert Kohn of the University of New South Wales, J. S. Maritz of La Trobe University, Joseph W. McKean of Western Michigan University, and William R. Schucany of Southern Methodist University. Finally, no textbook is complete without the labors of students; and the authors also wish to thank numerous students for their efforts on the exercises and their detailed comments and questions on the text. Special thanks to Lynda Chambers, Kathy Haskard and Andrew Mich.

We thank the successive editors Beatrice Shube and Kate Roach and the editorial staff at John Wiley & Sons for their assistance. We also thank the following sources of data sets for permission to reproduce them:

Table C.1; by permission of Dr. Neville G. White, Department of Genetics and Human Variation, La Trobe University

Table C.2; © by permission of the Bureau of Meteorology, Victoria, Australia

Table 7.9; © by permission of the Institute of Mathematical Statistics; see Atkinson's discussion of the paper by Chatterjee and Hadi (1986)

Example 2, p. 177 and Table 7.5; © by permission of John Wiley & Sons, Brownlee (1965)

Table C.5; by permission of the American Statistical Association; see Ruppert and Carroll (1980)

Problem 15, p. 195; Courtesy of Levi Strauss & Co., Albuquerque, New Mexico Production Facility.

Contents

CHAPTER ONE

The Field of Statistics

Simplification which retains the essential scientific essence of the problem is most likely to lead to useful answers but this requires understanding of, and interest in, scientific context.

Scientific investigation is an iterative process in which the model is not static but is continually evolving.

...I believe that scientific method employs and requires not one, but two kinds of inference—criticism and estimation; once this is understood the statistical advances made in recent years in Bayesian methods, data analysis, robust and shrinkage estimators can be seen as a cohesive whole.

G. E. P. Box (1983)

Robustness is a fundamental issue for all statistical analyses; in fact it might be argued that robustness is the subject of statistics.

J. B. Kadane (1984)

Scientists in this century devised statistical methods to help them objectively assess their subjective conjectures, and mathematicians and philosophers soon joined them in creating techniques, principles, and theories which are now known collectively as "the field of statistics." A better name might be "the battlefield of statistics," since its short history is marked by disagreement, controversy, and even feuding. Some statisticians have sought refuge from the wars by concentrating on the diverse arts of mathematical statistics and data analysis. Growth in these areas followed significant developments in measure theoretic probability and computer technology, respectively. During the last twenty years statisticians of all persuasions have been rewriting much of the literature under the banner of *robustness*, a pragmatic response to the overemphasis on parametric modeling of the past. While these changes make the field of statistics dynamic and exciting, there is now considerable confusion, bordering on the chaotic.

In this book we explain some of these recent developments in the simplifying language of descriptive measures. We will concentrate on the routine problems, such as the two-sample problem, because of their importance. Before doing so, we will elaborate on some of the above statements and especially review the relationship between the scientific method and statistical inference. In so doing we will argue the case for applied statistics. While this may seem a strange way to begin a book of mathematical statistics, it will continually remind us in future chapters that our original and continuing source of ideas is experimental science.

1.1 THE ROLE OF STATISTICS IN SCIENTIFIC INFERENCE

1.1.1 The Scientific Method

C. R. Blyth (1972) describes the scientific method as a C-V-C-V-··· sequence with some mixing, which consists of construction (C) and verification (V) of models for reality. In the construction phases the scientist examines data and the literature, discusses possibilities with colleagues, invokes hunches and other subjective criteria, and finally constructs new models for reality. These models may range from narrow hypotheses to global schemes.

In the verification phases the scientist tries to find fault with these models by objectively testing them against reality, which is usually in the form of experimental data. If the models survive these attempts to falsify them, the scientist may then be willing to publish them for other scientists' scrutiny. Those models that survive widespread objective testing become known as scientific laws. A good example are Mendel's laws of heredity, which he conjectured on the basis of insight and experimental observation long before the mechanism of chromosome replication was understood.

1.1.2 Statistical Support for the Scientific Method

Statistical methods are helpful in both the construction and the verification phases of the scientific method. For example, from the Mendelian hypotheses one derives expected phenotype frequencies in an experimental cross. If the observed frequencies are too far from the expected frequencies, then doubts arise about the hypotheses. The chi-squared test of Karl Pearson

(1900) helps quantify the vague words "too far" and thus assists in this verifying, objective phase. Of course Mendel's conjectures have proven consistent with a multitude of observations and have now achieved the status of laws. Now they are used in a constructive phase to help estimate or test hypotheses about the frequencies of recessive traits, for example. Statistical methods such as maximum likelihood assist in this construction. Note that mathematical statistics may play an important supporting role in either the construction or the verification phases of the scientific method: it is nontrivial to prove that the chi-squared distribution is a useful approximation to the discrete chi-squared statistic in the goodness-of-fit tests, and sufficiency reductions, standard errors, and asymptotic approximations are all part of parametric estimation.

Many controversies in statistics arise because the problems under dispute are isolated from their scientific context. Those who argue in favor of reasoning in the presence of data are clearly participating in a subjective, construction phase of the scientific method. Those who insist on specifying critical regions before looking at the data are participating in objective testing of well-defined statistical hypotheses and alternatives. If these statistical hypotheses concern a generally accepted probability model for the scientific problem, then the test only results in a slight refinement of the model; this is part of the construction phase of the scientific process since there is no doubting the underlying scientific laws. However, if these statistical hypotheses are derived from scientific conjectures, then rejection of the statistical null hypothesis will also cast doubt on the corresponding scientific conjecture.

1.1.3 The Significance of a Result

We may illustrate these ideas by considering the sex of children conceived by the in vitro fertilization (IVF) program at the Queen Victoria Medical Centre in Melbourne. Letting p denote the probability of a female birth we may speculate that p differs from the well-known frequency .47 of females among naturally conceived births. If X is the number of females among the first n births in the IVF program, then it is natural to reject the null hypothesis $H : p = .47$ in favor of the alternatives $K : p \neq .47$ if $|X - .47n| > c$, where c is to be determined by the desired significance level.

A statistician could back up this natural test by assuming that the sex of different births are independent, reducing the ordered sequence of outcomes to X by sufficiency, and showing that the test is uniformly most powerful among unbiased tests for the given hypotheses. But the important point here is that for a given n the statistician may calculate, *before looking at the data*, the chance of rejecting H when X has the binomial $(n, .47)$

distribution. Thus by choice of c he may adjust the size of the test (the maximum probability of a type I error) so that it does not exceed his own subjective level of significance, the chance he is willing to take of making a type I error.

Furthermore he may judge in advance that a change in p of .02 or more is scientifically significant, and protect himself from making the type II error of erroneously accepting the null hypothesis when there is such a change in p. It is an elementary exercise (see Problem 1) to choose n and c so that the chances of making either type of error are kept to his own acceptable subjective levels. In theory this can be done, and is sometimes done. But what happens in practice is often different, because many hypotheses (such as a change in p) are not anticipated, but are suggested by the data after observation. In the IVF context, the sex of the newborn is of minor importance to the fact of the newborn, so the scientists in the program are mainly concerned with the pregnancy rate. Setting up a simple binomial test for change in the proportion of males is only one of many other hypotheses that could be listed a priori.

Six of the first seven children born were girls, which led to speculation in the press that there was affirmative action in favor of girls. We may retrospectively test the null hypothesis $H : p = .47$ and calculate the P-value (probability value, descriptive level, observed significance level) of the result "six out of seven"; it is the probability under H of the two-sided test which just barely rejects H, namely, $P\{X = 0, 1, 6, 7\}$. Assuming independence of sex for different births, this P-value is roughly one in eight, which is not significant at the usual levels. Nevertheless, these data may lead us to conjecture that the sex of children conceived by this method may strongly favor girls. Or at least we may now be willing to entertain the conjecture that the method may favor one sex or the other, compared to the naturally conceived ratio of .47. Let us adopt the latter. To objectively test the null hypothesis in favor of the alternatives $K : p \neq .47$ at level .05, we must await further evidence. Now the next 140 births included only 56 girls, which yields a P-value of .03 and rejection of the null hypothesis. Further examination of the data reveals that roughly half the cases of single births are boys, but that among twins (nonidentical) there are 34 boys and 14 girls! This leads to conjectures regarding the preponderance of boys among twins. It is clearly impossible to anticipate all hypotheses of interest. The main point is that the P-value of data with regard to certain hypotheses should not be calculated for hypotheses suggested by the same data, since we are limited only by our imagination in finding null hypotheses that the data appear to refute.

Another important question is whether the *statistical* significance of a result is the same as *scientific* significance. A change in p of .01 may have little scientific significance, yet the above statistical test of $p = .5$ (or .47, or whatever), which is based on more than 10,000 observations, will have a test statistic with standard error less than .005. In this case the scientifically meaningless change of .01 is likely to be declared significant by the statistical test.

Let us summarize the above points:

1. Statistical tests are mathematical constructs which may assist the scientist in construction or verification phases of the scientific method.
2. When there are so many data that the standard error of the test statistic is of smaller order than the scientifically meaningful alternatives, statistical significance is not necessarily scientific significance.
3. When the Neyman-Pearson theory is used to set up an objective test of hypotheses, and the P-value of the data is subsequently calculated relative to these hypotheses, it carries objective value in the following sense: in a large number of significance tests of true null hypotheses with size equal to this P-value, the proportion of rejections will be approximately equal to it. However:
4. When the hypotheses are inferred from the data (from among many possibilities), then the P-value loses its objectivity, since it is the anomaly in the data that sparked our interest in the hypothesis. The power of such tests is limited only by our imagination in finding a hypothesis to reject on the basis of the data.

1.1.4 The Challenge to Statisticians

E. Bright Wilson issued a challenge to statisticians in 1952: "There is a great need for further work on the subject of scientific inference. To be fruitful, it should be carried out by critical, original minds who are not only versed in philosophy but also familiar with the way scientists actually work (and not just the way some of them say they work!). Unfortunately, the practical nonexistence of such people almost suggests that the qualities of mind required by a good philosopher and those needed by a working scientist are incompatible."

We exaggerate slightly in saying that scientists are unabashed seekers of the truths which they formulate; they are masters of observation who employ every possible technique (even statistical inference) to resolve a conjecture. Statisticians are more prone to speculation and bogging down in mathematical mire in the hope of finding an elegant approach to the problem at hand. Nevertheless we think that statisticians should continue to wear the three-cornered hat of mathematician, data analyst, and philosopher, since these roles have combined so productively in the past.

During this century statisticians have concentrated their efforts on mathematical research into sample-to-population inference. Such studies assume that the population is well defined and usually are supportive of the construction phases of the scientific method in which there is no questioning the underlying scientific laws. Comparatively little effort has been spent by statisticians in helping scientists formulate their theories and derive statistical hypotheses from them that will be falsifiable when subjected to objective tests.

Geoff Dolby (1982) elaborates on these themes and summarizes as follows:

> In conclusion, the application of statistics to the life sciences is likely to become more fruitful if greater emphasis is placed on the articulation of global conjectures by the specialist most intimately involved; if statistics is used to deduce probability statements from such conjectures; if emphasis is placed not on "validating" the probability statement involved but on FALSIFYING it; and if the inductivist attitude of the profession is softened. We badly need a brand of applied statistics which in temperament is inspirationist, global, deductivist, and falsificationist rather than empiricist, local, inductivist, and verificationist in spirit. Hypotheses are like nets: only he who casts can catch.

An example from our experience of the interaction between scientists and statisticians on a hypothesis of some general interest is in the recent paper by Staudte, Woodward, Fincher, and Stone (1983). The biochemists wanted to test whether two types of residues in barley sugar macromolecules had been biosynthesized independently or whether some pattern could be detected. The scientific null hypothesis is that the residues are independently synthesized with some scientists suggesting that local dependence may well exist. After lengthy discussion it was decided that a first-order Markov chain model with the residues as states was appropriate. Moreover, stationarity of the chain seemed a reasonable assumption. In such a chain dependence could be measured by the autocorrelation, and the statistician planned to use a sample version to carry out a test of independence. However, sequences of states could not be observed. The only data available were in the form of the relative proportions of the two states, obtained by weighing the fragmented residues among millions remaining after an enzymic hydrolysis. It was suggested that the test could be carried out if the proportions of two-state segments were estimated; this led to a new experiment of hydrolysis, which yielded the necessary information. The null hypothesis was not rejected; in fact confidence intervals for the autocorrelation tended to confirm the null hypothesis, much to the surprise of those looking for dependence. This example illustrates that knowledge of traditional parametric models may be useful in framing statistical hypotheses in new areas of scientific enquiry, and that the traditional sampling methods may be irrelevant because data are only available in a previously unseen form.

1.2 RECENT TRENDS IN STATISTICS

There are five major developments in statistics this century. First, the conceptual contributions by R. A. Fisher (1921) and Jerzey Neyman and Egon Pearson (1933) which grew out of their scientific work; second, the influence of measure theoretic probability begun by Kolmogorov (1933a); third, the technological development of high-speed computers; fourth, the discovery of highly efficient distribution-free procedures; and finally, the robustification of many classical procedures. There are many contributors to these developments and a book could easily be devoted to each of them. For those interested in the controversies in statistics alluded to earlier we suggest the lucid introduction by Efron (1978). Two excellent collections of history and one of biography have recently appeared, edited, respectively, by Owen (1976), David and David (1984), and Gani (1982); we heartily recommend them to the reader. We will mention only the important developments which help to place this book in context.

1.2.1 Mathematical Statistics

Statisticians prefer abstract settings for the same reasons other mathematicians do: to gain clarity through precision, to obtain generality, to discover new links between concepts, to unify large areas of knowledge, and to find simple approximations to complex entities. Elegance is not only a virtue, but a necessity.

An example of progression up the ladder of abstraction is the concept of a sufficient statistic first conceived by R. A. Fisher (1920). The notion could not be made precise until regular conditional probability was defined, but later it was extended to the notion of sufficient subfield. Various forms of the factorization theorem were rigorously proved. In combination with the notion of completeness and Jensen's inequality a form of the Rao-Blackwell theorem could be proved for convex loss functions. Also, the interchange of order in sufficiency and invariance reductions was shown to be justified in certain situations [see Hall, Wijsman, and Ghosh (1965)].

In addition to abstracting and relating one or more concepts, some statisticians tried to find a suitable framework which included as many statistical methods as possible. In fact, the decision theorists built an all-encompassing model which includes the Neyman-Pearson theory of hypothesis testing and much of classical estimation theory. [See Ferguson (1967) for a clear and comprehensive account.] The richness of decision theory reflects in part the potential inherent in studying functions on a family of measure theoretic probability models. Measure theoretic probability also formed the basis for convergence theorems for sequences of independent random variables and martingales, which has led to the many results on consistency and asymptotic efficiency for estimators and test statistics. Ideas from functional analysis have become increasingly useful; we mention only projection results from the L_2 theory employed in linear models and U-statistics, and the von Mises expansions.

Of course statisticians have not only benefited from probabilists by climbing up the abstraction ladder, but they have taken difficult problems with them which have stimulated a great deal of probabilistic research. The goodness-of-fit tests of Cramer (1928) and Kolmogorov (1933b) stimulated research into invariance theorems in weak convergence theory, for example.

1.2.2 The Impact of Computers

While the emergence of mathematical statistics led to greater abstraction, computers opened opportunities in the direction of greater quantification. The positive results for statisticians include the following:

1. Large quantities of data are instantly accessible for visual display in a variety of summarizing and detail-revealing forms. The instant interaction has greatly enhanced the power of data analysts.
2. Traditional estimators and tests are almost instantly calculable, and computationally intensive methods such as density estimation, adaptive procedures, and permutation tests are now feasible alternatives. New techniques such as the bootstrap will continue to emerge.
3. Monte Carlo studies may reveal the relative frequency performance of various statistics at finite sample sizes to determine the accuracy of asymptotic approximations. (The potential for solving difficult problems by combinations of computational power and analytical skills has been illustrated by solution of the four-color problem.)
4. Beautiful camera-ready plots and printouts of well packaged output are conveniently provided.
5. We may develop intuition for central limit theorems, sampling distributions, projections in linear models, etc., by means of good teaching software.

On the negative side, computer technology has led to a deluge of information and crunched numbers. And some abuses:

1. A proliferation of statistical packages which provides slick output for those who use statistics for "sanctification rather than elucidation," as Hampel (1973) so aptly puts it. There is an old saying in the computer world, "garbage in, garbage out." Now it is possible to put good raw data in and to get nicely packaged garbage out.
2. An explosion of jargon which in many cases amounts to a renaming of previous well established concepts.
3. The assumption that statistical computing can completely replace analytical thinking.

1.2.3 Robust Statistics

Robust statistics have been in use for hundreds of years [Stigler (1973)] but not seriously studied by mathematical statisticians until quite recently. Parametric results began to be questioned in the 1950s, when asymptotic methods developed to the point where the inefficiency of the t-test to the Wilcoxon test could be demonstrated [Pitman (1949), Hodges and Lehmann (1956)]. Then the sensitivity of some classical estimators was documented by Tukey (1960). Other factors also contributed to the growing interest in *robustness*. Hampel (1973) has given a comprehensive list of them and a provocative introduction to robust methods; the reader is encouraged to read his essay.

We only note that while statisticians have been long aware of the sensitivity of some statistics to slight changes in their assumptions, it is only recently that they have had the tools to describe these problems mathematically. In particular more statisticians are now familiar with the rudiments of functional analysis and computing. Huber (1964) and Hampel (1968) both have employed ideas from functional analysis in their pioneering work; and the Princeton robustness study [Andrews et al. (1972)] is typical of the computational approach to gaining insight into robust proposals.

There have been no new journals dedicated to papers on robust procedures, probably because of the widespread acceptance of the idea that every new statistical study is incomplete without some serious attention to the effect of slight changes in the assumptions. Rather, nearly all the traditional journals have included numerous robustness studies. One reason for the continued proliferation of such work is the almost universal applicability and intuitive appeal of Hampel's influence function and breakdown point (1971, 1973).

Let us consider a simple example which illustrates the need for robustness studies. In it the lack of robustness of the P-value for the t-test is clearly demonstrated.

Ridge Count Example

The dermal ridges on the fingers are largely an inherited characteristic, and geneticists have studied them to test theories of genetic inheritance. In particular the total ridge count (TRC) has been found to be a useful genetic marker. Appendix C.1 lists the TRCs for an Australian aboriginal tribe and for a first-year statistics class at La Trobe University. The former group is composed of relatives, and can be expected to show less variability than the latter, which is composed of people from many different ethnic backgrounds. In addition, it is hypothesized that the ridges evolved to aid in grasping, so that tribal members whose survival depended on hunting skills may be expected to have more ridges than people of European, Middle Eastern, and Oriental background. Assume that the TRCs for the tribe are normally distributed with mean μ and variance $\sigma^2 > 0$, and that we know that for the population at large the mean TRC is 138. To test $H : \mu = 138$ versus $K : \mu > 138$ we take the first six observations from data set 1, Appendix C. (All 27 observations are analyzed later.) The six observations are 166, 229, 142, 141, 136, and 153. This yields $\overline{x} = 161.2$, $s = 34.9$, and a t-statistic P-value for the above hypothesis of .08. By way of comparison, the Wilcoxon signed-rank test yields a P-value of .03. Thus the Wilcoxon test would be significant at level .05 while the t-test is not, and it clearly matters which test is chosen.

Of course one may object to making comparisons based on a normal model when the data do not appear to come from a normal distribution. However, the normal scores test would just barely reject these data as coming from a normal distribution at level .05, and if the outlier 229 were somewhat smaller, it would not do so.

Now let us see what happens to the P-values of the two tests as we vary the second observation over a wide range of values. From 153 to 166, x_2 is the second largest observation, and from 166 on x_2 is the largest of the six observations. The P-values are plotted in Figure 1.1 as a function of x_2.

The graph shows that *both* tests would reject H at level .05 if x_2 were only 160. But as x_2 increases to 189, and more evidence is given against H, the t-test fails to reject H. The P-value of the t-test is *very* sensitive to one outlier, while the Wilcoxon test remains unaffected by it.

For much larger values of n, the central limit theorem comes into effect and the P-value of the t-test does become less sensitive to outliers, as discussed in Problem 6. However, it is also shown there that even this robustness occurs only because both the numerator and the denominator

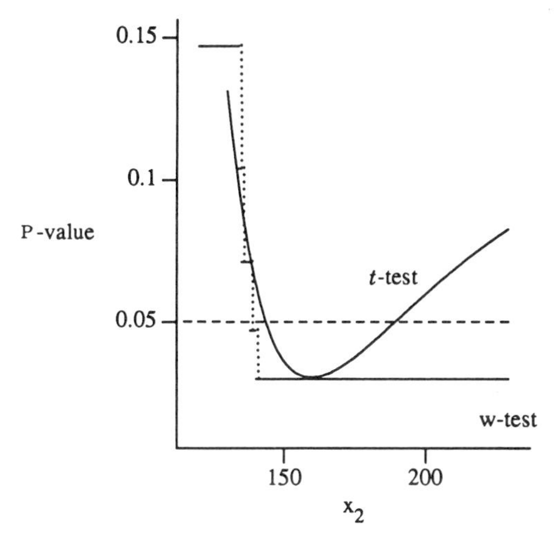

FIGURE 1.1 The effect of an outlier on the P-value.

of the t-statistic are so sensitive to the outlying observation, the denominator more so than the numerator. We can hardly expect such good fortune in general. The t-test P-value still begins to increase for one large outlier so that H will still not be rejected by it if too much evidence of this sort against H is presented! The robustness of the P-value and the power of various tests are considered in Chapters 5 and 6.

1.3 THE CASE FOR DESCRIPTIVE MEASURES

There are many reasons for collecting and examining data, but we will restrict attention to situations where the data may be considered a sample from a population of numerical values. The population will be represented in some idealized form as a probability distribution F.† We realize that there is no "true" F, and that even if we could examine the entire population, many possible F's would be a suitable model for it. Indeed, one reason for looking at data is to find an F, or a family $\mathcal{F}$ of F's, which could represent the population from which the data arise, and which are convenient to manipulate mathematically. Moreover, we would like the unknown parameters of this family to have meaning in the scientific context of the population.

†In a convenient abuse of notation we shall use F to represent both the cumulative distribution function and the associated probability distribution (probability measure); the context will make the meaning clear.

There is a tendency to head for a parametric family of models with which we are familiar, before trying to enunciate clearly the scientific parameters of interest. Often they are treated as though they were of secondary importance, and only to be considered as functions of the parameters of our favorite parametric family. We think that the order of importance should be reversed, that we should first say *what* we are trying to estimate or test hypotheses about. Calling this unknown $T(F)$, we want to specify T independently of the unknown underlying F, which is of secondary importance. Only then should we begin to restrict the possibilities for F to a nonparametric family or smaller parametric family. For example, in estimating the ore yield from a particular site, only the mean amount of ore per unit volume is of importance, and this is the quantity of interest, whether or not the distribution is skewed. On the other hand, in estimating a "typical" value of a natural characteristic such as height, weight, or lifetime, the distribution may be roughly bell-shaped but skewed to the right. In these cases the mean may well be very atypical and uninformative, while the meaning of the median is quite clear. Any T which maps a large class $\mathcal{F}$ of F's into the real line will be called a *descriptive measure.* By "large class $\mathcal{F}$" we mean one that contains the plausible models F for the unknown population as well as the empirical distributions F_n defined below.

There are other reasons for concentrating on the descriptive measure $T(F)$. It induces a natural nonparametric estimator $T(F_n)$, where F_n is the empirical distribution, and the properties of $T(F_n)$ can be deduced by studying the behavior of $T(G)$ for G in a neighborhood of F. We will look at such behaviour in subsequent chapters. Since T assigns a real number to each member of a class of functions, including the empirical distribution functions F_n, it is often referred to as a *statistical functional*, or simply a *functional*.

Let us define a simple descriptive measure and the corresponding "induced" estimator. For any real-valued function g define the functional

$$T(F) = \int g(x) dF(x)$$

for all distributions F for which the integral exists.† We will primarily be concerned with two distinct cases. First, when F has density f with respect to Lebesgue measure, this becomes $T(F) = \int g(x)f(x)dx$. Second, when F is discrete with mass concentrated on the set of points $\{x_i\}$,

$$T(F) = \sum_i g(x_i)p_i$$

†By "integral" we mean Lebesgue-Stieltjes integral [see Ash (1972)].

where $p_i = F(x_i) - F(x_i^-)$ is the probability on the point x_i.

Letting $g(x) \equiv x$ be the identity function we obtain the *mean* functional, which is often denoted by μ_X or $E_F[X]$, when X has distribution F. Here we continue to denote it by $T(F)$ to emphasize its functional dependence on the underlying F. Now if $F_n(x)$ denotes the usual empirical distribution of $X_1, \ldots, X_n$, then F_n is a discrete distribution which puts mass $p_i = 1/n$ on each of the n observations. So by the above definition,

$$T(F_n) = \frac{X_1 + \cdots + X_n}{n} = \overline{X}_n.$$

The reader may easily find the induced estimator $T(F_n)$ of the functional $T(F) = E_F[X^2]$, which arises from the above integral formulation when $g(x) = x^2$. More examples are worked out in the problems.

1.3.1 Nonparametric Neighborhoods of Parametric Models

Given $X_1, \ldots, X_n$ independent and identically distributed as F, where F belongs to a known family $\mathcal{F}$, the problem is to estimate some descriptive measure of the distribution, call it $\theta = T(F)$. In order to delineate the types of "robustness" we shall be concerned with, it is necessary to introduce some terminology. The model $\mathcal{F}$ is called *parametric* if it is indexed by a real vector or vector-valued quantity (called a *parameter*); otherwise it is nonparametric. For example, the class of all normal distributions $\{N(\mu, \sigma^2)\}$ is parametric, and the class of all continuous distributions is nonparametric. In either of these two cases the statistician may be interested in the median $T(F) = F^{-1}(\frac{1}{2})$, which is often called a *parameter*. A better term than parameter, which is gaining widespread usage is *estimand*. It is preferable because, strictly speaking, the median does not parameterize, i.e., determine F in the class of continuous distributions. The median does characterize what the statistician wants to know. Thus nonparametric models are qualitatively larger than parametric models, and a parameter may be of interest in either case.

Why shouldn't we use parametric models? The great bulk of research has been concerned with these simple and seemingly adequate models. Partly because of their simplicity it is possible to superimpose additional structure such as prior distributions, to extract powerful sufficiency reductions, and to prove asymptotic results, say for maximum likelihood estimators. The problem with parametric models is that what works well for one model often works poorly for neighboring models, and *in practice one cannot tell which model is closest to the data*. For example, if one hypothesizes the normal $\{N(\theta, 1)\}$ distributions and the data (with $n = 20$) are actually from a double-exponential distribution centered at θ, the estimator $\overline{X}$, which is thought to be 100% efficient, is only 70% efficient.

Moreover so called goodness-of-fit tests are unlikely to reject double-exponential data as unfit for normal consumption; a chi-squared goodness-of-fit test at level .05 has approximate power .17 of detecting double-exponential alternatives to the normal family. Users of such tests typically reject the normal model only if the fit is atrocious, not just bad, possibly because there is no clear method of solution if one rejects the normal model. And when there is no control over type II errors, as there cannot be in goodness-of-fit testing, accepting null models is statistically reprehensible. In other words, the fact that we know what to do when the normal model is assumed and do not know what to do otherwise is no justification for the assumption. A goodness-of-fit test may reject the normal model, but cannot confirm it.

If parametric models are inadequate, why not use nonparametric models? If one knows almost nothing about the data source it seems appropriate to use the empirical F_n to estimate the unknown F. This suggests using $T(F_n)$ to estimate $T(F)$. For example, in estimating a location parameter we first decide what descriptive measure is of interest, the mean or median, or whatever. Then we apply this chosen T to the empirical F_n to obtain the sample mean, or sample median, or whatever. In this way we obtain naturally induced estimators in the nonparametric setting.

If estimators derived for parametric models are inadequate, and if estimation in a completely nonparametric setting is only a matter of definition of $\theta = T(F)$ (substitute F_n for F), then how do we attack a problem when we are convinced that the "true" distribution is near some nice parametric one? One answer is to consider nonparametric neighborhoods of parametric models. For example in 1964 Huber considered the class of contaminated normal models $(1-\varepsilon)N + \varepsilon H$, where H is an arbitrary symmetric distribution. The objective of many robustness studies since then has been to find estimators which are highly efficient at the parametric model, and whose distributions change little in a small neighborhood of it.

1.3.2 Descriptive Measures

One way to study the stability of an estimator over a neighborhood of the parametric model is to simulate its performance for different F's in the neighborhood. In the Princeton robustness study [Andrews et al. (1972)] some 68 estimators of location were repeatedly evaluated on samples from a dozen "nearly normal" models. This approach is unsatisfying, even when it is informative. It may convince one to reject the sample mean as too unreliable, but it does not guarantee that the alternative estimators, such as the trimmed mean, will be reliable when tested on models not included in the study. The numbers may not lie, but the truth is not all out, either.

Another way to study the stability of an estimator T_n is to first write it $T(F_n)$, where $T(G)$ is a descriptive measure defined for all G in a neighborhood of F. Then the behavior of T on the neighborhood can provide insight into the behavior of $T(F_n)$, at least if F_n has high probability of lying in the neighborhood. This procedure separates the probabalistic convergence problem $F_n \to F$ from the analytic problem of determining the smoothness of the mapping $G \to T(G)$. "Smoothness" is defined as continuity with respect to appropriate topologies, see Hampel (1971).

It is not necessary to look at appropriate topologies to gain insight into the behavior of T at F. Hampel's influence function $\mathrm{IF}(x)$, a directional derivative of T at F, is often easy to derive with elementary calculus. It gives an approximation to the relative influence on T of small departures from F. Moreover, it is often the case that $T(F_n) - T(F)$ is approximately the average of the n independent terms $\mathrm{IF}(X_1), \ldots, \mathrm{IF}(X_n)$, so that the asymptotic normality of the estimator sequence $\{T(F_n)\}$ can be studied via the influence function.

To summarize, descriptive measures are useful not only for saying precisely what it is we want to estimate, regardless of the exact form of F, but they are also basic to a study of the robustness and asymptotic normality of their induced estimators.

1.4 THE DOMAIN AND RANGE OF THIS BOOK

This book is intended to guide a student with a background in undergraduate statistics through the basic material on robust statistics, so that he or she understands the effects of model inadequacy on standard procedures and can confidently apply robust methods which are known to be almost as efficient as optimal procedures for the model.

By a "background in undergraduate statistics," we mean that the reader should be familiar with the material covered in standard texts such as *Introduction to Mathematical Statistics* by Hogg and Craig (1978), and *Introduction to Linear Regression Analysis* by Montgomery and Peck (1982); or the more advanced text *Mathematical Statistics: Basic Ideas and Selected Topics* by Bickel and Doksum (1977). The mathematical level of this material is kept to the minimum required to understand the basic concepts. A good text for asymptotic theory is *Approximation Theorems of Mathematical Statistics* by Serfling (1980); some of the results proved there are listed in the Appendix of this text for quick reference.

The topics covered here are all basic: estimation of scale, location, dispersion, and regression parameters; one- and two-sample tests. Various methods for estimating standard errors are explained in detail, especially the bootstrap. Other methods which are presented are influence function estimates and density estimates. In the Appendix we compare Hajek's projection method with the delta method for proving asymptotic normality, and the reasons for the ubiquitous influence curve are given.

While this text is mainly concerned with methods for handling possible outliers or wider tailed distributions than normally assumed, there is also material on the effects of dependent observations on estimates and tests (when they are assumed to be independent). In addition, the effects of unequal variances on two-sample tests when equal variances are assumed are presented.

In each of these cases we have sought to find and present *qualitative* results, which are supplemented by quantitative results. Many qualitative results emerge from simple asymptotic theory, when the dependence on the sample size does not complicate the main ideas, so the student will be guided through the necessary limit results. The implications of the limiting results for finite samples are illustrated by applications to real data sets.

The representation of estimators as descriptive measures applied to an empirical distribution is central to the book. While this approach is more abstract, it is becoming very widely used and is straightforward for the i.i.d. case. In more structured problems, such as the regression models, it requires more mathematical sophistication but continues to yield statistical insight. In the case of dependence models, such as the autoregressive processes, the functional representation may not be obvious, as illustrated by the disagreement among current researchers [see the paper and discussion of Martin and Yohai (1986)].

1.5 PROBLEMS

Section 1.1

1. Let X have the binomial (n, p) distribution. Reject $H : p = .5$ in favor of $K : p \neq .5$ when $|X - .5n| > c$. The P-value of an outcome $\{X = x\}$ relative to H is the probability, under H, of all outcomes which would lead to rejection of H if $\{X = x\}$ does; it is the size of the test whose critical region just includes the outcome $\{X = x\}$. In this context it is $P\{|X - .5n| \geq |x - .5n|\}$.
   (a) Show that when $n = 7$, the P-value of $\{X = 6\}$ relative to the above hypothesis is exactly $\frac{1}{8}$.

(b) Show that when $n = 140$, the P-value of $\{X = 56\}$ is approximately .02.

(c) Choose n and c so that both (i) the size of the test is .05 and (ii) when $|p - .5| \geq .02$, the probability of accepting H is no more than .1.

(d) Calculate the probability that a size .05 test based on $n = 10{,}000$ observations will reject $H_0 : p = .5$ when in fact $p = .49$.

Section 1.2

2. Use the ridge count data of Section 1.2.3 to test the hypothesis that the standard deviation of ridge counts σ for the aboriginal tribe is equal to 25 against the alternative that it is greater than 25.

(a) Assume normality, and carry out a traditional chi-squared test of $H : \sigma = 25$ versus $K : \sigma > 25$, assuming the mean is unknown.

(b) Repeat part (a) several times, with the sample modified by an error in the largest observation. Plot the P-values as a function of x_2, as was done for the t-test of the mean in Figure 1.1. (For $x_2 = 209$, 229, and 249, the P-values should be approximately .3, .08, and .01.) This example shows how sensitive the P-value of this test is to one outlier.

Section 1.3

3. Assume that the TRCs for an Australian aboriginal tribe are normally distributed with unknown mean μ and standard deviation σ. We will use the TRCs for members of the tribe (see Appendix C, data set 1) to test whether $H : \mu = 144$ versus $K : \mu > 144$. The data are (after ordering) 98, 115, 119, 124, 128, 136, 136, 137, 141, 142, 144, 150, 152, 153, 154, 155, 156, 159, 160, 160, 166, 174, 174, 182, 204, 210, 229.

The Shapiro-Wilk (or normal scores) goodness-of-fit test for normality will not reject normality at level .1, so we may decide to proceed in the traditional fashion to carry out a t-test. The sample mean is 154 and the standard deviation is 29.04.

(a) Show that the P-value is .043.

(b) Show that the P-value of the z-test of H versus K which assumes $\sigma = 29$ is .037. In either case the hypothesis H is rejected in favor of K at level .05.

(c) Show that the Wilcoxon signed rank test (see Chapter 5), which only assumes symmetry and continuity of the underlying distribution, yields the P-value .047.

(d) Now assume that the largest observation includes a recording error ε which varies from -20 to $+80$. Plot the effect of such an error on the basic statistics $\overline{x}$ and s, and on the P-value of the various tests. Note that the z-test, which assumes knowledge of the variance, is much more sensitive to the outlier than is the t-test. Thus the apparent robustness of the P-value of the t-test for large sample sizes is revealed to be due to a fortunate cancellation of two highly unreliable statistics, the sample mean and the sample standard deviation.

4. In this problem we find the "asymptotic relative efficiency" of the mean to the median, where efficiency is measured in terms of the large sample variance of the estimator. Let $X_1, \ldots, X_n$ be i.i.d. with common symmetric density f. Under certain regularity conditions the asymptotic relative efficiency of the mean $\overline{X}_n$ to the median $X_{(n/2)}$ can be shown to be the limit of the ratio

$$\frac{\mathrm{Var}[X_{(n/2)}]}{\mathrm{Var}[\overline{X}_n]}$$

as $n \to \infty$. For explicit definitions and motivation, see Chapter 3. Note that this expression does not depend on the common variance of the observations, or the location of the center of symmetry. It may be shown that

$$\mathrm{Var}[X_{(n/2)}] \sim \frac{1}{4nf^2(F^{-1}(\frac{1}{2}))}.$$

See for example Theorem B.9 of the Appendix.

(a) Show that if X_1 has the double exponential distribution $[f(y) = \exp(-|y|)/2]$, then X_1 has variance 2 and the asymptotic relative efficiency of of the mean to the median is .5.

(b) Find the asymptotic relative efficiency of the mean to the median for normal data.

5. Let $X \sim F$, where F is an unknown distribution on the real line. It is desired to estimate $\theta = P\{X \leq x_0\}$ for some fixed point x_0.

(a) Express θ as a function $T(F)$, where F represents the cumulative distribution function of X.

(b) Describe the induced estimator $T(F_n)$ of $T(F)$ in words and also express it in terms of the indicator random variables $I_1, \ldots, I_n$, where $I_j = I\{X_j \leq x_0\}$.

(c) Find the mean and the variance of the induced estimator in terms of F, x_0, and n.

6. Define the variance functional $T(F) = \mathrm{E}_F[X^2] - \mathrm{E}_F^2[X]$, for all F for which second moments exist.

(a) Find the induced estimator $T(F_n)$ in terms of the n observations which give rise to the empirical distribution F_n. [Solution: $T(F_n) = (n-1)s_n^2/n$, where s_n^2 is the usual sample variance.]

(b) Express $\mathrm{E}[T(F_n)]$ in terms of $T(F)$.

7. Let $F^{-1}(u) = \inf\{y : F(y) \geq u\}$, $0 < u < 1$, define the *inverse* function of F. This inverse is the usual function inverse when F is a one-to-one mapping. It is also well defined when F is a *cumulative distribution function*, i.e., a right-continuous, nondecreasing function from the real line to the unit interval with $\lim_{x \to -\infty} F(x) = 0$ and $\lim_{x \to +\infty} F(x) = 1$.

(a) Sketch the inverse cumulative distribution function for the binomial $(3, \frac{1}{2})$ distribution, and the inverse of the normal approximation to it.

(b) Define the median functional by $T(F) = F^{-1}(\frac{1}{2})$. What is $T(F)$ when F is the distribution of part (a)? What is $T(F_n)$ in terms of $X_1, \ldots, X_n$?

(c) How could one define $T(F)$ for all F in part (b) so that $T(F_n)$ yielded the average of the two central order statistics when n was even?

1.6 COMPLEMENTS

1.6.1 Other Approaches to Robust Statistics

There are a variety of other approaches to overcoming the inadequacies of parametric models, other than those discussed in this text, but we will only be able to give the reader a few references to follow up. Whether these approaches lead to procedures which are regarded as robust or nonparametric depends on how broadly one interprets these terms.

First there are *adaptive* procedures, which adjust to the unknown underlying distribution; see Hogg (1974, 1979) and Welsh (1984, 1986).

Second, the robustness of *Bayesian* procedures is developed in the two collections *Scientific Inference, Data Analysis, and Robustness*, edited by Box, Leonard, and Wu (1983), and the *Robustness of Bayesian Analysis*, edited by Kadane (1984), in particular Part II by J. O. Berger.

Third, there are asymptotic *minimax* estimators (which minimize the maximum asymptotic variance as the true distribution varies over a contamination neighborhood of the model); and finite sample minimax interval estimators (which minimize the maximum probability of overshooting or undershooting the parameter by a fixed amount). See Chapters 4 and 10 of Huber (1981).

Fourth, there are estimators based on *minimum distance* methods which choose a member of a parametric family which comes closest in the chosen metric to an estimate of the distribution. See Beran (1977a,b), Heathcote (1979), Parr and Schucany (1980), Heathcote and Silvapulle (1981); Stather (1981); and Donoho and Liu (1988).

Fifth, there are methods to minimize the asymptotic variance or mean-squared error of an estimator as the underlying distribution varies over a *shrinking neighborhood* of a parametric model. See Huber-Carol (1970), Bickel (1978), Rieder (1978, 1980).

1.6.2 Significance of an Experimental Result

Controversy continues over the use and interpretation of significance testing, whether in the scientific domain, or more recently in the legal process. For a discussion of the issues and further references, see Leonard (1983), Edwards, Lindman, and Savage (1963), and Hall and Selinger (1986).

CHAPTER TWO

Estimating Scale—Finite Sample Results

In this chapter we will review some traditional ideas on scale parameter estimation, and in particular consider equivariant and unbiased estimators of the exponential scale parameter. Our goals are to illustrate the robustness or lack thereof of some of the classical estimators. We will introduce mixture models for contamination and the finite sample breakdown point to assist us in our analysis, and finally consider some different methods of estimating the standard error of our estimators, including the bootstrap. The difference between the location parameter and the scale parameter estimation problems is that in the former a *shift* of the distribution is possible, while in the latter a *fractional or multiplicative* change is possible. Mathematically, location is modeled by adding the same unknown constant to each observation from a fixed distribution, while scale is modeled by multiplying each observation by an unknown positive constant. The reader should be wary that the sample mean, median, and some other estimators commonly used as *location* parameter estimators may also be reasonable *scale* parameter estimators in the scale context, at least when multiplied by suitable constants. The location problem is treated in Chapter 4.

2.1 EXAMPLES

To begin we give examples where a scale parameter is of interest. In the first example an exponential model may be appropriate, and many characteristics of this population can be estimated, starting with an estimate of the scale parameter. In the second example the shape of the distribution is unclear, but nevertheless an estimate of the scale (in the sense of magnitude) of the population is desired.

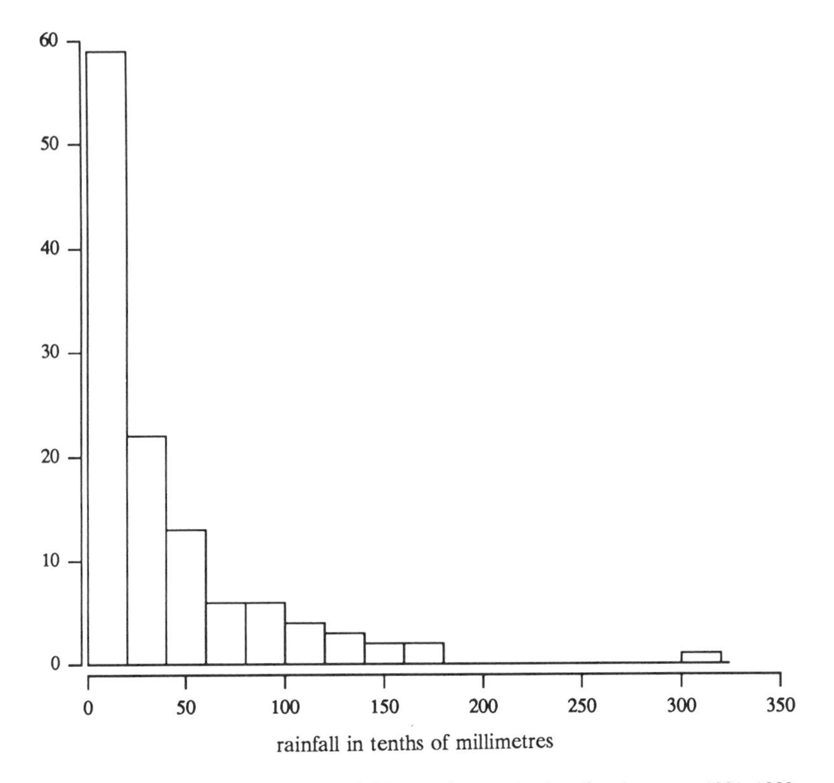

FIGURE 2.1 Melbourne's daily rainfall on winter rain days for the years 1981–1983.

Example 1: Melbourne's Daily Rainfall. Rainfall varies with the seasons in Melbourne, Australia, so for the sake of time homogeneity we restrict attention to the winter months of June, July, and August. During this rainy season roughly half the days have no measurable rainfall, and we will hereafter restrict attention to "rain days," those in which there is at least one millimeter of measured rainfall. The distribution of daily rainfall for the winter months of 1981–1983 is shown in Figure 2.1 in the form of a histogram. It can be approximated by the exponential density $f_\theta(x) = (1/\theta) \cdot e^{-x/\theta}$, $x > 0$, for some parameter value $\theta > 0$, where θ is the mean of this distribution. Since there is some day-to-day dependence, a Markov model is more appropriate if one wants to use all the information. However, as a first approximation we will select every fourth rain day observation from Table C.2 of the Appendix and assume independence. The results in millimeters

Table 2.1 Comparison of MLEs with Descriptive Estimates for the Melbourne Rainfall Data

Case	Characteristic of Interest	Maximum Likelihood Estimate	Descriptive Estimate
(1)	θ	4.50	4.50
(2)	$(\ln[2])\theta$	3.12	2.80
(3)	$(1-\ln[.9])\theta$	4.97	4.96
(4)	$.9\theta$	4.05	3.56
(5)	$(1-\ln[.1])\theta$	14.85	19.47

are:

1981: 6.4, 4.0, 3.2, 3.2, 8.2, 11.8, 6.0, .2, 4.2, 2.8, .6, 2.0, 16.4
1982: .4, 8.4, 1.0, 7.4, .2, 4.6, .2
1983: .2, .2, .8, .2, 9.8, 1.2, 1.0, .2, 30.2, 1.4, 3.0

There are several characteristics of the population which are of interest:

1. The mean or average rainfall θ
2. The median or "typical" rainfall
3. The mean absorbed rainfall after truncation of the lowest decile (which assumes that the lowest decile is lost to evaporation)
4. The mean absorbed rainfall after "Winsorizing" the highest decile (which assumes that rainfall above the 90th percentile is lost to run-off and hence only amounts up to the 90th percentile are absorbed, even when there is a larger rainfall)
5. The mean amount in the highest decile (which is of interest in flood control).

Each of the characteristics of interest can be expressed in terms of the scale parameter θ, and thence estimated with θ. For example, the maximum likelihood estimate (MLE) of θ is $\overline{X} = 4.50$. The median is $F_{\theta}^{-1}(\frac{1}{2}) = [\ln 2]\theta$, and the MLE of the median is therefore $[\ln 2]\overline{X} = 3.12$. We leave further details to be worked out in the problems, but summarize the results for all characteristics in Table 2.1.

In addition to the MLEs we have listed the descriptive estimates for comparison. *Descriptive estimates* are obtained by taking the sample version of the characteristic of interest: for example, the sample mean to estimate the population mean, the sample median to estimate the population median, etc. Thus rather than estimate the median $F_{\theta}^{-1}(\frac{1}{2})$ by the MLE $[\ln 2]\overline{X} = 3.12$, we could estimate it by the descriptive estimate, which is the sample median $X_{(16)} = 2.80$. To obtain the 10% Winsorized estimate, replace the three largest observations by the value of the fourth largest observation and take the mean of the 31 observations in the resulting Winsorized sample.

Note that the apparent agreement in case (3) is a coincidence, and that the descriptive estimates may be considerably smaller or larger than the MLEs for the exponential model. Hence it matters greatly whether we approach this problem through the parametric exponential model, or whether we use the nonparametic descriptive approach. The descriptive approach is developed in Chapter 3. □

Example 2: Lifetimes of Mammalian Cells. Mammalian cells in culture have lifetimes (cell cycle times) typically varying from 8 to 24 hours, so it is difficult to obtain observations on them in a fixed environment. In the famous "transition probability" model, a cell in the G1 phase awaits a specific (and still undiscovered) event which has a fixed probability per unit time of occurring. After the event the cell completes the processes leading to division in a nearly constant time γ. Thus lifetimes are assumed to follow a displaced exponential distribution with minimum γ and unknown scale parameter θ (see Problem 12). In this model it is also assumed that sister cell lifetimes are independent, so that the scale parameter θ can be estimated from absolute differences of sister cell lifetimes [Problems 12(b)]. Despite this appealing methodology, cell lifetime data rarely resemble the displaced exponential distribution; rather the displaced gamma, inverse normal, and log-normal all seem to be much more appropriate.

In Figure 2.2 a typical pedigree of EMT6 cell lifetimes is displayed. This data set illustrates the difficulties facing a statistician who analyzes cell lineage data: there are outliers (quiescent cells which enter a resting state before re-entering the cell cycle), missing data (due to cell death or cells wandering from the field of vision), and strong dependence between daughter cells. Thus model building is a complex process in this context; more details are available in Staudte, Guiguet, and Collyn D'Hooghe (1984) and Cowan and Staudte (1987). Despite these difficulties, a measure of scale is of great interest to the biologists, who want to compare the effects of nutrient depletion, radiation, etc., on cell lifetimes. The reader may try the role of exploratory data analyst and examine the data without further scientific details; the goal is to propose a summary statistic for the cell lifetimes and to find an accompanying standard error of the estimate. Such an analysis will lead to a series of questions for the scientist conducting the experiment, regarding time stationarity, independence, reasons for missing data, and outliers, among possibly many others. □

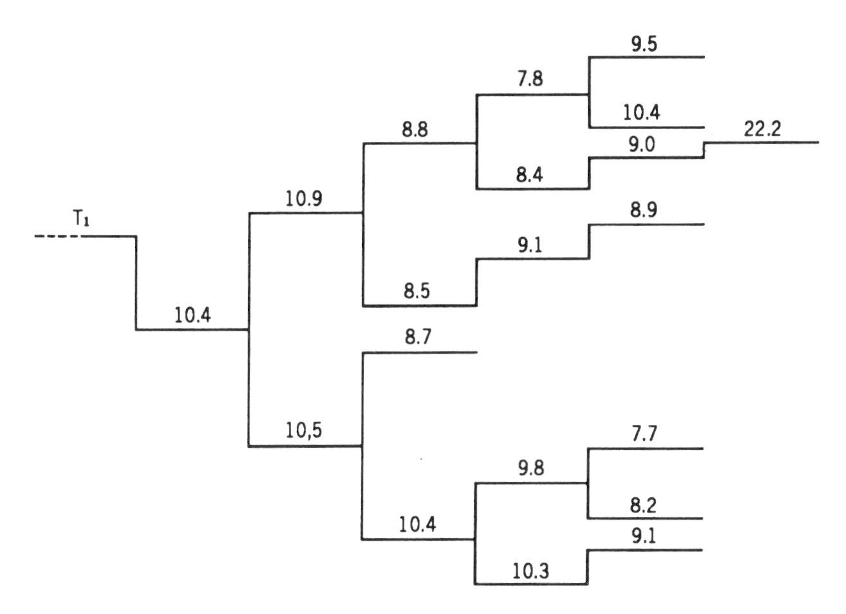

FIGURE 2.2 Pedigree of EMT6 cell lifetimes in hours.

2.2 SCALE PARAMETER FAMILIES

2.2.1 Definitions and Properties

Given any distribution F which is concentrated on the positive axis and $\theta > 0$ define

$$F_\theta(x) = F\left(\frac{x}{\theta}\right), \qquad x > 0.$$

Then $\{F_\theta;\ \theta > 0\}$ is called the scale parameter family generated by F, and θ is the *scale parameter*. The reader may verify that:

1. If $X \sim F_1 = F$, then $\theta X \sim F_\theta$, so that larger values of θ tend to render larger observations. Thus a scale parameter gives an indication of the *magnitude*, relative to the point 0, of the observations with the distribution F_θ.
2. If $X \sim F_\theta$, where $\{F_\theta,\ \theta > 0\}$ is a scale parameter family, then clearly X/θ has distribution which is free of θ. Conversely, given a positive random variable X with distribution indexed by a positive parameter θ and for which the distribution of X/θ is free of θ, it follows that the distributions of X form a scale parameter family and that θ is a scale parameter.

3. If F has density f, then F_θ has density

$$f_\theta(x) = \frac{1}{\theta} f\left(\frac{x}{\theta}\right), \qquad x > 0.$$

4. If $X \sim F_\theta$, we write $\mathrm{E}_\theta[g(X)]$ for $\int_0^\infty g(x)\,dF_\theta(x)$. To find the moments of F_θ it suffices to find them for F_1, since $\mathrm{E}_\theta[X^r] = \theta^r \mathrm{E}_1[X^r]$. In particular, the coefficient of variation of $X \sim F_\theta$ is free of θ:

$$\mathrm{CV}_\theta[X] = \frac{\sqrt{\mathrm{Var}_\theta[X]}}{\mathrm{E}_\theta[X]} = \frac{\sqrt{\mathrm{Var}_1[X]}}{\mathrm{E}_1[X]}.$$

2.2.2 Examples of Continuous Scale Parameter Families

Name	Mnemonic	Generating f	CV
Uniform	$U(0,\theta)$	$I_{(0,1)}(x)$	.577
Exponential	$\mathrm{Exp}(\theta)$	e^{-x}	1.0
Pareto, index r	$\mathrm{Par}(r,\theta)$	$\dfrac{r}{(1+x)^{r+1}}$	$\left(\dfrac{r}{r-2}\right)^{1/2}$, $r > 2$
Gamma, shape α	$G(\alpha,\theta)$	$\dfrac{x^{\alpha-1}e^{-x}}{\Gamma(\alpha)}$	$\alpha^{-1/2}$

2.3 FINITE SAMPLE PROPERTIES OF ESTIMATORS

In this section we will quickly review some standard criteria used to assess estimators. Our context is that of estimating the scale parameter of an exponential distribution. We will also introduce mixture models for contamination, and define the finite sample breakdown point.

2.3.1 Unbiasedness, Scale Equivariance, and Mean Squared Error

In the scale parameter problems we often assume that the observations $X_1, X_2, \ldots, X_n$ are i.i.d., each according to the distribution $F_\theta(x) = F(x/\theta)$, $x > 0$, where F is known and $\theta > 0$ is unknown. The problem is to choose an estimator $T_n = T_n(X_1, \ldots, X_n)$ which is close to θ in some sense, such as minimising the *mean squared error*:

$$\mathrm{MSE}_\theta[T_n] = \mathrm{E}_\theta[(T_n - \theta)^2].$$

This cannot be achieved for all θ by any one T_n, so the problem is often further simplified by restricting attention to estimators that are *unbiased* in that $E_\theta[T_n] = \theta$ for all θ. For such unbiased estimators the bias term $E_\theta T_n - \theta$ drops out of the following expression for the MSE, and the problem reduces to minimizing the variance by choice of unbiased estimator, since

$$\text{MSE}_\theta[T_n] = \text{Var}_\theta[T_n] + (E_\theta[T_n] - \theta)^2. \tag{2.3.1}$$

Another common way in which we simplify the study of scale estimators is by requiring that the estimator yield an equivalent result, regardless of the unit of measurement. That is, if each observation is multiplied by a constant c, the effect is to multiply T_n by c. Symbolically, if each $X_i \to cX_i$, for some constant c, then $T_n \to cT_n$. This property of *scale equivariance* implies that $E_\theta[T_n] = \theta E_1[T_n]$ and $\text{Var}_\theta[T_n] = \theta^2 \text{Var}_1[T_n]$. Hence any such equivariant estimator may be multiplied by a suitable constant, namely $1/E_1[T_n]$, to make it an unbiased estimator for θ.

Moreover, since scale equivariance is such an intuitively appealing property, nearly all proposed scale estimators have this property. For each of them $\text{MSE}_\theta[T] = \theta^2 \text{MSE}_1[T]$ and thus only the case $\theta = 1$ need be considered in comparing such estimators.

One could argue that scale estimators should be compared on the basis of *mean squared relative error* $E_\theta(T - \theta)^2/\theta^2$ rather than mean squared error since more error is tolerable when the parameter itself is larger. However, when we restrict attention to equivariant estimators then the relative mean squared error of T is precisely $\text{MSE}_1[T]$. Thus for equivariant estimators, comparisons made with either criterion will lead to the same conclusions.

2.3.2 Estimators of an Exponential Scale Parameter

In this section we will look closely at five estimators of an exponential scale parameter. These estimators have been chosen mainly because they illustrate concepts which we will need throughout the text, and not because they are recommended for the exponential scale problem. As we shall see in due course, the trimmed mean is the most favored of these five estimators due to robustness and efficiency considerations.

Assume that $X_1, \ldots, X_n$ are i.i.d. with common exponential density

$$f_\theta(x) = \theta^{-1} e^{-x/\theta}, \qquad x > 0$$

where $\theta > 0$ is the scale parameter. Five estimators of $\theta = E_\theta[X]$ are as follows:

1. The Sample Mean

$$\overline{X}_n = \frac{1}{n}\sum_{i=1}^{n} X_i$$

which is well known to be the maximum likelihood estimator as well as the minimum variance unbiased estimator (MVUE) of θ.

2. The Sample Median

$cX_{(m+1)}$ when $n = 2m + 1$ is odd, and $c(X_{(m)} + X_{(m+1)})$ when $n = 2m$ is even. Here and hereafter the notation $X_{(m)}$ is used to denote the mth smallest observation, or mth *order statistic*. The constant c is usually chosen to make the estimator unbiased for some parameter.

The median is not the best quantile estimator of scale for the exponential scale parameter, as we shall see. Define *the qth quantile estimator* $c_m X_{(m)}$, where $X_{(m)}$ is the mth order statistic, $m = [qn]$, and c_m is chosen to make the estimator unbiased. Siddiqui (1963) showed that the MVUE of θ among the quantile estimators is obtained by taking m to be the nearest integer to $.79681(n + 1) - .39841 + 1.16312(n + 1)^{-1}$, or $m \approx .8n$.

3. The β-Trimmed Mean

$$\overline{X}_{n,\beta} = \frac{1}{r}\sum_{m=1}^{r} X_{(m)}$$

where $r = n - [n\beta]$ is the number of observations remaining after the $[n\beta]$ largest have been "censored." Such censoring is common in life-testing experiments where we might not want to wait for the longest lived $[n\beta]$ components to fail.

4. The β-Winsorized Mean

This estimator was shown by Epstein and Sobel (1953) to be the MVUE of θ based on the first $r = n - [n\beta]$ order statistics $X_{(1)}, \ldots, X_{(r)}$:

$$W_{n,\beta} = \frac{1}{n}\left[\sum_{m=1}^{r} X_{(m)} + (n - r)X_{(r)}\right].$$

While the β-trimmed mean censors the largest $[\beta n]$ observations, the β-Winsorized version replaces each of them by the value of the largest uncensored one. Note that

$$W_{n,\beta} = \left(\frac{r}{n}\right)\overline{X}_{n,\beta} + \left(1 - \frac{r}{n}\right)X_{(r)}.$$

5. *Outlier Rejection Followed by Mean*

A rejection rule is applied to one or more of the largest observations, which may then be removed from the sample; the average of the possibly modified sample is denoted by $\overline{X}_{OR}$. Here we choose a rejection rule suggested by Lîkes (1966); namely, to reject the largest observation if its distance from the rest of the sample is large compared to its own magnitude. That is, reject $X_{(n)}$ if

$$\frac{X_{(n)} - X_{(n-1)}}{X_{(n)}} \geq c$$

for some predetermined c. This kind of estimator (rejection rule followed by mean) is probably the most commonly employed and the least understood, because of its complicated nature.

We note in passing that our list does not contain the Pitman estimator $n\overline{X}_n/(n+1)$, which minimizes the MSE among all equivariant scale estimators for the exponential scale parameter. [This follows from material in Ferguson (1967), Section 4.7, in particular Problem 4.] Nor are some of the estimators very robust to contamination, as we shall see.

2.3.3 Mixture Models for Contamination

Mixture models arise naturally in many everyday situations. For example, winter rainfall in Melbourne (see Example 1 of Section 2.1) may reasonably be modeled by a random variable which is zero with probability one-half (since it rains about half of the days), and exponentially distributed with probability one-half. The cumulative distribution function for this model may be written

$$G(x) = \tfrac{1}{2}\Delta_0(x) + \tfrac{1}{2}(1 - e^{-x/\theta})$$

where θ is the mean rainfall on rain days, and Δ_0 is the point mass distribution at 0. More generally, $\Delta_y(x)$ is defined to be 1 if $x \geq y$ and 0 otherwise. A *mixture model* is one that is a weighted average of other probability models: $G = (1-p)F + pH$, with $0 < p < 1$. It is convenient to think of $X \sim G$ as arising from a two-stage process. First, either F or G is selected, with weights $1-p$ and p, respectively. Then an observation is made on the selected distribution.

Following Tukey (1960) we introduce mixture models to represent chance contamination. Each observation X is with high probability $1-\varepsilon$ from the presumed model F but with small probability from the contaminating distribution H. Thus we write $X \sim (1-\varepsilon)F + \varepsilon H$. For large n roughly proportion ε of the observations $X_1, \ldots, X_n$ will be contaminants.

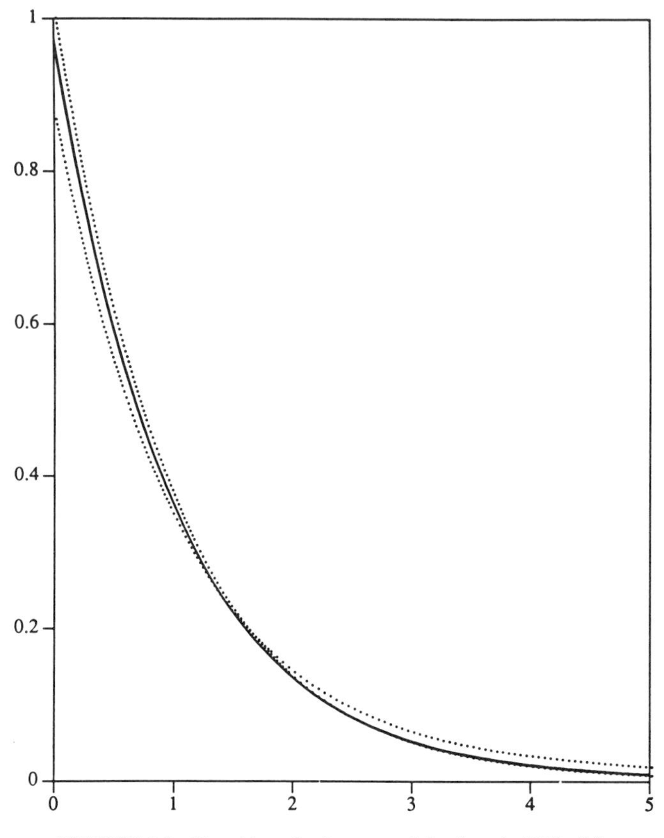

FIGURE 2.3 Densities of mixture models given in Table 2.2.

Frequently we will simply take H to be the point mass distribution at y, and write $H = \Delta_y$. Then for large y, $(1-\varepsilon)F + \varepsilon\Delta_y$ is a simple model for a proportion ε of outliers at y. When $F_\theta(x) = 1 - e^{-x/\theta}$, another common choice for H is $H(x) = 1 - e^{-x/c\theta}$, where $c > 1$. Then contamination is due to a small proportion of observations coming from the same family, but with a larger scale. If c is not too large, such contamination is difficult to detect . We may express this contamination model

$$(1-\varepsilon)F_1\left(\frac{x}{\theta}\right) + \varepsilon F_1\left(\frac{x}{c\theta}\right).$$

We will compare the MSE of the above listed estimators for samples of size 20 from the exponential distribution and three similar mixture models;

the densities of these models are illustrated in Figure 2.3. Note that there is hardly any visual difference. One might ask whether goodness-of-fit tests such as the chi-squared, Kolmogorov, and Anderson-Darling are able to distinguish such mixtures from the exponential model, but it is shown in Staudte (1980) that they have power less than .15 at level .05 of doing so. Thus we would be unlikely to detect such mixture models with a goodness-of-fit test. More important, if such a test rejects the exponential model, the statistician must face the question, What am I trying to estimate? We will dispense with such tests and address the important question in Chapter 3.

2.3.4 Simulation Results

For equivariant estimators the MSE satisfies

$$\text{MSE}_\theta[T_n] = \theta^2 \text{MSE}_1[T_n].$$

Thus when making comparisons between such estimators we assume without loss of generality that $\theta = 1$. For some mixture models and estimators it is easy to derive the MSE (see the problems), but in other cases we rely on simulation studies. 4000 samples of size 20 were generated from each of the above mixture models and the bias and variance were estimated by averaging the sample bias and sample variance. The results are summarized in Table 2.2.

Remarks

1. The multiplicative constants are chosen to make the estimators unbiased for the exponential scale parameter.
2. For the exponential model the sample mean is only slightly more efficient than the trimmed mean and outlier rejection plus mean; however, its MSE is two-thirds that of the quantile estimator.
3. For the three contamination models, the sample mean is slightly worse than the quantile estimator, but much worse on average than the other estimators.
4. Even in model 3, where the proportion of outliers is double the amount of trimming, the trimmed mean performs well relative to the other estimators.
5. For $n = 20$, $(\text{bias})^2$ is typically a much smaller contributor to MSE than the variance, although as $n \to \infty$ the variance will become negligible relative to $(\text{bias})^2$. Thus we cannot claim that any estimator is superior for all n on the basis of such limited results.

Table 2.2 $20 \times (\text{bias})^2$, $20 \times$ **Var, and** $20 \times$ **MSE for Estimators of the Exponential Scale Parameter**

Model	$\overline{X}_{20}$	$c_{16}X_{(16)}$	$k\overline{X}_{.1}$	$\overline{X}_{OR}$
	.000	0.001	0.000	0.002
1 $E(x) = 1 - e^{-x}$	.971	1.430	1.116	1.013
	.971	1.431	1.116	1.015
	0.201	0.070	0.069	0.077
2 $.95E(x) + .05E(x/3)$	1.544	1.831	1.362	1.476
	1.745	1.901	1.431	1.553
	3.228	1.805	1.487	2.204
3 $.8E(x) + .2E(x/3)$	3.167	3.640	2.423	3.201
	6.395	5.445	3.910	5.405
4 $.95E(x) + .05H(x/6)$	0.191	0.036	0.039	0.034
where $[H(x) = 1 - (x+1)^{-3}]$	2.070	1.720	1.299	1.374
	2.261	1.756	1.338	1.408

6. The β-Winsorized mean has been omitted in this table since its performance has been found to differ little from that of the β-trimmed mean.

2.3.5 Finite Sample Breakdown Point

So far we have only been concerned with the efficiency of estimators, but now let us examine the effect of outliers on their behavior. By an *outlier* in this scale parameter context we mean an observation which is so much larger than the bulk of the observations that it stands out, and that there is doubt about it being from the proposed model. This vague definition can be replaced, for example, by the rejection rule given in Example (5) of Section 2.3.2; or, better, by the rule, which designates an observation as outlying if it is more than three times the interquartile range from the median.

One method of assessing the effect of gross contamination by outliers on the estimators would be to include a mixture model in the simulation study of the last section which gives rise to many outliers (such as model 4 with $H(x) = 1 - (1 + x)^{-1}$.) It is more instructive, however, to calculate the *finite sample breakdown point* ε_n^*, first introduced by Hodges (1967), and generalized by Hampel (1968). It is the smallest proportion of the n observations which can render the estimator meaningless. For example, the sample mean has $\varepsilon_n^* = 1/n$, since only one sufficiently large outlier can carry it beyond any given value. More than $[\beta n]$ of the observations would need to be very large to similarly affect the β-trimmed mean, so it has breakdown point $\varepsilon_n^* = ([\beta n] + 1)/n$.

Often the limiting value $\lim_{n\to\infty} \varepsilon_n^*$ is quite adequate for summarizing what is a crude but useful measure of the global stability of the estimator. The limiting values for the five estimators of Section 2.3.2 are, respectively, 0, .5, β, β, and 0. In more complex estimation problems the process of calculating the breakdown point may be nontrivial, but it often leads to insight into the behavior of the estimator for "least favorable" data sets.

2.4 STANDARD ERRORS, THE BOOTSTRAP

2.4.1 Traditional Estimates of Standard Error

An estimator $T_n = T_n(X_1, \ldots, X_n)$ of a parameter θ may have been chosen on the basis of its efficiency, computational, and robustness properties, but its realization for a particular set of data is of little value unless there is an accompanying statistic which indicates its accuracy. The standard deviation of the estimator, $\sqrt{\mathrm{Var}_\theta[T_n]}$, is a measure of the spread or dispersion of the distribution of the estimator, so it indicates the long-run accuracy of the estimator. The standard deviation of the estimator will here be called the *standard error* of the estimator, and it will be denoted by $\mathrm{SE}[T_n]$.† The standard error itself often depends on the unknown parameter, and in many cases the form of dependence is unknown. Thus the standard error of the estimator must itself be estimated; the result will be denoted by $\widehat{\mathrm{SE}}[T_n]$.

Let us look at two of the traditional ways an estimate of the standard error may be found in the context of estimating an exponential scale parameter θ by the sample mean. In this case the standard error $\sigma_{\overline{X}_n} = \theta/\sqrt{n}$.

1. Maximum Likelihood Estimate

The MLE $\hat{\theta}$ of θ is known to be the sample mean, and since the MLE of any function of θ is given by the same function of the MLE, the MLE of the standard error is

$$\widehat{\mathrm{SE}}_{\mathrm{MLE}}[\overline{X}_n] = \frac{\overline{X}_n}{\sqrt{n}}.$$

The next technique may be employed in a wider context than parametric estimation: the estimator T_n need only estimate some population characteristic $T(F)$, where F possibly belongs to a nonparametric class; and the standard deviation of T_n must be a known function of F.

†The term *standard error* is frequently used in a vague sense to denote *any* estimate of the standard deviation. It is also used to denote an estimate of the standard deviation of an approximating normal distribution; see Section 3.5.

2. Descriptive Estimate

When $T_n = \overline{X}_n$, the standard deviation is $\sqrt{\mathrm{Var}_F[X]/n}$. The descriptive estimate of population variance $\mathrm{Var}_F[X] = \mathrm{E}_F[X - \mathrm{E}_F[X]]^2$ is the variance of the sample distribution F_n, namely, $\mathrm{Var}_{F_n}[X] = \{\sum_{i=1}^{n}(X_i - \overline{X}_n)^2\}/n$. See Problem 6 of Chapter 1. Hence the descriptive estimate of the standard error is

$$\widehat{\mathrm{SE}}_{\mathrm{desc}}[\overline{X}_n] = \frac{\{\sum_{i=1}^{n}(X_i - \overline{X}_n)^2\}^{1/2}}{n}.$$

2.4.2 Bootstrap Estimates of Standard Error

The word *bootstrap* is used to indicate that the observed data are used not only to obtain an estimate of the parameter but also to generate new samples from which many more estimates may be obtained, and hence an idea of the variability of the estimate.

1. Parametric Bootstrap Estimate

If the formula for the standard deviation of $\hat{\theta}$ in terms of θ could not be derived analytically, then an estimate could be obtained by means of the following algorithm with the aid of a device that can generate $\mathrm{Exp}(\theta)$ "observations":

(i) For each $b = 1,\ldots,B$ generate n independent observations from the $\mathrm{Exp}(\hat{\theta})$ distribution, where $\hat{\theta}$ is the estimate based on the original data, and form a bootstrap estimate $\hat{\theta}_b$.

(ii) Find the mean and the variance of the bootstrap estimates:

$$\overline{\theta}_B = \frac{1}{B}\sum_{b=1}^{B}\hat{\theta}_b, \qquad s_B^2 = \frac{1}{B-1}\sum_{b=1}^{B}(\hat{\theta}_b - \overline{\theta}_B)^2.$$

Then the parametric bootstrap estimate of the standard error of $\hat{\theta} = \overline{X}_n$ is

$$\widehat{\mathrm{SE}}_{\mathrm{boot}}[\overline{X}_n] = s_B.$$

The parametric bootstrap estimate is simply s_B, the standard deviation of the B estimates obtained by sampling from the distribution suggested by the exponential model and the original estimate of its parameter. Note that the bootstrap estimate of the standard deviation depends on the number of replications B, as well as the random number generator.

2. Nonparametric Bootstrap Estimate

This is obtained in exactly the same way as the parametric bootstrap in 1, except that sampling in step (a) is from the empirical distribution F_n

Table 2.3 "Convergence" of Parametric Bootstrap Estimates

B	$\bar{\theta}_B$	s_B	$\hat{\theta}_{(1)}$	$\hat{\theta}_{(B/4)}$	$\hat{\theta}_{(B/2)}$	$\hat{\theta}_{(3B/4)}$	$\hat{\theta}_{(B)}$
10	.71	.21	.41	.47	.77	.86	1.03
25	.68	.22	.25	.49	.68	.83	1.14
100	.67	.22	.27	.52	.64	.78	1.29

which places mass $1/n$ on each of the original observations. Thus no knowledge at all of the underlying F is required. This method is most frequently employed and often simply referred to as "the bootstrap."

Bootstrap techniques have been analyzed and shown to be useful in a wide variety of contexts by Efron (1982). In the next chapter we will look at other methods for finding standard errors which are suggested by asymptotic results; and we will examine the bootstrap in more depth.

2.4.3 An Illustration of Bootstrap Calculations

To illustrate the computation of estimates of standard error, we refer the reader to the data of Figure 2.2. *If* these data were realizations of independent, exponentially distributed lifetimes, then the absolute differences of sister lifetimes would be i.i.d. Exp(θ); see Problem 12 for details. The seven observations are, after ordering, .3, .4, .5, .5, .6, .9, 1.7. The sample mean and the standard deviation are .7 and .48, respectively. Hence the traditional MLE estimate of the standard error of $\hat{\theta}$ is $.7/\sqrt{7}$. The descriptive estimate of standard error is $\sqrt{6} \times .48/7 \approx .168$; see Section 2.4.1.

For the sake of comparison, we now calculate the bootstrap estimates of the standard error obtained by the resampling methods found in Section 2.4.2. (All that is required is a half-dozen Minitab instructions; these are given in Appendix D.1.1.) Table 2.3 shows the results of sampling repeatedly from the Exp(.7) distribution to obtain new samples of size 7 from which the sample mean can be calculated. It is mainly the third column which is of interest here, for it gives the standard deviation of our B bootstrap estimates. The table shows that the estimates change little with B. As $B \to \infty$, the entries will approach the corresponding moments and quantiles of a gamma distribution with parameters (7, .7), which has been rescaled by dividing by 7. Why? The variance of this distribution is $7 \times .7^2/7^2$, so the standard deviation is .2645. This is the limit of the entries in column 3. Of course if the initial estimate of θ is badly off, so will be the parametric bootstrap estimate of the standard error.

Table 2.4 "Convergence" of Nonparametric Bootstrap Estimates

B	$\overline{\theta}_B$	s_B	$\hat{\theta}_{(1)}$	$\hat{\theta}_{(B/4)}$	$\hat{\theta}_{(B/2)}$	$\hat{\theta}_{(3B/4)}$	$\hat{\theta}_{(B)}$
10	.71	.197	.49	.53	.69	.84	1.11
25	.68	.165	.44	.53	.64	.81	.98
100	.69	.177	.40	.56	.66	.81	1.17
∞	.7	.168	.30	—	—	—	1.7

Now consider the nonparametric bootstrap method of Section 2.4.2. Repeated samples of size 7 are selected from the empirical distribution defined by the original data. The sample mean is calculated for each of the B samples, and the standard deviation of these B estimates is the estimated standard error. (This is carried out effortlessly by the few Minitab instructions in Appendix D.1.2.) The results for different choices of B are given in Table 2.4. Notice that the limiting distribution is discrete in this case, and that again the limiting values are approached rapidly.

2.4.4 Evaluating the Standard Error Estimates

The reader may object to the use of standard error for a measure of accuracy of an estimate of scale, since the standard error of T_n may always be reduced by multiplication of T_n by a constant less than one. What is important is that the *relative* standard error be small. For example, in measuring cell lifetimes in hours, an estimate of 20 hours with standard error of 2 hours is considered reliable, while an estimate for a different cell line of 3 hours with standard error of 2 hours is almost useless. We would like the standard error to be small, relative to the size of the estimate itself. The *relative standard error* of T_n is the coefficient of variation of T_n:

$$\mathrm{CV}[T_n] = \frac{\sqrt{\mathrm{Var}[T_n]}}{\mathrm{E}[T_n]}.$$

In practice the numerator of this expression is estimated by one of the four methods described earlier, and the denominator by T_n itself.

Returning to our example of $\overline{X}_n$ in the Exp(θ) problem, we find that $\mathrm{CV}_\theta[\overline{X}_n] = 1/\sqrt{n}$. This quantity is also the minimum CV achievable by any scale equivariant estimator T_n. To see this, note that for such estimators $\mathrm{CV}_\theta[T_n] = \mathrm{CV}_\theta[cT_n]$ for any constant c, including that which makes cT_n unbiased for θ. For this choice of c, $\mathrm{CV}_\theta[T_n] = \mathrm{Var}_1[cT_n]$, and the problem of minimizing the variance by choice of unbiased estimator has the well-known solution $T_n = \overline{X}_n$.

How do we evaluate the estimates of standard error given in Sections 2.4.1 and 2.4.2? Intuitively speaking, the more that is assumed about F

and that can be exploited, the more accurate the estimate of standard error should be. For those rare parametric situations where several methods apply, such as the exponential scale problem, we have a choice; in other problems the bootstrap may be the only method available. We are trying to estimate a positive quantity σ_{T_n}, the standard error of an estimator, by some estimator $\hat{\sigma}_{T_n}$, using the realized observations $X_1, \ldots, X_n$. When these observations are themselves considered as random variables, the estimator $\hat{\sigma}_{T_n}$ has a variability which we want to be small relative to what it is estimating. Thus the coefficient of variation of the standard error estimator itself $\mathrm{CV}[\hat{\sigma}_{T_n}]$ is a guide to its effectiveness. For method 1 of Section 2.4.1, the CV of the estimator is clearly $1/\sqrt{n}$. For other methods, the CV may be calculated by simulation of many random samples from $\mathrm{Exp}(\theta)$, and computation of a bootstrap estimate of the CV of each of the bootstrap estimates of the standard deviation of the original estimator. This process takes us far enough along the road of an infinite computational progression to make us return to analysis with delight! The accuracy of the bootstrap estimators is the subject of intensive analytical research at the moment, since enough computations have been carried out to suggest that they work "reasonably well" in a large variety of statistical problems.

2.5 PROBLEMS

Section 2.1

1. In this problem we will introduce trimmed and Winsorized means and verify the entries in Table 2.1.

   (a) For any distribution function F the qth quantile x_q satisfies the relation $F(x_q) = q$. Show that if $F(x) = F_\theta(x) = 1 - \exp(-x/\theta)$, then $x_q = -\theta \ln(1-q)$.

   (b) If $X \sim F$ define the *left q-trimmed mean* by

   $$\mathrm{E}[X \mid X > x_q] = \frac{1}{1-q} \int_{x_q}^{\infty} x \, dF(x).$$

   Show using integration by parts (or the memoryless property) that, for the exponential distribution with mean θ, $\mathrm{E}_\theta[X \mid X > x_q] = x_q + \theta$. When $q = .1$ this yields the characteristic (3) of interest in Example 1, namely, the mean amount of rainfall after truncation of the lowest decile. Also define a *right* q-trimmed mean and find it for $\mathrm{Exp}(\theta)$, the exponential distribution with mean θ.

(c) Define a *right q-Winsorized mean* by $\mathrm{E}[W_q]$, where

$$W_q = XI\{X \leq x_{1-q}\} + x_{1-q}I\{X > x_{1-q}\}$$

is the right Winsorized variable. Show that the right q-Winsorized mean may be written as $\mathrm{E}[W_q] = (1-q)\mathrm{E}[X \mid X < x_{1-q}] + qx_{1-q}$, a probability combination of a right q-trimmed mean and the quantile x_{1-q}. Show that for the $\mathrm{Exp}(\theta)$ distribution, the right q-Winsorized mean is $\mathrm{E}[W_q] = (1-q)\theta$.

(d) Express each of characteristics 1–5 of interest of Example 1 in terms of the mean (scale parameter θ) of the exponential distribution, and compare your results with those in Table 2.1.

(e) Also find the maximum likelihood estimates of these characteristics 1–5 of interest by substituting $\overline{X}$ for θ in the expressions obtained in part (d), and verify the numerical estimates in Table 2.1.

(f) Verify the descriptive estimates given in Table 2.1.

Section 2.2

2. Verify the properties of scale parameter families listed in Section 2.2.1.

3. For each of the following models calculate the mean, the variance, and the coefficient of variation. In all cases $\theta > 0$. (Reminder: first find the results for $\theta = 1$, and then employ the properties of scale parameter families found in Section 2.2.1.)

(a) $f_X(x) = \theta^{-1}$, $0 < x < \theta$

(b) $f_X(x) = \theta^{-1}e^{-x/\theta}$, $0 < x$

(c) $f_X(x) = r\theta^r(x+\theta)^{-r-1}$, $0 < x$, $r \geq 1$

(d) $f_X(x) = \theta^{-\alpha}x^{\alpha-1}e^{-x/\theta}/\Gamma(\alpha)$

(e) $f_X(x) = (1-\varepsilon)\theta^{-1}e^{-x/\theta} + \varepsilon(c\theta)^{-1}e^{-x/c\theta}$, $x > 0$, $c > 1$

In case (c) you may use the fact that for positive random variables $\mathrm{E}[X] = \int_0^\infty [1-F(x)]\,dx$ and $\mathrm{E}[X^2] = \int_0^\infty 2x[1-F(x)]\,dx$; these results are derived by integration by parts.

Section 2.3

4. For any random variable X with $\mathrm{E}[X^2] < \infty$ and any constant c show that

$$\mathrm{E}[X-c]^2 = \mathrm{Var}[X] + (\mathrm{E}[X]-c)^2$$

and hence confirm (2.3.1).

5. Let $X_1,\ldots,X_n$ be i.i.d. with

$$F_\theta(x) = F\left(\frac{x}{\theta}\right), \qquad x > 0; \quad \theta > 0.$$

Assume that $T_n = T_n(X_1,\ldots,X_n)$ is scale equivariant. Show that

$$\mathrm{E}_\theta[T_n] = \theta \mathrm{E}_1[T_n]$$

and

$$\mathrm{Var}_\theta[T_n] = \theta^2 \mathrm{Var}_1[T_n].$$

6. Verify that Examples 1–5 of Section 2.3.2 are scale equivariant estimators.

7. Let $X_1,\ldots,X_n$ be i.i.d. with

$$F_\theta(x) = F\left(\frac{x}{\theta}\right), \qquad x > 0$$

where $\theta > 0$ is unknown. Assume that F has density f. The *maximum likelihood estimator* (MLE) of θ is by definition the value $\hat{\theta}$ which maximizes

$$L(\theta) = \frac{1}{\theta^n} \prod_{i=1}^{n} f\left(\frac{x_i}{\theta}\right).$$

(a) Show that if f has derivative f', the MLE is the solution to

$$n^{-1} \sum_{i=1}^{n} \psi\left(\frac{X_i}{\theta}\right) = 0$$

where $\psi(y) = yf'(y)/f(y) + 1$. In the special case of the exponential distribution, $\hat{\theta} = \overline{X}$.

(b) If attention is restricted to the first r order statistics in a sample of size n from the exponential distribution, the MLE of θ is n/r times the β-Winsorized mean, where $\beta = 1 - r/n$.

8. Let $X_1,\ldots,X_n$ be i.i.d. Exp(θ). Define $X_{(0)} = 0$ and for $j = 1,\ldots,n$ let $Y_j = X_{(j)} - X_{(j-1)}$.

(a) Show that $Y_1,\ldots,Y_n$ are independent with $Y_j \sim \mathrm{Exp}(\gamma_j)$, with $\gamma_j = \theta/(n - j + 1)$.

(b) Show that

$$\mathrm{E}_\theta[X_{(r)}] = \theta \sum_{j=1}^{r} (n - j + 1)^{-1}$$

and that

$$\mathrm{Cov}_\theta[X_{(r)}, X_{(s)}] = \theta^2 \sum_{j=1}^{r} (n-j+1)^{-2}, \qquad r \le s.$$

9. Continuing with the assumptions of Problem 8, choose the constant c_m to make $c_m X_{(m)}$ unbiased for θ. In particular for $n = 20$, $m = 16$ find the constant and show that

$$20\,\mathrm{Var}[c_{16} X_{(16)}] \approx 1.5\theta^2$$

in agreement with the empirical entry in Table 2.2.

10. Let $X_1, \ldots, X_n$ be i.i.d.,

$$(1-\varepsilon)\mathrm{Exp}(\theta) + \varepsilon\mathrm{Exp}(3\theta).$$

Find the MSE of $\overline{X}_{20}$ as a function of ε.

11. Determine ε_n^* and $\lim_{n\to\infty} \varepsilon_n^*$ for each of the five estimators discussed in Section 2.3.2.

12. Define the displaced exponential distribution by the density

$$g(x) = \theta^{-1} e^{-(x-\gamma)/\theta}, \qquad x \ge \gamma$$

where $\theta \ge 0$, and γ is unrestricted.

(a) Show that θ is not a scale parameter for any family as defined in Section 2.2 unless $\gamma = 0$.

(b) Show that if X, Y are i.i.d. with common density g, then $|X - Y|$ has the $\mathrm{Exp}(\theta)$ distribution.

(c) Assuming that sister cell lifetimes satisfy the assumptions of part (b), estimate θ using the data in Example 2 of Section 2.1. The ordered absolute differences of available pairs are .3, .4, .5, .5, .6, .9, 1.7.

Section 2.4

13. **(a)** Continuing with the data of Problem 12(c), find the four suggested estimates of the standard error of the sample mean from Section 2.4. (Use B = 10, 25, and 100 bootstrap replications in cases 1 and 2 of the resampling methods of Section 2.4.2, and then conjecture what happens as B approaches infinity.)

(b) Use Problem 8 to find the appropriate constant c_6 which makes $c_6 X_{(6)}$ an unbiased estimator of the exponential parameter. Evaluate the estimator for the seven observations used in part (a), and find a bootstrap estimate (parametric or nonparametric) based on $B = 100$ bootstrap replications of the standard error.

(c) What are the relative advantages and disadvantages of the estimators in parts (a) and (b)?

14. Maximum likelihood and descriptive estimates of five characteristics of Melbourne's rainfall are given in Table 2.1. For each of these 10 estimates find an estimate of the associated standard error, by either traditional or bootstrap methods.

2.6 COMPLEMENTS

2.6.1 The Breakdown Point

The concept of a finite sample breakdown point was first suggested by Hodges (1967) as "tolerance of extreme values" in the context of the location parameter problem. It was generalized to statistical functionals and a variety of contexts by Hampel (1968); see also Hampel et al. (1986), Section 2.2. The definition used in this chapter is essentially that of Donoho and Huber (1983). It does not apply to contexts where the range of the parameter being estimated is bounded, but it can and will be extended to such contexts as needed. It seems to be more useful to change the definition to suit the context, rather than to employ an omnibus definition.

If one were concerned that a measure of scale might be taken to zero by a configuration of "inliers," then a different definition of breakdown point would be needed in the scale context; it would allow for implosion as well as explosion caused by outliers. It would be defined as the minimum proportion of observations which can take the estimator to 0 or to $+\infty$.

2.6.2 Further Developments on the Bootstrap

Because bootstrap methods are relatively easy to carry out, given the accessibility of machine computing power, they are becoming very popular. There are a number of bootstrap methods other than the basic ones presented in this text, including a variety of methods for obtaining confidence intervals. The analysis of how well they work is a difficult topic, which is currently receiving much attention. The reader who wishes to pursue this subject may begin with the following articles and the references contained within them: Beran (1982, 1987), Diciccio and Romano (1988), Efron and Tibshirani (1986), Efron (1982, 1987), Hall (1986, 1988), Hinkley (1988), Tibshirani (1988).

CHAPTER THREE

Estimating Scale—Asymptotic Results

In statistics, some people worry about not seeing the forest for the trees. I like to look at the bark.

C. R. Blyth (c. 1967)

This is a rather lengthy chapter since it not only contains the basic ideas on robustness which are used in subsequent chapters, but also contains sections which flow naturally on from it into various areas of research. These more abstract sections, which are starred, may be skipped at first reading of the text without loss of continuity.

It seems worthwhile to begin this material with a long preamble so that the student will begin to have an overview and a directed path through it. It may be wise to return to this introductory map if direction is lost.

In Section 3.1, "Consistency, Asymptotic Normality, and Efficiency," we review the standard material on consistency, asymptotic normality, and asymptotic relative efficiency of a sequence of estimators, which we assume the student has some acquaintance with. (In this connection, we note that Appendix B.1 contains some basic convergence theorems which are useful in statistics, plus a few simple exercises at the end to help the reader review them.) These asymptotic properties of estimators are presented here in the language of descriptive measures, so that they may be utilized in either parametric or nonparametric settings. They will be illustrated for the estimators of scale discussed in Chapter 2, where finite sample results were obtained. One of the objectives of the section is to demonstrate that in some contexts asymptotic efficiency can be a useful guide to the behavior of an estimator even for sample sizes as low as 20.

In Section 3.2, "Robustness Concepts," we consider three different ways of assessing the robustness of an estimator $T(F_n)$ of $T(F)$, namely, the breakdown point, influence function, and continuity of the descriptive measure T which defines it. These concepts are often described as *quantitative* robustness, *infinitesimal* robustness, and *qualitative* robustness. All can be derived from the descriptive measure T which induces the estimator sequence through $T_n = T(F_n)$. These concepts were introduced by Hampel (1968) in his Ph.D. thesis.

The breakdown point of T is essentially a measure of how far G can be from F before $T(G)$ is arbitrarily far from $T(F)$, and it is thus a measure of the global stability of T at F. Surprisingly the breakdown point does not usually depend on F, and in this way also it is a global property of T. Moreover it is often the limit of the finite sample breakdown points ε_n^* studied in Chapter 2, and is therefore an indication of how much gross contamination the associated estimators $T(F_n)$ can tolerate.

The influence function of T at F summarizes the infinitesimal robustness properties of T; it depends very much on F, and is thus a summary of *local* behavior. For each x it gives the relative effect on $T(F)$ of a very small (in fact, infinitesimal) amount of contamination at x. When this relative influence is considered as a function of x, it is called the *influence function* of T at F.

The influence function of T at F is also intimately tied up with the asymptotic normality of the associated sequence $\{T(F_n)\}$, as we shall see. It suffices now to say that when a strong form of derivative of T exists at F, the associated sequence $\{T(F_n)\}$ is asymptotically normal. Moreover this strong form of derivative coincides with the influence function of T at F.

Furthermore, it turns out that the asymptotic variance of the estimators $T(F_n)$ can usually be defined in terms of the influence function. This in turn leads to "influence function" estimates of the standard error of $T(F_n)$. These standard error estimates are presented in Section 3.5, for comparison with the bootstrap estimates of the standard error. Thus Sections 3.1, 3.2.1, 3.2.2, and 3.5 are essential reading for a basic understanding of the applications of the influence function.

The theory of robustness developed by Huber (1964) and Hampel (1968) is basically centered on parametric models. That is, their methods recognize that the parametric model may not be the "true" model, but nevertheless they make inferences about its parameters with robust and efficient methods.

The material in Section 3.3, "Descriptive Measures of Scale," provides a contrasting philosophical approach. This alternative approach [introduced by Bickel and Lehmann (1975)], places emphasis on selecting a descriptive measure that satisfies certain invariance properties for a nonparametric class of distributions. The properties effectively define what is meant by a measure of location, scale, dispersion, or association in a nonparametric context. Among all the measures of scale, say, it is desired to choose one that is easy to estimate for the large nonparametric class. By "easy" we mean that in terms of some measure of efficiency, usually related to the asymptotic variance of the induced estimators, the chosen descriptive measure is more efficient and hence requires smaller samples than its competitors. Robustness considerations may also be important in this choice.

Now we will describe the more abstract portions of the chapter which are starred and may be omitted at first reading. Section 3.2.3 discusses the properties of quantitative and infinitesimal robustness for the class of L-estimators, which are merely linear combinations of order statistics. Nearly all the examples considered in Chapter 2 are of this form. Then we consider the property of *qualitative* robustness, which is a strong form of continuity of T. In Section 3.2.4 we briefly discuss this notion and give references for those readers seeking more information on this topic.

In Section 3.4, "Stability of Estimators on Neighborhoods of the Exponential Scale Family,*" two approaches are given which measure the efficiency of estimators as the underlying distribution G of the observations varies over an ε-neighborhood of F, where F belongs to a known parametric family. The details are elementary but somewhat lengthy. The material provides a further comparison of the two philosophical approaches to robust statistics mentioned above at their intersection of application, and an introduction to further areas of research.

Finally in Section 3.5, "Estimates of Standard Error," we present the influence function estimates of standard errors of the mean and trimmed mean. Then we discuss the exact bootstrap estimates of standard error of the mean and of the median, for comparison with some Monte Carlo bootstrap estimates.

3.1 CONSISTENCY, ASYMPTOTIC NORMALITY, AND EFFICIENCY

Traditionally the three main concerns of asymptotic theory are the properties listed below of a sequence of estimators $\{\hat{\theta}_n\}$ of θ. It is usually hoped that the limiting results (as $n \to \infty$) will say something about the behavior of $\hat{\theta}_n$ for a particular sample size n of interest, or perhaps, more optimistically, for all moderate to large sample sizes.

1. *Consistency*. A sequence of estimators $\{\hat{\theta}_n\}$ is *consistent* for a parameter θ if $\hat{\theta}_n \xrightarrow{P} \theta$.
2. *Asymptotic Normality*. We say that $\{\hat{\theta}_n\}$ is *asymptotically* normal with parameters $\theta, V(\theta)$ if $\sqrt{n}(\hat{\theta}_n - \theta) \xrightarrow{d} N(0, V(\theta))$.

3. *Relative Efficiency*. Given two sequences of estimators, say $\{\hat{\theta}_{1,n}\}$ and $\{\hat{\theta}_{2,n}\}$, which are each consistent for θ and also asymptotically normal, the *relative efficiency* of the first to the second is given by the ratio $V_2(\theta)/V_1(\theta)$, which happily often does not depend on the parameter θ.

We want to define and review these properties in the language of descriptive measures, so that we may use them more easily in both parametric and nonparametric settings. We have already looked at some descriptive measures in Chapter 1, especially in the problems. Now we will systematically extend those estimators introduced in Chapter 2 for the exponential scale problem to a nonparametric setting.

3.1.1 Representing Estimators by Descriptive Measures

In Chapter 2 we studied the stability of various estimators on mixture models which introduced small amounts of contamination into the desired model. We found that for estimating an exponential scale parameter the sample mean is unreliable, and that there are reliable alternatives. This approach does not guarantee that the alternative estimators will be stable when tested on further models, or that the results carry over to different values of n. In this section we introduce another approach which begins by expressing previously discussed estimators (Section 2.3.2) as functionals applied to the empirical distribution function F_n. The functionals, or *descriptive measures*, as we shall often call them, assign real values to probability distributions. They will enable us to make precise what we are estimating, and to study more easily the efficiency and robustness of the associated estimators.

In the following examples we take F_n to be the usual *empirical distribution function*:

$$F_n(x) = \frac{\#\{X_i \le x\}}{n} = \frac{1}{n}\sum_{i=1}^{n} I\{X_i \le x\}$$

where I is the indicator function. The graph of an empirical distribution is sketched in Figure 3.1. The student should find and sketch the graph of the probability function corresponding to this distribution function. It is defined by $p_{F_n}(x) = F_n(x) - F_n(x^-)$ for all x, and may be expressed in terms of n and the order statistics $x_{(1)} \le x_{(2)} \le \cdots \le x_{(n)}$. It is also a useful exercise to sketch the graph of the inverse distribution function $F_n^{-1}(u)$, $0 < u < 1$.

Now we are ready to express the estimators of Section 2.3.3 in functional form.

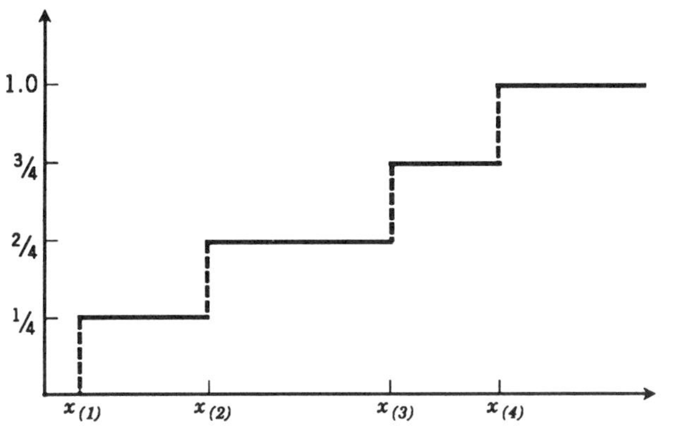

FIGURE 3.1 An empirical distribution function.

1. The Mean

Let $T(G) = \int x\,dG(x)$ for any G for which the integral is defined and finite. Recall that if G has a density g this may be written $T(G) = \int xg(x)\,dx$. For example, if $g(x) = \theta^{-1}e^{-x/\theta}$, then $T(G) = \theta$.

For a second example let G be discrete with probability function $p_G(x) = G(x) - G(x^-)$, then $T(G) = \int x\,dG(x) = \sum_x xp_G(x)$. For example, if G is the Poisson (θ) distribution, then $p_G(x) = e^{-\theta}\theta^x/x!$, $x = 0, 1, 2, \ldots$, and $T(G) = \theta$.

Whether or not F is discrete or absolutely continuous or otherwise, let F_n be the empirical distribution of n observations from F. Then

$$T(F_n) = \int x\,dF_n(x) = \sum_x xp_{F_n}(x) = \sum_{i=1}^{n} \frac{X_i}{n} = \overline{X}_n$$

since F_n assigns probability $1/n$ to each obervation X_i.

2. The qth Quantile Estimator

Let G^{-1} denote the inverse function to G, namely, $G^{-1}(q) = \inf\{y : G(y) \geq q\}$. For example, if $G(x) = 1 - e^{-x/\theta}$, $x > 0$, then setting $q = G(x)$ and solving for x yields $x = -\theta \ln(1-q)$. Hence the inverse function is $G^{-1}(q) = -\theta \ln(1-q)$, $0 < q < 1$. The exponential distribution and its inverse are plotted in Figure 3.2.

For a second example the reader may find the inverse when G is the Poisson (θ) distribution function. This is more difficult to find because of the discrete nature of the distribution, but it is a useful exercise.

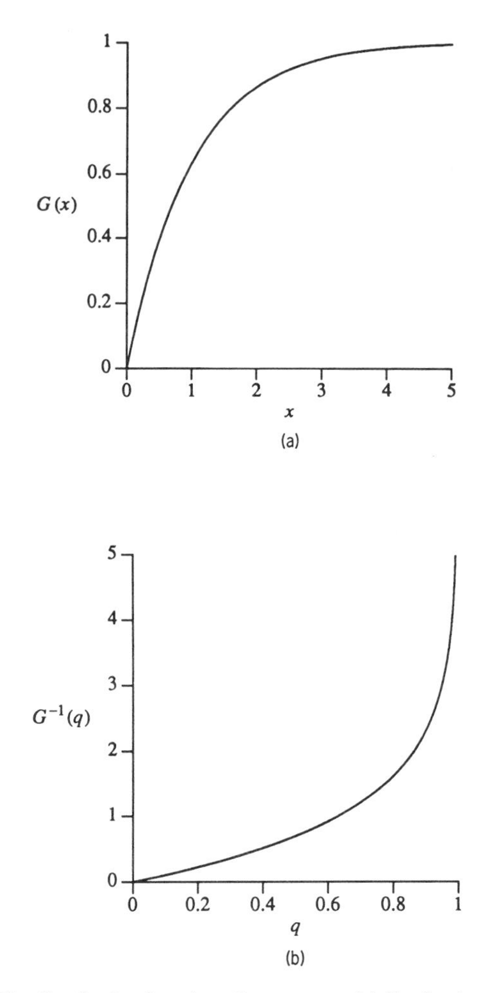

FIGURE 3.2 (a) The distribution function of an exponential distribution. (b) Its inverse function.

Now fix q and define the functional T for each G by $T(G) = G^{-1}(q)$. The functional T assigns to F_n the value

$$T(F_n) = F_n^{-1}(q) = X_{(m)}$$

where m is the smallest integer $\geq nq$; see Figure 3.1. Thus $F_n^{-1}(q)$ is either the $[nq]$th-order statistic $X_{([nq])}$, *or* the next largest one. When F_n is based on a sample from F we think of $X_{([nq])}$ as the natural estimator of the quantile $x_q = F^{-1}(q)$.

3. The β-Trimmed Mean

Let

$$T_\beta(G) = \frac{1}{1-\beta}\int_0^{G^{-1}(1-\beta)} x\,dG(x).$$

Note that we are only considering distributions on the nonnegative real axis [with $G(0^-) = 0$], and are only concerned with trimming on the right. If $X \sim G$, this $T_\beta(G)$ is just the expected value of X, given that $X < x_{1-\beta} = G^{-1}(1-\beta)$. In symbols, it is $\mathrm{E}_G[X \mid X < x_{1-\beta}]$. In words, it is the mean of the conditional distribution of X, given that X is less than its $(1-\beta)$th quantile. The reader is asked to study this conditional distribution for the exponential model in the problems.

From part 2 above we see that $F_n^{-1}(1-\beta)$ is essentially the $n-[n\beta]$th-order statistic. Thus a β-trimmed empirical distribution will have the largest $[\beta n]$ observations removed, and the average of those remaining will be the β-trimmed sample mean. In symbols,

$$T_\beta(F_n) = \frac{1}{n-n\beta}\sum_{i=1}^{n-[n\beta]} X_{(i)} \approx \frac{1}{r}\sum_{i=1}^{r} X_{(i)} = \overline{X}_{n,\beta}$$

where $r = n - [n\beta]$, and $[j]$ is the largest integer $\leq j$.†

4. The β-Winsorized Mean

Define a probability mixture of the previous two examples:

$$W_\beta(G) = (1-\beta)T_\beta(G) + \beta G^{-1}(1-\beta).$$

This is a weighted average of the β-trimmed mean and the $(1-\beta)$th quantile. Then the sample version is

$$W_\beta(F_n) = (1-\beta)\overline{X}_{n,\beta} + \beta X_{(n-[n\beta])} \approx \overline{W}_{\beta,n}.$$

5. A Rejection Rule Followed by the Mean

This example illustrates the main shortcoming of the functional approach, for the rejection rule $(X_{(n)} - X_{(n-1)})/X_{(n)} > c$ depends on n as well as F_n and hence cannot be written in the form $T(F_n)$ for some T that is free of n. (But see the problems for a "rejection rule plus mean" which can be so expressed.)

†The approximation symbol is used when two estimators differ by an amount which is negligible in that the absolute difference, multiplied by the square root of the sample size, converges to zero in probability as the sample size increases without bound.

6. Linear Functionals

We will see that many functionals $T(G)$ may be expressed in the form $T(G) = \int k(x)dG(x)$ for some function $k(x)$ which is called the *kernel* of T. Such functionals are called *linear functionals* because $T(aF + bG) = aT(F) + bT(G)$. They have the desirable property that

$$T(F_n) = \int k(x)dF_n(x) = \frac{1}{n}\sum_{i=1}^{n} k(X_i).$$

This is desirable because it means that the estimator is a sum of i.i.d. random variables, and hence it will have a simple asymptotic theory.

In the above examples the functionals $T(G)$ are defined for G lying in nonparametric classes. When F_n is based on a sample from F, we usually find that $\{T(F_n)\}$ is consistent for $T(F)$. Thus it is natural to associate the sequence of estimators $\{T(F_n)\}$ with the characteristic $T(F)$, which is also often called the *estimand*. We may therefore sometimes refer to the sequence $\{\overline{X}_n\}$ as "the mean" and use the notation $T(F) = \int x\,dF(x)$.

3.1.2 Consistency, Asymptotic Normality, and Relative Efficiency

A descriptive measure T generates a natural sequence of estimators $\{T(F_n)\}$ for the estimand $T(F)$, assuming F_n is based on a sample of size n from F. We have a statistical problem if F belongs to a a family of distributions $\mathcal{F}$ which includes as few as two members but which may be a large nonparametric class. The least that we expect from the estimator is *consistency*:

$$T(F_n) \xrightarrow{p} T(F), \qquad F \in \mathcal{F} \tag{3.1.1}$$

i.e., $T(F_n)$ converging to $T(F)$ in *probability* as n approaches infinity, for every F in the family. The qualifier $F \in \mathcal{F}$ will always be understood, although often omitted for convenience. This type of consistency is also called *weak* consistency, to distinguish it from *strong* consistency which is based on almost sure convergence (see Appendix B for further discussion).

The most common example is the case when $T(F)$ is the mean of F, and then the weak law of large numbers guarantees that $\overline{X}_n \xrightarrow{p} \mu$ for every F that has finite variance. In general consistency is guaranteed if T is continuous at F in a very weak sense (see Section 3.2.4).

Despite the fact that a functional $T(F)$ is defined for a large nonparametric class of distributions, we will usually be interested in its behavior for F in the neighborhood of an F_θ belonging to a particular parametric model $\{F_\theta : \theta \in \Theta\}$. Then another kind of consistency is useful, Fisher consistency of T for the model. We say that $T(F)$ is *Fisher consistent* for θ if $T(F_\theta) = \theta$

for all $\theta \in \Theta$. For an example, the mean is Fisher consistent for the scale parameter θ of an exponential model. The median is not Fisher consistent for θ, but $T(F) = (\ln 2)F^{-1}(1/2)$ is Fisher consistent for it.

Under certain conditions we also have *asymptotic normality*:

$$n^{1/2}[T(F_n) - T(F)] \xrightarrow{d} N(0, V). \tag{3.1.2}$$

In other words, the sequence defined by

$$\left(\frac{n}{V}\right)^{1/2} [T(F_n) - T(F)]$$

converges in *distribution* to the standard normal distribution Φ. The variance of the limit distribution $V = V(T,F)$ will often be called the *asymptotic variance* (although, strictly speaking, it is the sequence $\{V(T,F)/n\}$ which is the asymptotic variance; for further discussion see the problems). Furthermore, in many cases we have

$$n \operatorname{Var}[T(F_n)] \to V(T,F) \tag{3.1.3}$$

and then $V(T,F)/n$ gives an approximation to the variance of $T(F_n)$. In the problems we will examine examples which satisfy some of equations (3.1.1) to (3.1.3), but not the others.

Now suppose we are given two descriptive measures S,T which agree at F, $S(F) = T(F)$, and which each satisfy (3.1.2). Then approximately

$$S(F_m) \sim N\left(S(F), \frac{V(S,F)}{m}\right)$$

and

$$T(F_n) \sim N\left(T(F), \frac{V(T,F)}{n}\right).$$

Thus the sample size n required by $T(F_n)$ to achieve the same approximating normal distribution as $S(F_m)$ is

$$n = \frac{mV(T,F)}{V(S,F)}.$$

The *asymptotic relative efficiency* (ARE) of S to T at F is therefore defined to be the factor

$$E(S,T;F) = \frac{V(T,F)}{V(S,F)}. \tag{3.1.4}$$

This concept is particularly useful in many parametric problems in which the efficiency $E(S,T;F_\theta)$ is free of θ, for then it summarizes in one number the relative (large) sample sizes required by S and T to achieve the same accuracy in estimating θ.

For an example consider the exponential scale family $F_\theta(x) = 1 - e^{-x/\theta}$, $x > 0$, where $\theta > 0$ is unknown. We will find the ARE of several previously mentioned estimators.

1. The Mean

The sample mean of Example 1 satisfies $T(F_\theta) = \theta$ and (3.1.2) becomes:

$$n^{1/2}[\overline{X}_n - \theta] \xrightarrow{d} N(0, \theta^2).$$

2. The qth Quantile Estimator

For the qth quantile of Example 2 and any F that has a continuous, nonzero density at $x_q = F^{-1}(q)$, (3.1.2) becomes

$$n^{1/2}[X_{(nq)} - x_q] \xrightarrow{d} N(0, V) \tag{3.1.5}$$

where

$$V = \frac{q(1-q)}{f^2(x_q)}.$$

See David (1981) or Appendix B.2. When F is exponential and we employ the .8 quantile,

$$X_{(.8n)} \xrightarrow{d} F_\theta^{-1}(.8) = -\theta \ln(1 - .8) \approx 1.61\theta.$$

To obtain a descriptive measure which is consistent for θ let

$$S(F) = \frac{F^{-1}(.8)}{1.61} \approx .62 F^{-1}(.8).$$

Then $S(F_\theta) = \theta$ and by (3.1.5)

$$m^{1/2}[S(F_m) - \theta] \xrightarrow{d} N(0, V(S, F_\theta))$$

where

$$V(S, F_\theta) = \frac{(.62)^2 \theta^2 .8(1 - .8)}{(1 - .8)^2} \approx 1.54\theta^2.$$

Hence the asymptotic relative efficiency of the unbiased quantile estimator S to the sample mean T at F_θ is $E(S, T; F_\theta) \approx .65$. The simulation estimates of Table 2.1 for $20 \times \text{Var}$ yield the comparable ratio

$$\frac{20 \times \text{Var}[\overline{X}_{20}]}{20 \times \text{Var}[c_{16} X_{(16)}]} \approx .68.$$

Table 3.1 ARE of Robust Estimators of an Exponential Scale Parameter for Comparison with Finite Sample Results of Table 2.2

Estimator	ARE	Ratio of Variances	Simulation Result
$c_{16}X_{(16)}$	.648	.665	.679
$k\overline{X}_{20,.1}$	.847	.867	.870

3. The β-Trimmed Mean

For β-trimmed means one may show [see Stigler (1969)] that if F is continuous at the unique quantile $x_{1-\beta} = F^{-1}(1-\beta)$, then asymptotic normality (3.1.2) holds with

$$V = \frac{1}{1-\beta}[\sigma_\beta^2 + \beta(x_{1-\beta} - \mu_\beta)^2] \tag{3.1.6}$$

where μ_β and σ_β^2 are the mean and the variance of the truncated distribution:

$$\begin{aligned} \mu_\beta &= \int_0^{x_{1-\beta}} \frac{x\,dF(x)}{1-\beta} \\ \sigma_\beta^2 &= \int_0^{x_{1-\beta}} \frac{(x-\mu_\beta)^2\,dF(x)}{1-\beta}. \end{aligned} \tag{3.1.7}$$

For the exponential model one may verify that $\mu_\beta = 1 + \beta \ln\beta/(1-\beta)$ and $\sigma_\beta^2 = 1 - \beta\ln^2\beta/(1-\beta)^2$, so that the asymptotic variance of the .1-trimmed mean (after modification to ensure Fisher consistency) is $1.179\theta^2$. Thus the asymptotic relative efficiency of the unbiased .1-trimmed mean to the mean is $1/1.179 \approx .85$. We leave the details to the problems and summarize the results in Table 3.1. The coefficients are chosen to make the estimators unbiased for the exponential scale parameter.

The main point of this section is that for a parametric family the asymptotic relative efficiency is a useful concept in that it allows us to make comparisons of competing estimators which are generally valid for sample sizes as low as 20. Such comparisons tacitly assume that convergence in (3.1.2) and (3.1.3) is more or less monotone for $n \geq 20$, so that the approximations are roughly increasing in accuracy with increasing n.

3.2 ROBUSTNESS CONCEPTS

Now we consider three different ways of assessing the robustness of an estimator $T(F_n)$, the breakdown point, influence function, and continuity of the

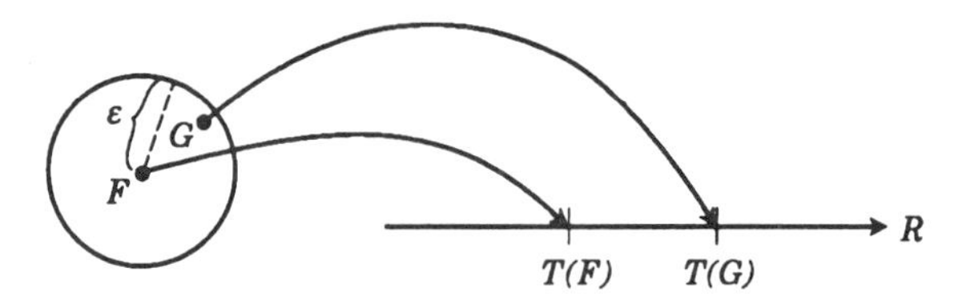

FIGURE 3.3 A two-dimensional representation of an ε-neighborhood of F.

descriptive measure T which induces it. These notions are often described as *quantitative* robustness, *infinitesimal* robustness, and *qualitative* robustness. All can be derived from the descriptive measure T, which induces the estimator sequence $T_n = T(F_n)$.

It is imperative at this point to think of the distribution F as a point, one of a large family $\mathcal{F}$ of distributions which includes the empirical distributions F_n as well as F. It is a helpful simplification to represent the members of $\mathcal{F}$ as points in the plane, as shown in Figure 3.3. The disk enclosing F represents all distributions G whose distance from F is less than ε, where "distance" is measured by $\sup_y |F(y) - G(y)|$, say.

Now as G varies over the ε-neighborhood of F, for small ε the descriptive measure $T(G)$ hopefully will not stray too far from $T(F)$; hence for $G = F_n$ the estimator $T(F_n)$ will say *something* about the unknown $T(F)$. This is why continuity of T at F is a desirable property. But as epsilon increases from zero to 1, a point will be reached where $T(G)$ [and hence $T(F_n)$] is arbitrarily far from $T(F)$, and we say that the estimator "breaks down." The epsilon for which this happens is called the *breakdown point*. The breakdown point is a quantitative measure of the robustness of the descriptive measure $T(F)$. It indicates the maximum proportion of gross outliers which the induced estimators $T(F_n)$ can tolerate.

The next important concept, the influence function, yields information about the rate of change of the induced estimators with respect to the proportion of observations which are not from the model F, but from a "contamination" point x. To make this precise, define the mixture distribution $F_{x,\varepsilon} = (1-\varepsilon)F + \varepsilon\Delta_x$ which yields observations from F with high probability $1-\varepsilon$ and from the point x with small probability ε. It is not hard to show that $F_{x,\varepsilon}$ lies in the ε-neighborhood of F.

In fact, $F_{x,\varepsilon}$ will lie on or near the boundary of the neighborhood if x is sufficiently large, because as $x \to \infty$, the supremum distance of $F_{x,\varepsilon}$ from F will increase to ε.

The value of $T(F_{x,\varepsilon})$ indicates how the estimator $T(F_n)$ will behave if the sample is mostly from F but contains a small proportion of observations from the bogus point x.

The difference $T(F_{x,\varepsilon}) - T(F)$, divided by ε, gives the relative influence of the contamination at x on the descriptive measure at the point F. The limit of this relative influence, as ε tends to zero, is the value of the influence function at x. By plotting the graph of the influence function one obtains a curve that reveals much about the descriptive measure and hence about the induced estimators. [Hampel (1968) at first dubbed his influence function the "influence curve."] Generally speaking, the influence function of a robust estimator should be continuous and bounded. Examples to follow in Section 3.2.2 will illustrate these qualities, or lack thereof.

We have completed a rather lengthy introduction to the concepts of breakdown point and influence function which the reader may have found a little too general; we will now be more specific about definitions and examples.

3.2.1 The Breakdown Point

In Section 2.3.5 we discussed the finite sample breakdown point and calculated it for several examples. Since this breakdown point often converges to a limit as $n \to \infty$, it may be approximated by the limiting value. Now that we have seen how to represent many estimators in terms of descriptive measures applied to the empirical distribution, we may use the descriptive measures to find the limiting or "asymptotic" breakdown point. We will continue to write ε_n^* for the finite sample breakdown point, and now also write ε^* for the (asymptotic) breakdown point which will be derived from the descriptive measure.

The simplest definition of *breakdown point* ε^* for a measure of scale is the minimum proportion ε of outlier contamination at x for which $T(F_{x,\varepsilon})$ is unbounded in x. When the minimum does not exist, ε^* is taken to be the infimum of such ε.

Now we consider again the examples from Section 3.1.1.

1. The Mean

The mean assigns to $F_{x,\varepsilon}$ the value $T(F_{x,\varepsilon}) = (1-\varepsilon)T(F) + \varepsilon x$. In other words, the contaminated model has a mean that is a weighted average of the mean of the distribution $T(F)$ (which is fixed) and the point x, which we are free to vary. Hence for *any* $\varepsilon > 0$, $T(F_{x,\varepsilon})$ is unbounded in x and $\varepsilon^* = 0$. This result corresponds to the limit of the finite sample results mentioned earlier (Section 2.3.5), where only proportion $\varepsilon_n = 1/n$ of the observations may carry $T(F_n) = \overline{X}_n$ beyond all bounds.

2. The qth Quantile

To determine ε^* for the quantile functional $T(F) = F^{-1}(q)$ we assume for simplicity that F is strictly increasing and sketch the graph of $F_{x,\varepsilon}$ and $F_{x,\varepsilon}^{-1}$; see Figure 3.4. We note that

$$F_{x,\varepsilon}^{-1}(u) = \begin{cases} F^{-1}\left(\dfrac{u}{1-\varepsilon}\right), & u < (1-\varepsilon)F(x) \\ x, & (1-\varepsilon)F(x) \le u < (1-\varepsilon)F(x) + \varepsilon \\ F^{-1}\left(\dfrac{u-\varepsilon}{1-\varepsilon}\right), & (1-\varepsilon)F(x) + \varepsilon \le u. \end{cases} \tag{3.2.1}$$

To find the breakdown point, fix $\varepsilon < 1-q$; then for all x sufficiently large, $(1-\varepsilon)F(x) > q$ and $F_{x,\varepsilon}^{-1}(q) = F^{-1}(q/(1-\varepsilon))$. It follows that $F_{x,\varepsilon}^{-1}(q)$ is bounded in x. However, if $\varepsilon > 1-q$, then $F_{x,\varepsilon}^{-1}(q) \ge x$, which can be made arbitrarily large. Thus $\varepsilon^* = 1-q$.

3. The β-Trimmed Mean; and 4. The β-Winsorized Mean

It is intuitively clear that the β-trimmed mean and β-Winsorized means have breakdown point $\varepsilon^* = \beta$; details are left to the reader.

5. Outlier Rejection Rule Followed by Mean

This cannot be represented as a functional, so we consider the finite sample definition. An unfavorable situation arises if there are two outliers close together, for then the criterion $[X_{(n)} - X_{(n-1)}]/X_{(n)} > k$ will fail to reject $X_{(n)}$ (the so-called masking effect). In this situation the resulting mean can be arbitrarily large. Since only proportion $\varepsilon_n^* = 2/n$ of the observations can break down the estimator, we summarize the result for large n by taking $\varepsilon^* = \lim_{n\to\infty} \varepsilon_n^* = 0$.

Finally we note that the breakdown point of an estimator T is a global measure of its stability, since it indicates how far the contamination model can be from the hypothesized model before rendering T totally worthless. It is also a global property in the sense that often it is independent of the model F. In the next section we derive the influence function, which reveals the sensitivity of T to peculiarities of F.

3.2.2 The Influence Function

The breakdown point gives us a rough idea of how much really bad contamination an estimator can handle before it gives nonsensical results; the influence function, by contrast, gives us a precise idea of how the estimator responds to a small amount of contamination at any point. Not surprisingly,

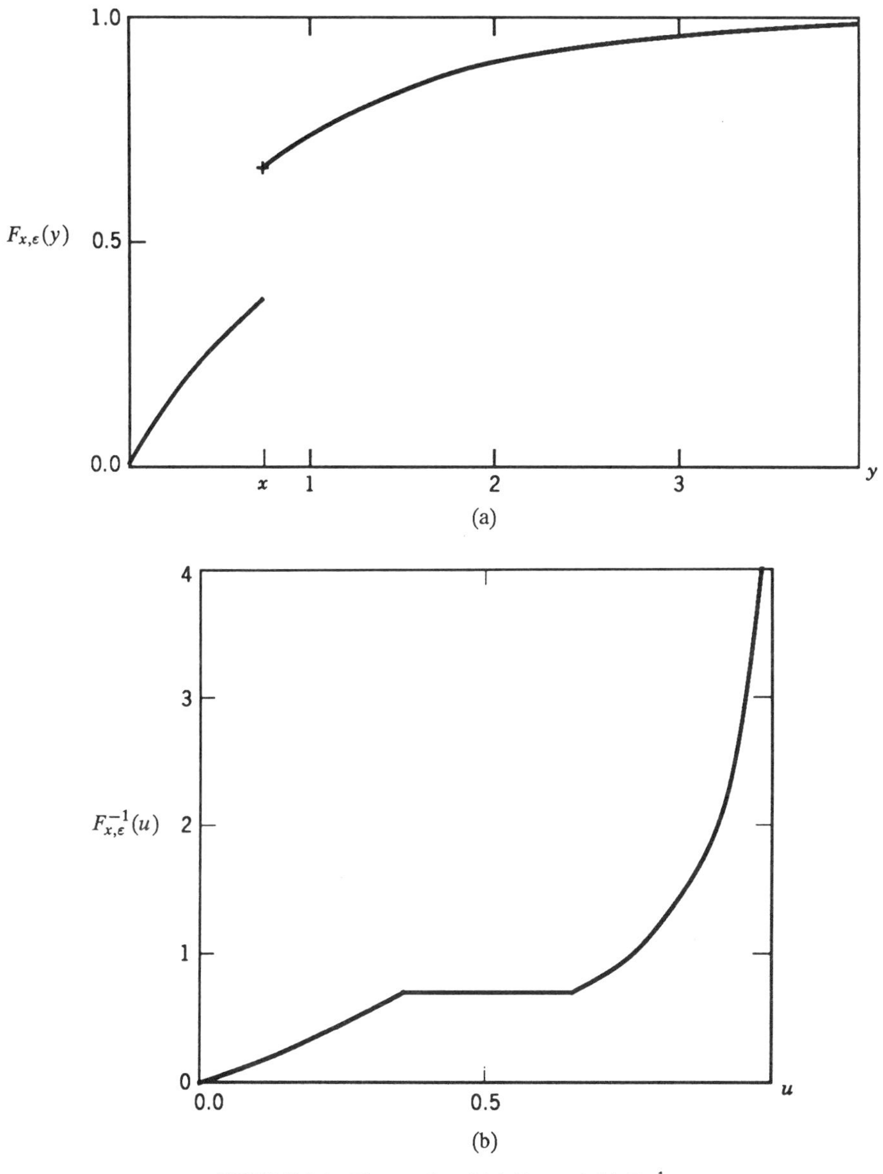

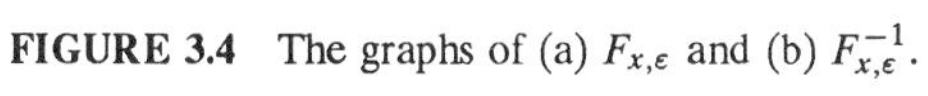
FIGURE 3.4 The graphs of (a) $F_{x,\varepsilon}$ and (b) $F_{x,\varepsilon}^{-1}$.

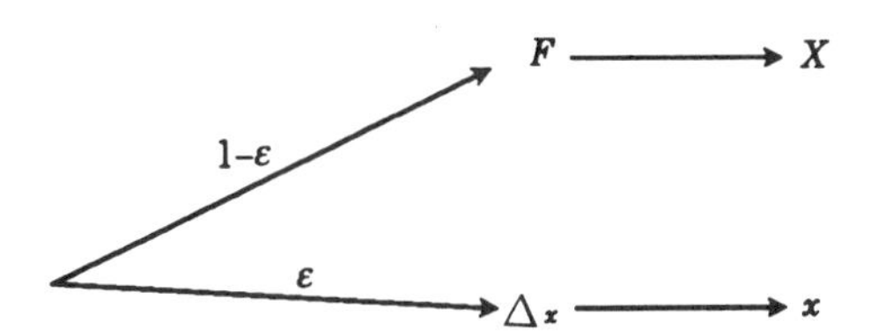

FIGURE 3.5 Diagram of two-step process involved in sampling from $F_{x,\varepsilon}$.

those estimators which are very sensitive to the form of F will be most influenced by small amounts of contamination. The influence function at x can be thought of as an approximation to the relative change in an estimate caused by the addition of a small proportion of spurious observations at x.

The mixture distribution $F_{x,\varepsilon} = (1-\varepsilon)F + \varepsilon\Delta_x$ is an appropriate model for contamination: sampling from $F_{x,\varepsilon}$ incurs the risk of obtaining a "bad" observation at x with small probability ε and a "good" observation from F otherwise. An example of $F_{x,\varepsilon}(y)$, sketched as a function of y, is shown in Figure 3.5. The following tree diagram illustrates the fact that sampling from $F_{x,\varepsilon}$ may be considered a two-step process. First select a distribution F or Δ_x with respective probabilities $1-\varepsilon$ and ε; and second select an observation $X \sim F$ or x from the chosen distribution.

The relative influence on $T(F)$ of proportion ε "bad" observations at x is formulated by

$$\frac{T(F_{x,\varepsilon}) - T(F)}{\varepsilon}.$$

The *influence function* of T at F is defined for each x by

$$\mathrm{IF}(x) = \lim_{\varepsilon \downarrow 0} \left[\frac{T(F_{x,\varepsilon}) - T(F)}{\varepsilon} \right] \tag{3.2.2}$$

provided the limit exists as a real number. Sometimes we write $\mathrm{IF}(x) = \mathrm{IF}_{T,F}(x)$ to emphasize the dependence on T and F. Other notations include $\mathrm{IF}(x;T,F)$, $\mathrm{IC}_{T,F}(x)$, $\Omega(x;T,F)$, and $T'_F(x)$.

The influence function $\mathrm{IF}(x)$ is a directional derivative of T at F in the direction of $\Delta_x - F$; see Figure 3.6. Some authors [including Huber (1980) and Hampel et al. (1986)] use the phrase "in the direction of Δ_x"; and if one is standing at F, this is true. However, to be consistent with the usual vector space terminology we will use the origin as the reference point for direction. The influence function is related to stronger derivatives of the functional T at F. (See Appendix B for further discussion and references.)

Let us reconsider our examples from Section 3.1.

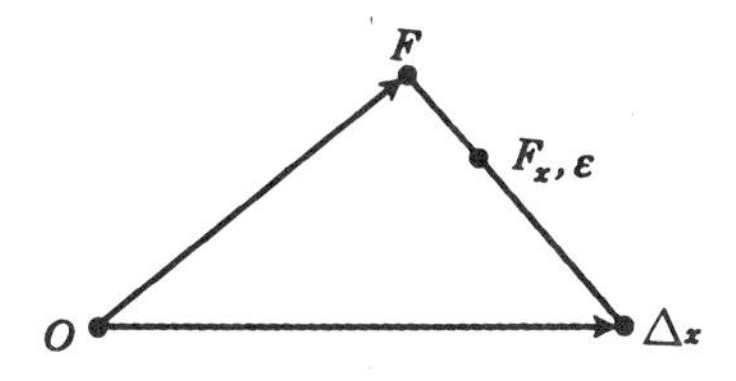

FIGURE 3.6 Vector diagram illustrating that the influence function is a directional derivative.

1. The Mean

For the mean it is easy to see that $T(F_{x,\varepsilon}) - T(F) = \varepsilon(x - T(F))$, so that the influence function is $\text{IF}(x) = x - T(F)$, $x > 0$; see Figure 3.7(a). Small values of x (for example, early failures in a life-testing experiment) will have a negative but bounded influence on the mean, while the influence of outliers is positive and unbounded.

2. The qth Quantile

To derive the influence function $\text{IF}_q(x)$ for the qth quantile functional $T(F) = F^{-1}(q)$, assume that F has a density f which is continuous and positive at $x_q = F^{-1}(q)$. Let $F_{x,\varepsilon} = (1-\varepsilon)F + \varepsilon\Delta_x$, and define

$$g(\varepsilon) = T(F_{x,\varepsilon}) = F_{x,\varepsilon}^{-1}(q).$$

We will find $\text{IF}_q(x) = (d/d\varepsilon)[g(\varepsilon)]_{\varepsilon=0}$ indirectly by first calculating $g'(\varepsilon)$ for $\varepsilon > 0$ and then taking $\lim_{\varepsilon\downarrow 0} g'(\varepsilon)$. (See Problem 10 in this regard.) Now for $(1-\varepsilon)F(x) + \varepsilon \le q$, we have by (3.2.1) that $F_{x,\varepsilon}^{-1}(q) = F^{-1}((q-\varepsilon)/(1-\varepsilon))$. Thus

$$g'(\varepsilon) = \frac{d}{d\varepsilon}F^{-1}\left(\frac{q-\varepsilon}{1-\varepsilon}\right) = \frac{(d/d\varepsilon)((q-\varepsilon)/(1-\varepsilon))}{f\{F^{-1}[(q-\varepsilon)/(1-\varepsilon)]\}}$$

so that

$$\text{IF}_q(x) = \lim_{\varepsilon\downarrow 0} g'(\varepsilon) = \frac{q-1}{f(x_q)}, \qquad x < x_q.$$

The other cases are left to the problems, wherein we show that

$$\text{IF}_q(x) = \begin{cases} \dfrac{(q-1)}{f(x_q)}, & x < x_q \\ 0, & x = x_q \; . \\ \dfrac{q}{f(x_q)}, & x > x_q \end{cases} \tag{3.2.3}$$

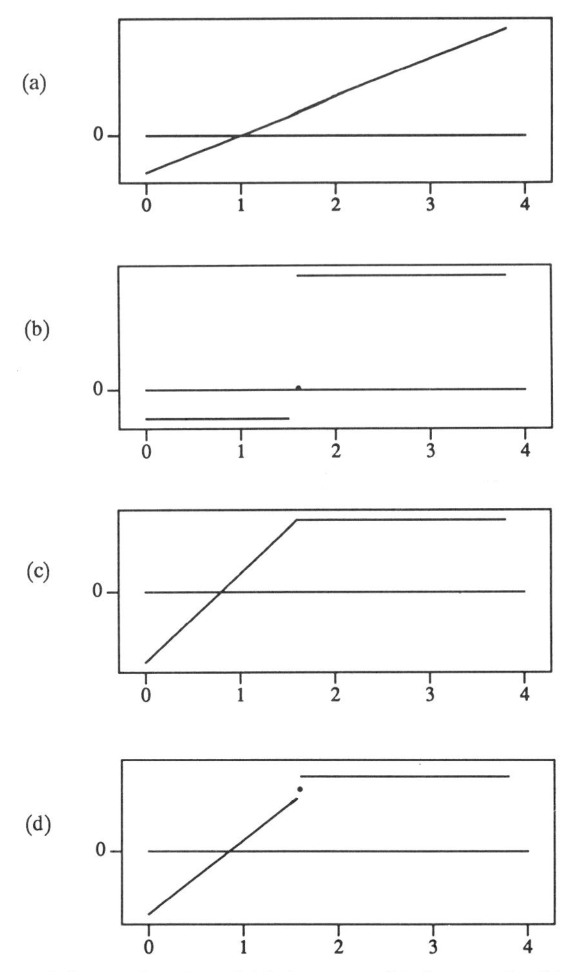

FIGURE 3.7 Influence function of (a) the mean, (b) the quantile, (c) the trimmed mean, and (d) the Winsorized mean.

This influence function is drawn in Figure 3.7(b). Note that contamination at x_q has zero influence on this quantile. Contamination at any $x > x_q$ has the same positive effect on the qth quantile of F, because for any ε the contamination shifts the quantile to the right [to $F^{-1}(q/(1-\varepsilon))$], regardless of the value of x. Similarly, contamination to the left of the qth quantile causes a fixed shift to the left, and hence a fixed negative influence. The sizes of these effects depend (inversely) upon the density f at x_q. When

$f(x_q)$ is near 0, IF_q will have a large discontinuity at x_q, indicating that x_q will be very difficult to estimate.

3. The β-Trimmed Mean

For the β-trimmed mean $T_\beta(G)$ we will need

$$x_{1-\beta} = F^{-1}(1-\beta); \qquad \mu_\beta = T_\beta(F) = \int_0^{x_{1-\beta}} \frac{y}{1-\beta}\, dF(y).$$

To derive the influence function for the β-trimmed mean, let $F_{x,\varepsilon}(y) = (1-\varepsilon)F(y) + \varepsilon\Delta_x(y)$, and define

$$\begin{aligned} g(\varepsilon) = T_\beta(F_{x,\varepsilon}) &= \int_0^{F_{x,\varepsilon}^{-1}(1-\beta)} \frac{y}{1-\beta}\, dF_{x,\varepsilon}(y) \\ &= \int_0^{F_{x,\varepsilon}^{-1}(1-\beta)} \frac{y}{1-\beta}\, dF(y) + \varepsilon \int_0^{F_{x,\varepsilon}^{-1}(1-\beta)} \frac{y}{1-\beta}\, d(\Delta_x - F)(y). \end{aligned}$$

Then using the chain rule on the first term

$$\begin{aligned} g'(\varepsilon) = {} & \frac{F_{x,\varepsilon}^{-1}(1-\beta)}{1-\beta} f(F_{x,\varepsilon}^{-1}(1-\beta)) \frac{\partial}{\partial\varepsilon}[F_{x,\varepsilon}^{-1}(1-\beta)] \\ & + \int_0^{F_{x,\varepsilon}^{-1}(1-\beta)} \frac{y}{1-\beta}\, d(\Delta_x - F)(y) \\ & + \varepsilon \frac{\partial}{\partial\varepsilon} \int_0^{F_{x,\varepsilon}^{-1}(1-\beta)} \frac{y}{1-\beta}\, d(\Delta_x - F)(y). \end{aligned}$$

Now

$$\begin{aligned} \mathrm{IF}_{T_\beta,F}(x) = \lim_{\varepsilon\downarrow 0} g'(\varepsilon) = {} & \frac{F^{-1}(1-\beta)}{1-\beta} f(x_{1-\beta})\mathrm{IF}_{1-\beta}(x) \\ & + \int_0^{F^{-1}(1-\beta)} \frac{y}{1-\beta}\, d(\Delta_x - F) \\ = {} & \frac{x_{1-\beta}}{1-\beta} f(x_{1-\beta})\mathrm{IF}_{1-\beta}(x) - \mu_\beta + \frac{x}{1-\beta} I\{x \le x_{1-\beta}\}. \end{aligned}$$

Substituting the influence function $\mathrm{IF}_{1-\beta}(x)$ given in (3.2.3) into this last expression yields:

$$\mathrm{IF}_{T_\beta,F}(x) = \begin{cases} \dfrac{x - \beta x_{1-\beta}}{1-\beta} - \mu_\beta, & 0 \le x \le x_{1-\beta} \\ x_{1-\beta} - \mu_\beta, & x_{1-\beta} < x. \end{cases} \qquad (3.2.4)$$

This influence function is plotted in Figure 3.7(c). Note that it is both continuous and bounded, so that the associated estimators have neither the local difficulties of the quantile estimator (as revealed by a discontinuity in its influence function), nor the problems with outliers of the untrimmed mean (as revealed by the unboundedness of its influence function).

4. The β-Winsorized Mean

The β-Winsorized mean W_β is a linear combination of the β-trimmed mean and the $(1-\beta)$th quantile, so its influence function is the same linear combination of the corresponding curves for the β-trimmed mean and the $(1-\beta)$th quantile. It is plotted in Figure 3.7(d). Note that for small β the influence of "bad" observations at 0 (early failures) is less than that of outliers, whether one trims or Winsorizes. Outliers have greater influence on the Winsorized mean; moveover, since the jump in IF_{W_β} at $x_{1-\beta}$ is inversely proportional to $f(x_{1-\beta})$, one would prefer the trimmed mean if gaps in the lifetime distribution are possible at $x_{1-\beta}$.

These examples suggest that continuous bounded influence functions correspond to estimators which are insensitive to outliers and small round-off errors.

More Properties of the Influence Function

The importance of the influence curve does not derive solely from its usefulness as a diagnostic tool. As illustrated in Figure 3.6 the influence function is really a directional derivative of T at F. When an even stronger derivative exists, that is, when a continuous linear functional exists which provides a good approximation to $T(G)-T(F)$ for all G near F, regardless of direction, then the kernel of the stronger derivative will agree with the influence function at $G = \Delta_x - F$.

Thus it is often the case that the influence function appears in a first derivative term in a series expansion of $T(G)$ for G in a neighborhood of F:

$$T(G) = T(F) + \int \mathrm{IF}_{T,F}(x)\,d(G-F)(x) + R \tag{3.2.5}$$

where R is a remainder term. While the term involving the integration may not appear to be a "derivative," we must remember that the derivative of a function at a point is the best linear approximation to it at the point. And a mapping of the form $L(H) = \int k(x)\,dH(x)$ is a linear functional (see Example 6 of Section 3.1.1). In this case the kernel of the linear functional is the influence function $k(x) = \mathrm{IF}_{T,F}(x)$ while $H = G - F$. It turns out that $\mathrm{E}_F[\mathrm{IF}_{T,F}(X)] = 0$, so that the integral in (3.2.5) of the influence function with respect to $dF(x)$ drops out. For further discussion of derivatives for functionals, see Chapter 2 of Huber (1981).

Now we are interested in a probabilistic version of (3.2.5), namely, when it is applied to the empirical distribution $G = F_n$. Then we have an expansion for $T(F_n)$ for F_n in a neighborhood of F:

$$T(F_n) = T(F) + \int \mathrm{IF}_{T,F}(x)d(F_n - F)(x) + R_n \qquad (3.2.6)$$

and often the remainder term satisfies $n^{1/2}R_n \xrightarrow{P} 0$. Hence the difference

$$n^{1/2}\left\{[T(F_n) - T(F)] - \int \mathrm{IF}_{T,F}(x)dF_n(x)\right\}$$

converges to zero in probability, and we have the approximation

$$n^{1/2}[T(F_n) - T(F)] \approx n^{-1/2}\sum_{i=1}^{n} \mathrm{IF}_{T,F}(X_i). \qquad (3.2.7)$$

The central limit theorem applied to the right-hand side of (3.2.7) shows that this average of i.i.d. random variables is asymptotically normal with parameters

$$\mathrm{E}_F[\mathrm{IF}_{T,F}(X)] = 0 \quad \text{and} \quad \mathrm{Var}_F[\mathrm{IF}_{T,F}(X)]. \qquad (3.2.8)$$

Hence the same limiting result is obtained for the left-hand side of (3.2.7), and $\{T(F_n)\}$ is also asymptotically normal. (The existence of the stronger derivative alluded to above is sufficient to guarantee asymptotic normality of the associated estimators, but the existence of the influence function is not. See Appendix B.2 for more details.)

For a simple example of (3.2.7) and (3.2.8) consider again the mean functional $\mu = T(F) = \int x\,dF(x)$ (Example 1 above). It has influence function $\mathrm{IF}_{T,F}(x) = x - \mu$, so $\mathrm{E}_F[\mathrm{IF}_{T,F}(X)] = 0$ and (3.2.6) becomes

$$\overline{X}_n = \mu + \frac{1}{n}\sum \mathrm{IF}_{T,F}(X_i) + R_n,$$

where $R_n = 0$. In this case the "approximation" is exact; and furthermore $n^{-1/2}[\overline{X}_n - \mu]$ is asymptotically normal with parameters 0 and $\mathrm{Var}_F[\mathrm{IF}_{T,F}(X)] = \mathrm{E}_F[X - \mu]^2$.

For a second example (the quantile example) we note that Bahadur (1966) has shown that if F has a continuous and positive density f at $x_q = F^{-1}(q)$, then

$$n^{1/2}[X_{(nq)} - x_q] = n^{-1/2}\sum_{i=1}^{n} \mathrm{IF}_q(X_i) + n^{1/2}R_n$$

where IF_q is given by (3.2.3) and $R_n = 0(n^{-3/4}\log n)$ with probability 1. Thus the approximation (3.2.7) holds and again the estimators are asymptotically normal. The reader may verify that the asymptotic variance of the quantile estimator is

$$\mathrm{Var}_F[\mathrm{IF}_q(X)] = \frac{q(1-q)}{[f(x_q)]^2}. \tag{3.2.9}$$

3.2.3* L-Estimators

The advantage of grouping together all estimators of a certain type is that it leads to an economy of thought, and an overview of the numerous proposals which have been put forward.

Any estimator which is a *linear* combination of the order statistics $T_n = \sum_{i=1}^{n} c_i X_{(i)}$ is called an *L-estimator*. Since examples 1–4 are of this form, it is prudent to collect here a few facts about the entire class. We may represent such estimators by the functional

$$T(F) = \int_0^1 F^{-1}(t)\,dK(t) \tag{3.2.10}$$

where K is a probability distribution on [0,1] with $K(0) = 0$. For then using the fact that $F_n^{-1}(t) = X_{(i)}$ is constant on the interval $(i-1)/n < t \le i/n$,

$$T(F_n) = \sum_{i=1}^{n} c_i X_{(i)}$$

where

$$c_i = K\left(\frac{i}{n}\right) - K\left(\frac{i-1}{n}\right), \qquad i = 1,2,\ldots,n.$$

An estimator which is based on only one or two order statistics can be expressed in this form by taking K to be a discrete distribution; for example, $K = \Delta_{1/2}$ yields the median. Another example, the β-trimmed mean, is obtained by taking K to be the continuous uniform distribution on $[0, 1-\beta]$. The Winsorized mean is obtained by taking K to be a mixture of a certain discrete and a certain continuous uniform distribution. What are they?

We may obtain the influence function of any L-estimator by exploiting the representation of it as a probability average of quantiles (3.2.10) via

$$\lim_{\varepsilon \downarrow 0} \int_0^1 \frac{F_{x,\varepsilon}^{-1}(t) - F^{-1}(t)}{\varepsilon}\,dK(t) = \int_0^1 \mathrm{IF}_t(x)\,dK(t)$$

where $\mathrm{IF}_t(x)$ is the influence function (3.2.3) of the quantile function

*Starred sections will generally require more mathematical background than the others, and may be skipped at first reading without loss of continuity.

$F^{-1}(t)$. It is shown in the problems that the interchange of limits is justified if $F^{-1}(t)$ is uniformly differentiable on the support of K and if the resulting derivative is K-integrable. By substitituting $\mathrm{IF}_t(x)$ in the integral and writing $u(t)$ for $\int_0^t dK(x)/f(F^{-1}(x))$ we obtain, after integration by parts,

$$\mathrm{IF}_T(x) = u(F(x)) - \int_0^1 u(t)\,dt. \tag{3.2.11}$$

To determine the breakdown point ε^* for L-estimators we substitute (3.2.1) into (3.2.10) to obtain

$$\begin{aligned} T(F_{x,\varepsilon}) = & \int_0^{(1-\varepsilon)F(x)} F^{-1}\left(\frac{t}{1-\varepsilon}\right) dK(t) \\ & + x[K((1-\varepsilon)F(x)+\varepsilon) - K((1-\varepsilon)F(x))] \\ & + \int_{(1-\varepsilon)F(x)+\varepsilon}^{1} F^{-1}\left(\frac{t-\varepsilon}{1-\varepsilon}\right) dK(t) \end{aligned}$$

which is clearly unbounded in $x > 0$ if and only if $K(1) - K(1-\varepsilon)$ is positive. Thus the breakdown point ε^* is the largest ε for which $K(1-\varepsilon) = 1$, namely, $\varepsilon^* = 1 - K^{-1}(1)$.

3.2.4* Qualitative Robustness

Next we consider the property of *qualitative* robustness, which cannot really be understood without a discussion of topologies on the class of distributions. The Kolmogorov or supremum metric mentioned above defines a topology with lots of open sets so that it is easy for a functional T to be continuous at a particular F in this strong topology. This kind of continuity is strong enough to guarantee consistency of the induced estimators, but it is not strong enough to guarantee *qualitative* robustness. Continuity with respect to a weaker topology (with fewer open sets) on the class of distribution functions is required.

Hampel (1968, 1971) introduced the concept of *qualitative robustness* of a sequence of estimators $\{T_n\}$. Roughly speaking, it requires that a slight change in the model F should result in only a small change in the distribution of T_n, uniformly in n. When $T_n = T(F_n)$ for some functional T, this notion may be formulated in terms of continuity of T with respect to an appropriate topology. Here we restrict attention to the strong topology generated by the Kolmogorov metric $D(F,G) = \sup_x |F(x) - G(x)|$, since some of the ideas are more easily explained with it. (We note that the appropriate topology suggested by Hampel is the weak-star topology; however, when F is continuous the neighborhood systems of F are the same for the two topologies, so that continuity at such an F is the same concept.) For further discussion of metrics and topologies on the class of distributions, and the applications to statistics, see Chapter 5, Staudte (1980) and Donoho and Liu (1988).

Define T to be *continuous* at F if for any sequence of distribution functions $\{G_n\}$ with $D(G_n,F) \to 0$ we also have $|T(G_n) - T(F)| \to 0$. Now by a well-known theorem of Kolmogorov, if F_n is the empirical distribution based on a sample from F, then with probability 1, $D(F_n,F) \to 0$ as $n \to \infty$. Hence if T is continuous at F, then we have consistency of $\{T(F_n)\}$ for $T(F)$. However, continuity of T at F alone does not guarantee stability of the estimator sequence $\{T(F_n)\}$. Consider the following examples.

1. Even if the mean functional T is restricted to G for which $T(G) = \int x\,dG(x)$ is finite, T fails to be continuous at any F. For given any fixed F, define $F_{x,\varepsilon} = (1-\varepsilon)F + \varepsilon\Delta_x$. Then $D(F, F_{x,\varepsilon}) \leq \varepsilon$ for all x, but $|T(F) - T(F_{x,\varepsilon})| = \varepsilon|x - T(F)|$, which can be made arbitrarily large, by choice of x.
2. Fix $0 < t < 1$. The quantile functional $T(G) = G^{-1}(t)$ is well defined at all G, and continuous at F if and only if F is strictly increasing at $F^{-1}(t)$. Note that even if T is continuous at F, it is not continuous at $F_{x,\varepsilon}$ which can be made arbitrarily close to F by choice of ε (see Figure 3.4). More details are given in the problems.
3. The β-trimmed mean $T(G) = [1/(1-\beta)]\int_0^{1-\beta} G^{-1}(t)\,dt$ is everywhere continuous.
4. The Winsorized mean inherits the problem of the quantile functional, namely, it is not continuous in any open set of distributions F.

Because qualitative robustness requires more mathematics to define, it is not often discussed in articles on robust statistics. However, for a simple class of estimators (the linear functionals defined in example 6 of Section 3.1.1), qualitative robustness is essentially equivalent to the continuity and boundedness of the influence function. Morover, for maximum likelihood estimators of location with *known* scale and nondecreasing influence function, qualitative robustness is equivalent to boundedness of the influence function. See Chapters 2 and 3, Huber (1981). Unfortunately, these estimators are not in general scale equivariant when the influence function is bounded. There are methods for modifying these estimators to make them scale equivariant (see Chapter 4), but then it is not clear whether continuity and boundedness of the influence function will guarantee qualitative robustness.

Despite the lack of simple-to-check criteria for determining whether a given estimator is qualitatively robust, it appears that boundedness of the influence function is quite generally a necessary condition. Let $\gamma^* = \sup_x \mathrm{IF}_{T,F}(x)$ denote the *gross error sensitivity* of the estimator T at F. This gives the maximum infinitesimal influence that contamination can have on $T(F)$. When γ^* is finite, the estimator T is called B-robust or *bias robust* at F, because then the asymptotic bias $[T(F_{x,\varepsilon}) - T(F)] \approx \varepsilon\gamma^*$ is bounded. A very fruitful approach to robust estimation due to Hampel (1968) and continued in Rousseeuw (1981) and Hampel et al. (1986), is to place a bound on γ^*, and subject to this bound find the T that is most efficient. A special case of this approach is given in Theorem 4.3 of Section 4.3.3.

3.2.5 Concluding Remarks

It is possible to represent many estimator sequences $\{T_n\}$ by $\{T(F_n)\}$, where F_n is the usual empirical distribution and T is a functional defined on a class of distributions containing the model F. It may happen that results derived for $T(G)$, G near F, do not carry over to $T(F_n)$ when n is small. However, there are many advantages in studying an estimator sequence via T.

First, we gain conceptual clarity. We know by virtue of defining T what we would estimate, namely $T(F)$, no matter what F happens to be. This frees us from estimating the parameter of a specific model, say θ of $\theta^{-1} \cdot e^{-x/\theta}$, when the data may well be insufficient to determine the goodness of fit.

Second, we may often derive the breakdown point $\varepsilon^*(T,F)$, the influence function $\mathrm{IF}_{T,F}(x)$, and the asymptotic variance $V(T,F) = \mathrm{E}_F[\mathrm{IF}_{T,F}(x)]^2$ directly from T, and these will be a guide to the behavior of $T(F_n)$, at least for larger n. The calculations with T are often simpler than finite sample manipulations with $T(F_n)$. For example in the exponential scale problem, it is messy to find the value of m that minimizes the variance of $c_m X_{(m)}$, but it is simple to show that $q \approx .8$ minimizes its asymptotic variance $q/(1-q)[\ln(1-q)]^2$. Similarly one could find the finite sample breakdown point ε_n^* of $T(F_n)$ for each n, and then "summarize" this sequence for large n by defining $\varepsilon^* = \lim \varepsilon_n^*$; however it is simpler to find ε^* directly from T. There is an interchange of limits involved in working with T, since we are effectively letting $F_n \to F$ first, then deriving limiting properties of $T(F)$, and finally drawing conclusions about $T(F_n)$. However, in our experience this interchange rarely leads to different conclusions.

3.3 DESCRIPTIVE MEASURES OF SCALE

So far we have only considered robust estimation in the parametric setting of a scale parameter for an exponential distribution, but now we want to change our orientation and emphasize the descriptive nature of measures of scale. What is a measure of scale for nonparametric families? What properties would we like it to have? We will see that in a certain asymptotic sense the classical MVUE criterion for estimators of scale parameters extends to nonparametric familes, provided one uses *standardized* variance as a measure of efficiency. Simulation studies in Section 3.3 show that these asymptotic approximations are informative for sample sizes as low as 20.

3.3.1 Measures of Scale

We use the notation $F \rightarrow T(F)$ when we want to think of T as a functional on a class of distributions including F, but it is awkward to express certain invariance properties with it. For example, we say that T is *scale equivariant* if $T(F(x/c)) = cT(F(x))$; but it is more natural to write $T(cX) = cT(X)$ with the understanding that $T(X)$ means $T(F)$ when $X \sim F$.

We define a *measure of scale* for positive† random variables as any descriptive measure T which satisfies the two conditions:

$$
\begin{aligned}
&\text{(i)} \qquad T(cX) = cT(X), \qquad c > 0 \\
&\text{(ii)} \qquad a \leq X \leq b \quad \text{implies} \quad a \leq T(X) \leq b.
\end{aligned}
\tag{3.3.1}
$$

Condition (i) ensures that T will serve as a scale parameter in the traditional sense; for if T is restricted to the family $\{F_\theta : \theta > 0\}$ where $F_\theta(x) = F_1(x/\theta)$, then $T(F_\theta) = \theta T(F_1)$, and any multiple of a scale parameter is a scale parameter; see property 2 of Section 2.2.1.

Condition (ii) ensures that T has some descriptive value across parametric families. The range $T(F) = F^{-1}(1) - F^{-1}(0)$, e.g., satisfies (i) but not (ii); the range tells us that $U(0,\theta)$ has half the magnitude of $U(0,2\theta)$; but it cannot distinguish between $U(0,\theta)$ and $U(\theta,2\theta)$.

Other conditions which further restrict the notion of scale measure may be added to (3.3.1), but we prefer to keep the list as brief as possible.

3.3.2 Efficiency in Terms of Standardized Variance

Suppose T satisfies conditions (3.3.1) for a measure of scale. Then the associated sequence of estimators $\{T(F_n)\}$ inherits the scale equivariance property; see Problem 14. Also assume that $\{T(F_n)\}$ is asymptotically normal

†By "positive" we mean $\Pr\{X > 0\} = 1$; and similarly, if we write $a \leq X \leq b$, it is understood that X lies within the closed interval $[a,b]$ with probability 1.

(3.1.2) with asymptotic variance $V(T,F)$. Define the *standardized variance* of T at F by

$$v(T,F) = \frac{V(T,F)}{T^2(F)}. \tag{3.3.2}$$

When F is restricted to a parametric scale family F_θ, $\theta > 0$, the standardized variance $v(T,F)$ is free of θ. Moreover, when (3.1.3) is satisfied, so that $E_\theta[T(F_n)] \to T(F_\theta)$ and $n\,\mathrm{Var}_\theta[T(F_n)] \to V(T,F_\theta)$, the standardized variance is the limiting variance of $n^{1/2}$ times the unbiased estimator $T(F_n)/E_1[T(F_n)]$. Thus trying to choose T to minimize $v(T,F_\theta)$ is the asymptotic problem associated with finding minimum variance unbiased estimators, as discussed in Section 2.3.1.

Leaving the parametric restriction on F aside, it will still often be the case that $T(F_n)/T(F)$ converges in probability to 1, and that

$$\mathrm{Var}\left[n^{1/2}\frac{T(F_n)}{T(F)}\right] \to v(T,F).$$

For such measures of scale T we want one that we may estimate with small standardized variance (and hence small relative standard error) for the class of F of interest. In Sections 3.3.3 and 3.4 we consider F near the exponential, but first we illustrate the properties of $v(T,F)$ for some parametric families, the uniform, exponential, and Pareto families. In each case we look at the descriptive measures (i) quantiles and (ii) trimmed means. The "pth norms" are treated in Problem 15.

Example 1. The uniform $U(0,\theta)$ family is generated by $F(x) = x$, $0 < x < 1$. Hence:

(i) For the tth quantile $T(F) = F^{-1}(t)$, we have by (3.1.5) that $v(T,F) = t^{-1} - 1$, which is decreasing in t. Thus larger quantiles are preferable for estimating scale, which is consistent with the classical results concerning $X_{(n)}$ for $U(0,\theta)$.

(ii) Using (3.1.6) we find that β-trimmed means have

$$v(\beta,F) = \left(\frac{1}{1-\beta}\right)\left(\frac{1}{3}+\beta\right)$$

which is increasing in β. The more we trim, the worse the estimator of scale. □

Example 2. The exponential scale family is generated by $F(x) = 1 - e^{-x}$.

(i) The quantile $F^{-1}(t) = -\ln[1-t]$ has $v(T,F) = t/(1-t)[\ln(1-t)]^2$, which is minimized by the solution of $-2t = \ln(1-t)$, or $t \approx .8$.

Table 3.2 Optimal Quantile Estimates of Scale θ for the Pareto (r,θ) Family and the Corresponding Values of the Standardized Variance

r	t_r = optimal t	$v(t_r, r)$
1	.5	4.00
2	.62	3.76
3	.66	2.36
$\vdots$	$\vdots$	$\vdots$
$+\infty$	.8	1.56

(ii) For β-trimmed means, (3.1.6) yields

$$\mu_\beta = 1 + \frac{\beta \ln \beta}{1-\beta}$$

$$\sigma_\beta^2 = 1 - \frac{\beta \ln^2 \beta}{1-\beta^2}$$

and

$$v(\beta, F) = \frac{1-\beta^2 + 2\beta \ln \beta}{(1-\beta+\beta \ln \beta)^2}.$$

We expect trimming to have a deleterious effect on the mean $\overline{X}_n$, and indeed $v(\beta, F)$ is increasing in β, albeit slowly from $v(0,F) = 1$ to $v(.1,F) \approx 1.18$ to $v(1,F) = \infty$. □

Example 3. For the Pareto family with $F_r(x) = 1-(x+1)^{-r}$, $x > 0$, the quantiles are of special interest.

(i) First we note that $T(F_r) = F_r^{-1}(t)$ has

$$v(T, F_r) = \frac{t}{r^2(1-t)[1-(1-t)^{1/r}]^2}.$$

The optimal choice of t for a given r is the solution of $(1-t)^{-1} = (1+2t/r)^r$. The solutions and the standardized variance of each are listed in Table 3.2 for some special cases.

This example illustrates the fact that in *very* heavy-tailed distributions the large quantiles are inefficient estimates of scale. For $r = 1$, the mean fails to exist, and the median is the best quantile estimator of scale. As r increases, so does the optimal quantile, with limiting value equal to the solution of $(1-t)^{-1} = e^{2t}$, or $t \approx .8$; which is the optimal quantile for the exponential distribution of Example 2.

(ii) For the special case $r = 1$, the β-trimmed mean yields an estimator with standardized variance $v(\beta)$ sketched in Figure 3.8. Note that

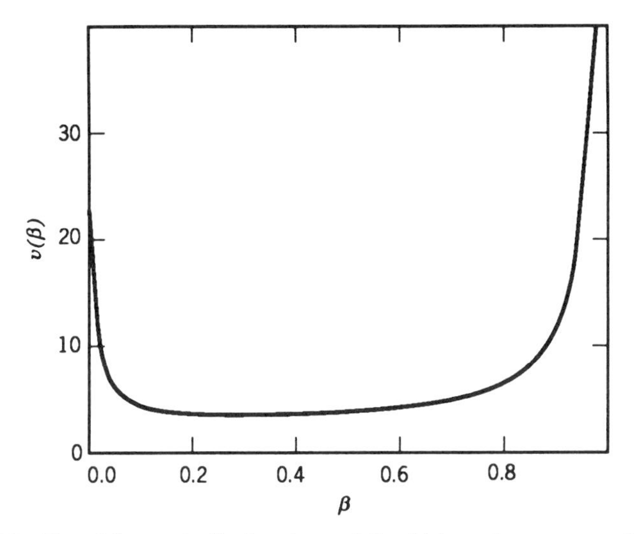

FIGURE 3.8 Plot of the standardized variance of the β-trimmed mean versus the trimming proportion.

30% trimming of outliers is recommended for this heavy-tailed distribution. For values of $r > 1$, a large proportion of the observations are very small, and trimming near 0 may be more efficient than trimming outliers. □

3.3.3 Simulation Results

We now review the simulation results obtained earlier in the mean square analysis of Section 2.3.4. For each estimator T_n and mixture distribution we list in Table 3.3 the estimates of the finite sample standardized variance $20 \times \text{Var}[T_{20}]/\text{E}^2[T_{20}]$, which is approximately the asymptotic standardized variance $v(T, F)$ defined in the last section.

Remarks

1. For the exponential model in Table 3.3 we have also listed in parentheses some exact values to two decimal places of the sample size 20 standardized variances. For the sake of comparison, the asymptotic values derived in Example 2 of Section 3.3.2 are included in square brackets. These results suggest that the asymptotics will be a useful guide to estimator performance for n as small as 20.

Table 3.3 Estimates of Standardized Variances of Classical Estimators of the Exponential Scale Parameter

Model	$\overline{X}$	$c_{16}X_{(16)}$	$k\overline{X}_{.1}$	$\overline{X}_{OR}$
$E(x) = 1 - e^{-x}$	0.97 (1.00) [1.00]	1.45 (1.51) [1.56]	1.12 (1.15) [1.18]	1.04
$.95E(x) + .05E\left(\frac{x}{3}\right)$	1.28	1.63	1.22	1.31
$.8E(x) + .2E\left(\frac{x}{3}\right)$	1.61	2.15	1.55	1.81
$.95E(x) + .05H\left(\frac{x}{6}\right)$	1.72	1.58	1.19	1.27

2. The sample mean compares favorably to the best quantile estimator using the criterion of standardized variance, but the outlier rejection plus mean and the trimmed mean perform even better. If standardized MSE were the criterion, the sample mean would fare considerably worse because of its large bias.
3. We mentioned earlier, Remark 5 of Section 2.2.4, that because $(\text{bias})^2$ and variance were changing with n, the $n = 20$ results may not carry over to other sample sizes. In particular the asymptotic variance is no guide to the $(\text{bias})^2$ and MSE of an estimator. From our newly adopted point of view, however, these problems disappear. As explained just after (3.3.2), the asymptotic standardized variance is a guide to the finite sample performance, no matter which model is correct.

3.3.4 Summary

We have seen in Sections 3.3.2 and 3.3.3 that the asymptotic standardized variance $v(T,F)$ of T at F is a meaningful guide to the performance of scale estimators $\{T(F_n)\}$ of $T(F)$. In particular:

1. $v(T,F)$ is constant on each scale parameter family

$$\left\{F_\theta : F_\theta(x) = F\left(\frac{x}{\theta}\right), x > 0;\ \theta > 0\right\}.$$

2. When F is restricted to a scale parameter family, the problem of finding the equivariant MVUE of the scale parameter θ is equivalent (asymptotically) to minimizing $v(T,F)$ by choice of a measure of scale T. Now we may meaningfully discuss minimizing $v(T,F)$ for F belonging to nonparametric families.

3. The *relative standard error* of $T(F_n)$ in estimating $T(F)$ may be approximated by $\sqrt{v(T,F_n)/n}$, often for sample sizes as small as $n = 20$.

3.4* STABILITY OF ESTIMATORS ON NEIGHBORHOODS OF THE EXPONENTIAL SCALE PARAMETER FAMILY

So far we have discussed the efficiency of various estimators for particular parametric families, mainly in terms of their asymptotic variances. Since the asymptotic variance depends on the influence function, the latter can give us insight into the effect of small changes on the asymptotic variance. Now in practical terms, the true distribution only lies within some neighborhood of our hypothesized model, so we would like to know how the asymptotic variance and hence how the efficiency varies over this neighborhood. Ideally it is stable, so that our claimed efficiency holds up, whether or not we have the true F. Solutions to this problem are difficult, but some progress has been made.

Let $\mathcal{F}_\delta$ be a neighborhood of the exponential scale parameter family. For example, in the sequel we assume that $\mathcal{F}_\delta$ is the class of distributions generated by scale transformations of $\{F : F$ has density of the form $f(x) = (1-\delta)e^{-x} + \delta h(x)$ where h is an arbitrary continuous density on the positve axis$\}$.

We want to choose a measure of scale T which in some sense minimizes $v(T,F)$ as F ranges over $\mathcal{F}_\delta$. In this section we look at two approaches to this problem, the relative efficiency approach and the infinitesimal approach. In the first, we restrict attention to two measures of scale S,T and compare $v(S,F)$ with $v(T,F)$ as F ranges over $\mathcal{F}_\delta$. In the second approach we fix T and try to gain insight by studying the rate of change in the asymptotic standardized variance $v(T,F_{x,\varepsilon})$ for small ε, where $F_{x,\varepsilon} = (1-\varepsilon)F + \varepsilon\Delta_x$.

The relative efficiency approach to choosing between different descriptive measures was introduced by Bickel and Lehmann (1975) in a series of papers advocating an emphasis on descriptive measures for various nonparametric families, including nonparametric neighborhoods of parametric models. The advantage of this approach is that the statistician is freed from reliance on a fixed parametric model. The disadvantage is that it is difficult to get tight quantitative results that show that a particular choice of descriptive measure leads to induced estimators which are uniformly efficient over a nonparametric class.

The infinitesimal (or change-of-variance) approach was developed mainly by Rousseeuw [see Chapter 2.5 of Hampel et al. (1986) and references therein] and enables one to approximate the asymptotic variance of an estimator on contamination neighborhoods of a parametric family. It has the advantage of a straightforward methodology which leads to numerous insights. The disadvantage is that calculations are often more cumbersome than for the derivation of the influence function.

3.4.1 The Relative Efficiency Approach

Given two measures of scale S,T with standardized variances $v(S,F)$ and $v(T,F)$, we define the *relative efficiency* of S to T at F by

$$e(S,T;F) = \frac{v(T,F)}{v(S,F)}.$$

This efficiency gives the ratio of sample sizes required to achieve the same relative standard error in estimating the scale of F, and clearly will be constant on each scale parameter family.

Example 1: Quantiles versus the Mean. Let $S(F)$ be the tth quantile of F, $S(F) = F^{-1}(t)$, for some fixed t, $0 < t < 1$, and let $T(F)$ be the mean of F. Then using earlier results,

$$e(F^{-1}(t),\mu(F);F) = \frac{\sigma^2}{\mu^2}\frac{f^2(F^{-1}(t))[F^{-1}(t)]^2}{t(1-t)} \tag{3.4.1}$$

where μ, σ^2, and f are the mean, variance, and density of F, respectively. We leave it to the problems to show that this expression is unbounded above as F varies over $\mathcal{F}_\delta$, no matter how small $\delta > 0$ is taken. Thus any quantile estimator will perform much better than the mean for some $F \in \mathcal{F}_\delta$. However, we will show in Theorem 3.1 below that the quantile $F^{-1}(t)$ never does too badly relative to the mean, provided the proportion of contamination δ satisfies $\delta < t < 1-\delta$. □

Example 2: Trimmed Means versus the (Untrimmed) Mean. Define $\mu_\beta = \mu_\beta(F)$ to be the β-trimmed mean introduced in example 3 of Section 3.1.1. The relative efficiency of μ_β to μ_0 at F is

$$e(\mu_\beta,\mu_0;F) = \frac{v(\mu_0,F)}{v(\mu_\beta,F)} \tag{3.4.2}$$

where

$$v(\mu_\beta,F) = \frac{V(\mu_\beta,F)}{[\mu_\beta(F)]^2} = \frac{\sigma_\beta^2 + \beta(x_\beta - \mu_\beta)^2}{(1-\beta)\mu_\beta^2}$$

and $\mu_\beta, \sigma_\beta^2$ are the mean and the variance of the β-trimmed distribution. Examples are given in the problems which show that (3.4.2) is unbounded over $\mathcal{F}_\delta$. However, the following theorem shows that the reciprocal of (3.4.2) is bounded. □

Theorem 3.1

(i) If $\delta < t < 1-\delta$, then

$$\inf_{F \in \mathcal{F}_\delta} e(F^{-1}(t), \mu_0; F) > 0.$$

(ii) If $\delta < \beta < 1-\delta$, then

$$\inf_{F \in \mathcal{F}_\delta} e(\mu_\beta, \mu_0; F) > 0.$$

Proof of (i). Let μ, σ^2 denote the mean and the variance of F. We need to show that

$$\frac{\sigma^2}{\mu^2} \frac{f^2[F^{-1}(t)][F^{-1}(t)]^2}{t(1-t)} > k > 0. \tag{3.4.3}$$

That is, the efficiency of the tth quantile to the mean is bounded below by a positive constant as F varies over $\mathcal{F}_\delta$. To minimize σ^2/μ^2, we must choose a density h with first two moments μ_1, μ_2, which minimizes

$$\frac{\sigma^2}{\mu^2} = \frac{(1-\delta)2 + \delta\mu_2}{[(1-\delta)+\delta\mu_1]^2} - 1 = \frac{1-\delta^2 + \delta[(\mu_2 - \mu_1^2) + (1-\delta)\mu_1(\mu_1 - 2)]}{[1-\delta+\delta\mu_1]^2}.$$

From this expression it is clear that for fixed μ_1, σ^2/μ^2 is minimized by taking $\mu_2 = \mu_1^2$, i.e., h degenerate at μ_1. It is easy to show that among all degenerate distributions the one concentrated at $\mu_1 = 2$ yields the minimum value $\sigma^2/\mu^2 \geq (1-\delta)/(1+\delta)$. Of course "degenerate" densities h are not allowed, but we may choose a sequence of proper densities such that σ^2/μ^2 approaches $(1-\delta)/(1+\delta)$.

To get a positive lower bound on $F^{-1}(t)f[F(t)]$, note that for $\delta < t < 1-\delta$, $t = (1-\delta)\varepsilon[F^{-1}(t)] + \delta H[F^{-1}(t)]$ implies

$$\ln\left(\frac{1-\delta}{1-t}\right) \leq \ln\left(\frac{1-\delta}{1-\delta-t}\right). \tag{3.4.4}$$

Now $f(x) = (1-\delta)e^{-x} + \delta h(x) \geq (1-\delta)e^{-x}$, xe^{-x} is increasing for $x < 1$, decreasing for $x > 1$. Thus

$$F^{-1}(t)f[F^{-1}(t)] \geq \min\left\{(1-t)\ln\left(\frac{1-\delta}{1-t}\right), (1-\delta-t)\ln\left(\frac{1-\delta}{1-\delta-t}\right)\right\} \tag{3.4.5}$$

Table 3.4 Value of t^* that Maximizes the Lower Bound on $e(F^{-1}(t), \mu_0; F)$ for Fixed δ and the Corresponding Relative Efficiency

δ	t^*	Lower Bound at t^*	$e(F^{-1}(t^*), \mu_0; F)$
.00	.80	.6476	.6476
.01	.77	.6060	.6452
.05	.69	.4801	.6162
.10	.62	.3712	.5738

for $\delta < t < 1-\delta$. Combining these lower bounds on $F^{-1}(t)f[F^{-1}(t)]$, σ^2/μ^2 yields (3.4.3).

Proof of (ii). Let $\nu_\beta = \nu(\mu_\beta, F)$. We learned in part (i) that $\nu_0 = \sigma^2/\mu^2 \geq (1-\delta)/(1+\delta)$. Since $\delta < \beta < 1-\delta$, we may also use (3.4.4) of part (i) to conclude that $F^{-1}(1-\beta)$ is bounded away from 0 and $+\infty$, no matter what the choice of h. It follows that μ_β is bounded away from 0 and $V(\mu_\beta, F)$ is bounded, so that ν_β is also bounded. This completes the proof. □

The above theorem is qualitative in nature; it only tells us that quantiles and trimmed means never do "too badly" relative to the mean whenever $F \in \mathcal{F}_\delta$. Ideally we would find the greatest lower bounds of Theorem 3.1 as functions of δ and t. It is clear from the proof that the lower bounds which we obtained are not tight; nevertheless we will examine them. In case (i), the lower bound on $e(F^{-1}(t), \mu_0; F)$ as F ranges over $\mathcal{F}_\delta$ is $[1/t(1-t)](1-\delta)/(1+\delta)$ times the square of the right-hand side of (3.4.5). For $\delta = .05$ this lower bound on $e(F^{-1}(t), \mu_0; F)$ as a function of t has its maximum value near $t = .7$; the value of the lower bound in this case is $\approx .48$. Some other values of δ and "optimal" choice of t are given in Table 3.4.

While the bounds on the efficiency in Table 3.4 may not seem too impressive, one must keep in mind that even the best quantile estimator is not very efficient relative to the mean for exponential data. Similar results for trimmed means would be much higher.

3.4.2 The Infinitesimal Approach

Another way of studying $e(S,T;F)$ as F varies from a parametric model would be to consider its behavior, for small t, at the mixture model, $F_{x,\varepsilon} = (1-\varepsilon)F + \varepsilon\Delta_x$. It turns out that it is easier to study the natural logarithm of the asymptotic variance, so define $w(T,F) = \ln[\nu(T,F)]$ and for each x let

$$\dot{w}_T(x) = \lim_{\varepsilon \downarrow 0} \frac{w(T, F_{x,\varepsilon}) - w(T,F)}{\varepsilon}.$$

This defines the *change-of-variance curve*. It is shown in the problems that

$$e(S,T;F_{x,\varepsilon}) \approx e(S,T;F)e^{\varepsilon[\dot{w}_T(x)-\dot{w}_S(x)]}. \tag{3.4.6}$$

Thus we may estimate lower and upper bounds on the efficiency of S to T as the distribution strays distance ε from F by finding

$$\sup_x[\dot{w}_T(x)-\dot{w}_S(x)] \quad \text{and} \quad \inf_x[\dot{w}_T(x)-\dot{w}_S(x)] \tag{3.4.7}$$

and substituting these quantities, respectively, for the bracketed term in the exponent of (3.4.6).

Rousseeuw (1981) has exploited this "change of variance" idea in a number of other contexts; his approach (called optimal V-robustness) is to place a bound on $\sup_x[\dot{w}_T(x)]$ and find the T that minimizes $v(T,F)$ subject to this constraint. Here we will only calculate and interpret the values of $w_T(x)$ for some examples.

Example 1. Let $T(F)=\mu(F)$, the mean functional. Then

$$w(T,F)=\ln[\sigma^2(F)]-\ln[\mu^2(F)]$$

and

$$\dot{w}_T(x)=\frac{\mathrm{IF}_{\sigma^2(F)}(x)}{\sigma^2(F)}-\frac{2\,\mathrm{IF}_{\mu(F)}(x)}{\mu(F)}=\left(\frac{x-\mu}{\sigma}\right)^2+1-\frac{2x}{\mu}$$

where $\mu=\mu(F)$, $\sigma^2=\sigma^2(F)$. Note that $w_T(x)$ is an unbounded convex function of x on $(0,\infty)$ with minimum at $x=\mu+\sigma/\mu$.

Example 2. Let $S(F)=F^{-1}(t)$, the tth quantile functional. Then

$$w(S,F)=\ln[t(1-t)]-2\ln[f(F^{-1}(t))]-2\ln[F^{-1}(t)]$$

and

$$\frac{-\dot{w}_S(x)}{2}=\frac{\mathrm{IF}_{f(F^{-1}(t))}(x)}{f(F^{-1}(t))}+\frac{\mathrm{IF}_{F^{-1}(t)}(x)}{F^{-1}(t).}$$

We have found

$$\mathrm{IF}_{F^{-1}(t)}(x)=\frac{t-I[x<F^{-1}(t)]}{F(F^{-1}(t))}$$

earlier and in the problems we show

$$\mathrm{IF}_{f(F^{-1}(t))}(x)=\delta_x(F^{-1}(t))-f(F^{-1}(t))+f'(F^{-1}(t))\mathrm{IF}(x)F(t)$$

Table 3.5 Lower Bounds on the Efficiency of the Best Quantile Estimator Relative to the Mean over ε-neigborhoods of the Model

ε	.00	.01	.05	.10
$e^{-.62\varepsilon}$	1.00	.99	.97	.94
$\inf_x e(S,T;F_{x,\varepsilon})$	.65	.64	.63	.61

where $\delta_x(y) = +\infty$ if $y = x$ and 0 otherwise. Therefore

$$\frac{-\dot{w}_S(x)}{2} = \delta_x(F^{-1}(t)) - 1 + \left\{ \frac{f'(F^{-1}(t))}{f(F^{-1}(t)} + \frac{1}{F^{-1}(t)} \right\} \times \left\{ \frac{t - I[x < F^{-1}(t)]}{f(F^{-1}(t))} \right\}. \quad \square$$

Now we may find approximate lower and upper bounds on the relative efficiency of the tth quantile estimator to the sample mean for ε-neighbourhoods of a particular parametric family. For example, we consider $F(y) = 1 - e^{-y}$ and $t = .8$, the optimal quantile estimator for the exponential scale parameter.

For $T(F) = \mu(F)$, we have from Example 1, $\dot{w}_T(x) = (x-2)^2$, and from Example 2, where $F^{-1}(.8) = \ln(5)$, we have

$$-\dot{w}_S(x) = 2\delta_x(\ln 5) - 1 + \{-1 + [\ln(5)]^{-1}\} \left\{ \frac{.8 - I\{x < \ln 5\}}{.2} \right\}.$$

Clearly

$$\sup_x [\dot{w}_T(x) - \dot{w}_S(x)] = +\infty$$

while

$$\inf_x [\dot{w}_T(x) - \dot{w}_S(x)] = -1 + \left\{ -1 + \frac{1}{\ln 5} \right\} \left(\frac{-.2}{.2} \right) = -\frac{1}{\ln 5} \approx -.62.$$

In view of (3.4.6) we see that the quantile estimator $X_{(.8n)} = F_n^{-1}(.8)$ has infinitely greater efficiency than the mean $\overline{X}_n$ as a scale estimator for any ε-neighborhood of the exponential scale family. Also, the relative efficiency on this neighborhood has lower bound (approximately) of

$$\inf_x e(S,T;F_{x,\varepsilon}) \approx .6476 e^{-.62\varepsilon}.$$

The quantity $e^{-.62\varepsilon}$ represents the proportion of efficiency of S to T retained as F varies over an ε-neighborhood of the model. Some values are given in Table 3.5. These lower bounds are only approximations, but they are significantly higher than the crude lower bounds obtained in Table 3.4.

To summarize, the infinitesimal approach shows that in a mixture model neighborhood of the exponential model the .8 quantile estimator never has efficiency less than .6 relative to the sample mean for mixing proportions as high as .1 contamination. Since the β-trimmed mean is also robust and generally more efficient than the quantile estimator, we would expect its performance on such mixture model neighborhoods to be even better.

3.5 ESTIMATES OF STANDARD ERROR

In this section we continue the discussion of estimates of standard error begun in Section 2.4. The asymptotic results suggest that standard errors may be estimated via formulas for asymptotic variances, and these in turn may be estimated by influence function estimates. Examples of such estimates are given in Section 3.5.1.

Another approach to standard errors is the bootstrap, which involves resampling from the empirical distribution. While this idea was briefly introduced in Section 2.4.2, it is more fully developed (in Section 3.5.2) and better understood here in the context of descriptive measures, since it is essentially an estimate of a functional (namely, the standard deviation of the estimator considered as a function of the underlying distribution F) evaluated at the empirical distribution.

3.5.1 Influence Function Estimates

In Sections 2.4.1 and 2.4.2 we discussed several methods of estimating the standard error $\sqrt{\mathrm{Var}[T_n]}$ of an estimator T_n. When $n\,\mathrm{Var}[T_n] \to V(T,F)$ holds, and n is sufficiently large, another method of estimating the standard error is by the square root of the asymptotic variance $V(T;F)/n$, evaluated at $F = F_n$. Furthermore, as pointed out in Section 3.2.2, it is often the case that $V(T,F) = \mathrm{E}_F[\mathrm{IF}_{T,F}(X)]^2$, so we may estimate $V(T,F)$ by

$$\mathrm{E}_{F_n}[\mathrm{IF}_{T,F_n}(X)]^2 = \frac{1}{n}\sum_{i=1}^{n} \mathrm{IF}^2_{T,F_n}(X_i). \tag{3.5.1}$$

Such an estimate requires only knowledge of the influence function for the functional T. The resulting estimated standard error $\widehat{\mathrm{SE}}[T_n]$ is obtained by taking the square root after dividing (3.5.1) by n; it is called the *influence function estimate* of standard error.

Before looking at some examples, we note that even if the finite sample variances do not converge [so (3.1.3) is *not* satisfied], it may still be the case that the asymptotic normality result (3.1.2) holds; and the statistician may be content to have an estimate of the variance of the limiting distribution.

For with it [and the standard error defined to be the square root of the asymptotic variance $\mathrm{SE}[T_n] = \sqrt{V(T,F)/n}$] one may construct approximate standard confidence intervals for the estimand $T(F)$ of the form

$$\left[T_n - k\widehat{\mathrm{SE}}[T_n],\ T_n + k\widehat{\mathrm{SE}}[T_n]\right].^\dagger$$

Such intervals will have nominal confidence level $2\Phi(k) - 1$. Any estimate of the standard error via an estimate of the asymptotic variance $V(T,F)$ should be judged on the actual coverage probability and expected length of the resulting interval, since it is the confidence interval which motivates the estimate. It will depend on the context whether the finite sample standard deviation of the estimator is desired, or whether an estimate of the standard deviation of the asymptotic distribution will suffice.

Example 1: The Mean. The influence function is $\mathrm{IF}_{T,F}(x) = x - T(F)$, so an estimate of the asymptotic variance is by (3.5.1) equal to

$$\frac{1}{n}\sum_{i=1}^{n}(X_i - \overline{X}_n)^2$$

and the influence function estimate of the standard error is

$$\widehat{\mathrm{SE}}[\overline{X}_n] = \frac{1}{n}\sqrt{\sum_{i=1}^{n}(X_i - \overline{X}_n)^2} = \sqrt{n-1}\frac{s_n}{n}.$$

Example 2: The qth Quantile. This population characteristic cannot be so easily estimated by appealing to asymptotic results since the influence function (3.2.3) depends on the density function of the unknown distribution F. A density estimate is required in this case. Such an estimate is presented in the location problem of Section 4.3.2.

Example 3: The β-Trimmed Mean. The influence function of the β-trimmed mean is given in (3.2.4) and may be rewritten as follows:

$$\mathrm{IF}_{\beta,F}(x) = \begin{cases} \dfrac{x - W_\beta(F)}{1-\beta}, & x \le x_{1-\beta} \\[2ex] \dfrac{x_{1-\beta} - W_\beta(F)}{1-\beta}, & x > x_{1-\beta} \end{cases}$$

where $x_{1-\beta} = F^{-1}(1-\beta)$, $\mu_\beta(F) = (1-\beta)^{-1}\int_0^{x_{1-\beta}} x\,dF(x)$, and $W_\beta(F) = (1-\beta)\mu_\beta(F) + \beta x_{1-\beta}$. Hence (3.5.1) becomes the sample variance of the β-Winsorized observations, times $(1-\beta)^{-2}$.

† For moderate n asymmetric intervals are more appropriate in the scale context.

It is particularly fitting that the β-trimmed mean should have such a simply derived standard error, for it is highly efficient and robust for entire neighborhoods of the exponential model, as seen is Sections 3.3 and 3.4. Thus we may apply it confidently in those problems where there is evidence that the underlying distribution is approximately of the exponential form.

Returning to the cell lifetime data of Section 2.1 which was analyzed in Section 2.4.3 and to Problems 12 and 13 of Chapter 2, we may estimate the Exp(θ) parameter by an appropriately modified trimmed mean using the seven absolute differences .3, .4, .5, .5, .6, .9, 1.7. With $\beta = \frac{1}{7}$ and $c_7 = 1.3614$ (see Problem 2 (c)), the trimmed mean

$$c_7 \overline{x}_{7,1/7} = .725$$

is unbiased for θ if the exponential model happens to be correct. Moreover it has known variance in this case of

$$\mathrm{Var}_\theta[c_7 \overline{x}_{7,1/7}] = \theta^2 c_7^2 \sum_{j=1}^{6} \left(1 - \frac{1}{8-j}\right)^2 = .17\theta^2.$$

Hence $\mathrm{SE}[c_7 \overline{x}_{7,1/7}] \approx .4\theta \approx .3$.

If we cannot justify the Exp(θ) model assumption, we may still estimate the scale of the unknown distribution with the trimmed mean, and its standard error via the sample variance of the Winsorized sample, as derived above in example 3. In this case the β-Winsorized sample is .3, .4, .5, .5, .6, .9, .9, which has standard deviation .234 . Hence the influence function estimate of the standard error is

$$\mathrm{SE}[c_7 \overline{x}_{7,1/7}] = \frac{.234 c_7}{(1-\beta)\sqrt{7}} \approx .14.$$

Of course we are not very content with this result; the relative standard error is roughly one in five, which indicates that more data (or a more sophisticated analysis) is required.

3.5.2 Bootstrap Estimates of Standard Error

Let $T_n = T_n(X_1, \ldots, X_n)$ be an estimator of some estimand $T(F)$, where $X_1, \ldots, X_n$ are i.i.d. F. The realized observations define the empirical distribution F_n, which is held fixed in this section. We continue the discussion of the bootstrap introduced in Section 2.4.2 by rewriting the standard error of an estimator $\mathrm{SE}_F[T_n]$ with subscript and variable interchanged to emphasize its dependence on the distribution F from which the observations are derived:

$$\mathrm{SE}_{T_n}(F) = \sqrt{\mathrm{Var}_F[T_n]}.$$

According to Efron (1982) the bootstrap estimate of $\mathrm{SE}_{T_n}(F)$ is defined by substituting the empirical distribution F_n for F in this expression. This procedure (which we called the descriptive estimate in Section 2.4.1) is possible only when $\mathrm{SE}_{T_n}(F)$ is a known function of F, and some examples are given below. In most cases, however, $\mathrm{SE}_{T_n}(F_n)$ itself must be estimated, and this is done by repeatedly sampling from F_n and finding the standard deviation of the estimates based on these samples. Thus the *bootstrap estimate* of standard error usually means an *estimate* of $\mathrm{SE}_{T_n}(F_n)$ [and hence of $\mathrm{SE}_{T_n}(F)$], derived by a special simulation experiment. This simulation experiment employs the following algorithm [Efron (1982)]:

(i) Fix F_n, the empirical distribution which puts mass $1/n$ on the n data points $x_1, x_2, \ldots, x_n$.
(ii) Draw a random sample *with replacement* of size n from the empirical distribution in (i). That is, let $X_1^*, X_2^*, \ldots, X_n^*$ be i.i.d. F_n. Call the n realized values the *bootstrap sample*, and calculate the bootstrap estimate $T_n^* = T_n(X_1^*, \ldots, X_n^*)$.
(iii) Independently repeat step (ii) a number of times B to obtain estimates $T_n^{*1}, \ldots, T_n^{*B}$ and use the standard deviation of these B estimates to estimate $\mathrm{SE}_{T_n}(F_n)$:

$$\mathrm{SE}_{T_n}(F_n) \approx \sqrt{\frac{\sum_{b=1}^{B}(T_n^{*b} - \overline{T_n^*}_B)^2}{B-1}} \tag{3.5.2}$$

where

$$\overline{T_n^*}_B = \frac{\sum_{b=1}^{B} T_n^{*b}}{B}.$$

A simple Minitab program for implementing this algorithm is given in Appendix D. As Efron (1982) pointed out, the number of bootstrap replications B need not be large (often $B = 100$ will suffice) because the error in (3.5.2) is then likely to be smaller than the error in approximating $\mathrm{SE}_{T_n}(F)$ by $\mathrm{SE}_{T_n}(F_n)$.

As $B \to \infty$, the right-hand side of (3.5.2) converges in probability to the left-hand side, but it is $\mathrm{SE}_{T_n}(F)$ that we want to estimate.

Example 1: The Mean. Estimate $T(F) = \int x\,dF(x)$ by $T(F_n) = \overline{X}_n$, which has variance $\mathrm{Var}_F[\overline{X}_n] = \sigma_F^2/n$. Hence $\mathrm{SE}_{\overline{X}_n}(F) = \sigma_F/\sqrt{n}$ is a known function of F and we need not use simulation to estimate it. We need only find the standard deviation σ_{F_n} of the fixed empirical distribution F_n. It is determined by the variance

$$\sigma_{F_n}^2 = \frac{1}{n}\sum_{i=1}^{n}(x_i - \overline{x}_n)^2. \tag{3.5.3}$$

Note that alternatively we may think of the variance in (3.5.3) as $\text{Var}^*[X^*]$, where $X^* \sim F_n$ has the bootstrap distribution which puts mass $1/n$ on each of the fixed data points $x_1,\ldots,x_n$. (We will indicate when the bootstrap distribution is employed by placing an asterisk on P, E, or Var as the case may be.) To derive $\text{Var}^*[X^*]$, note that

$$P^*\{X_i^* = x_j\} = \frac{1}{n}, \qquad 1 \leq i, \quad j \leq n.$$

Hence

$$\text{E}^*[X_i^*] = \sum_{j=1}^{n} x_j P^*\{X_i^* = x_j\} = \overline{x}$$

and similarly

$$\text{E}^*[(X_i^*)^2] = \frac{1}{n}\sum_{j=1}^{n} x_j^2.$$

Thus the sample mean $T_n = \overline{X}_n$ has a distribution which we approximate by the distribution of $T_n^* = \overline{X}_n^*$, and

$$\text{Var}^*[T_n^*] = \frac{1}{n}\text{Var}^*[X_i^*] = \frac{1}{n^2}\sum_{j=1}^{n}(x_j - \overline{x})^2.$$

This last expression is the variance of the bootstrap distribution of $T_n^* = \overline{X}_n^*$, and the square root of it is the standard error estimate.

Example 2: The Median. Let $T(F) = F^{-1}(1/2)$. Restrict attention to the case of odd sample sizes $n = 2m + 1$, so that $T_n = T(F_n) = X_{(m+1)}$. We will find an explicit expression for the variance of the median of the bootstrap distribution, and hence an explicit expression for the estimate of standard error of T_n. Let $x_{(i)}$ denote the ith smallest observation in the original sample. Let $X_{(i)}^*$ denote the ith smallest observation from a bootstrap sample $X_{(1)}^*,\ldots,X_{(n)}^*$. Let $N_i^* = \#\{j : X_j^* = x_{(i)}\}$, $i = 1,\ldots,n$. Then the reader may

Table 3.6 The Ordered Absolute Differences of Sister Cell Lifetimes

k	1	2	3	4	5	6	7
$x_{(k)}$	.3	.4	.5	.5	.6	.9	1.7

verify that $\sum_{i=1}^{k} N_i^* \sim B(n, k/n)$, so

$$P^*\{X^*_{(m+1)} > x_{(k)}\} = P^*\left\{\sum_{i=1}^{k} N_i^* \le n\right\}$$

$$= \sum_{j=0}^{n} \binom{n}{j} \left(\frac{k}{n}\right)^j \left(\frac{n-k}{n}\right)^{n-j}. \qquad (3.5.4)$$

Thus using a well-known relation [(1.2.6) of David (1981)] between binomial sums and the incomplete beta function, we find

$$w_k = P^*\{X^*_{(m)} = x_{(k)}\} = \frac{n!}{(m!)^2} \int_{(k-1)/n}^{k/n} (1-y)^m y^m \, dy$$

and hence that

$$\text{Var}^*[X^*_{(m+1)}] = \sum_{k=1}^{n} w_k x_{(k)}^2 - \left(\sum_{k=1}^{n} w_k x_{(k)}\right)^2. \qquad (3.5.5)$$

This result was obtained independently by Maritz and Jarrett (1978) and Efron (1979). Thus again in this case it is not necessary to use simulation, for the standard error of the median is estimated by the standard deviation of the bootstrap distribution of $X^*_{(m+1)}$:

$$\text{SE}[X_{(m+1)}] \approx \sqrt{\text{Var}^*[X^*_{(m+1)}]}.$$

To illustrate this method we will find the exact bootstrap distribution of the median for the cell lifetime data which was analyzed in Problems 12 and 13 of Chapter 2 (Table 3.6).

Let $p_k = P^*(\hat{\theta}^* = x_{(k)})$, $k = 1, 2, \ldots, 7$, that is, p_k is the bootstrap probability that the median of a bootstrap sample of size n is equal to the kth-order statistic. Let $\hat{p}_k$ be the estimate of p_k based on 500 bootstrap samples. Using Table 1 of Maritz and Jarrett (1978), we obtain the comparison of p_k and $\hat{p}_k$ for this data set as given in Table 3.7.

Table 3.7 Comparison of Exact and Mont Carlo Bootstrap Distributions for the Ordered Absolute Differences of Sister Cell Lifetimes

k	1	2	3	4	5	6	7
p_k	.0102	.0981	.2386	.3062	.2386	.0981	.0102
$\hat{p}_k$	.01	.128	.548		.208	.098	.008
$x_{(k)}$	.3	.4	.5	.5	.6	.9	1.7

The bootstrap standard error estimate of the median can be found from

$$\mathrm{SE}_{\mathrm{BOOT}}^2 = \sum_{k=1}^{n} p_k x_{(k)}^2 - \left\{ \sum_{k=1}^{n} p_k x_{(k)} \right\}^2$$

and its obvious approximation by replacing p_k by $\hat{p}_k$. In this case $\mathrm{SE}_{\mathrm{BOOT}}^2 = .3475377 - (0.563499)^2 = .03006$; therefore $\mathrm{SE}_{\mathrm{BOOT}} = 0.173$. The approximation, obtained by replacing p_k by $\hat{p}_k$, gives $\mathrm{SE}_{\mathrm{BOOT}} \approx 0.167$.

Sheather (1986a) proposed an estimate of the variance of the sample median which like the above exact bootstrap estimate does not require resampling but which has the advantage of not using the incomplete Beta function.

3.6 PROBLEMS

Section 3.1

1. In this problem we calculate the mean and variance of a conditional distribution for use with the trimmed mean (example 3 of Sections 3.1 and 3.2).
   - **(a)** Find the conditional distribution of X, given that $X < x_{1-\beta}$, when $X \sim \mathrm{Exp}(\theta)$, and plot its density.
   - **(b)** Then find the mean and variance of this distribution (3.1.7).
   - **(c)** Evaluate these quantities for $\beta = .1$, and find the asymptotic variance (3.1.6) of the .1-trimmed mean.
   - **(d)** What constant does the .1-trimmed mean need to be multiplied by in order to make it Fisher-consistent for the exponential scale parameter θ?
   - **(e)** Determine the asymptotic relative efficiency of the best Fisher-consistent quantile estimator of the exponential scale parameter relative to the .1-trimmed mean.

2.* Assume $X_1, \ldots, X_n$ i.i.d. Exp(θ), $\theta > 0$.

(a) Show that if m increases with n so that $m/n \to t$, for some $0 < t < 1$, then the finite sample variance of the unbiased $c_m X_{(m)}$ converges in the sense of (3.1.3):

$$n \times \mathrm{Var}[c_m X_{(m)}] \to \frac{\theta^2 t}{(1-t)[\ln(1-t)]^2}.$$

Hint: Use results from the problems in Chapter 2, and approximate partial sums by integrals.

(b) Show that the above expression is minimized by the solution of $-2t = \ln(1-t)$ or $t \approx .8$.

(c) Let $r = n - [n\beta]$ for some $0 < \beta < 1$. Find c_n such that $c_n \overline{X}_{n,\beta}$ is unbiased for θ. *Hint*:

$$\sum_{m=1}^{r}\sum_{j=1}^{m}\frac{1}{n-j+1} = \sum_{j=1}^{r}\sum_{m=j}^{r}\frac{1}{n-j+1}$$
$$= \sum_{j=1}^{r}\left(1 - \frac{n-r}{n-j+1}\right) = r - [n\beta]\sum_{j=1}^{r}\frac{1}{n-j+1}.$$

(d) Show that for c_n of part (c)

$$n\,\mathrm{Var}_\theta[c_n \overline{X}_{n,\beta}] \to \theta^2\left[\frac{1-\beta^2+2\beta\ln\beta}{(1-\beta+\beta\ln\beta)^2}\right]$$

and that the limit is increasing in β.

3. Consider the class of linear functionals defined in example 6, Section 3.1. Using theorems in Appendix B or elsewhere, give conditions on the kernel k for which the corresponding sequence of estimators satisfies

(a) (3.1.1);

(b) (3.1.2);

(c) (3.1.3).

4. Let $S(F) = 2*\mu(F)$, where $\mu(F)$ is the mean functional defined in example 1 of Section 3.1.1. When $F \sim U(0,\theta)$, show that $S(F_n)$ satisfies (3.1.1) to (3.1.3).

5. In this problem we examine an estimator that does not possess all the desirable properties (3.1.1) to (3.1.3).

(a) Show that $T(F) = F^{-1}(1)$ defines a measure of scale. For which distributions is it defined? What is $T(F_n)$?

*Starred problems require more mathematical background than the others.

(b) Let F be a member of the family of uniform distributions $U(0,\theta)$, $\theta > 0$. Find the breakdown point for T at F and show that the influence function for T does not exist. (Since the influence function does not exist, we expect that the corresponding estimator sequence will have unusual asymptotic behavior.)

(c) Assume $X_1,\ldots,X_n$ i.i.d. $U(0,\theta)$, $\theta > 0$, and show that $T_n = X_{(n)}$ is the MLE of θ. Show that this estimator sequence satisfies (3.1.1), but not (3.1.2), for all $\theta > 0$.

(d) Is there a sequence of real numbers $\{s_n\}$ and limit distribution H such that

$$\frac{T(F_n) - T(F)}{s_n} \xrightarrow{d} H$$

as $n \to \infty$? *Hint*: $(1 + c/n)^n \to e^c$ as $n \to \infty$.

6.* **(a)** Suppose that S,T agree at F and that each satisfies (3.1.2) and (3.1.3). For each n let m_n be the least integer for which

$$\text{Var}[S(F_m)] \leq \text{Var}[T(F_n)].$$

Show that $\lim_{n \geq \infty}[m_n/n]$ exists and equals $E(T,S;F)$.

(b) Show that (3.1.2) implies that

$$\lim_{n\to\infty} n\,\text{Var}[T(F_n)] \geq V(T,F).$$

(c) Give an example for which the inequality in (b) is strict.

7. Let $X_1,\ldots,X_n$ be i.i.d. $(1 - \varepsilon/\sqrt{n})\text{Exp}(x/\theta) + (\varepsilon/\sqrt{n})\text{Exp}(x/c\theta)$ where $c > 1$. Determine the bias for θ and the variance of $\overline{X}_n$ and show that the ratio

$$\frac{(\text{bias})^2(\overline{X}_n)}{\text{Var}(\overline{X}_n)} \to (c-1)^2\varepsilon \qquad \text{as} \quad n \to \infty.$$

Thus with this model for contamination the $(\text{bias})^2$ and variance are of the same order in their contribution to mean squared error as $n \to \infty$.

Section 3.2

8. Consider the Dixon type discordance test [see Likes (1966)] for an exponential sample which rejects a proportion β of outliers if

$$\frac{X_{(n)} - X_{(r)}}{X_{(n)}} > c$$

where $r = n - [n\beta]$. If the sample mean of the (possibly) censored sample is the estimator of the exponential scale parameter,

(a) Show that this estimator may be expressed in the form $T(F_n)$ where

$$T(F) = [1 - I_{OR}(F)]\mu(F) + I_{OR}(F)\mu_\beta(F).$$

Here μ_β denotes the trimmed mean functional and $I_{OR}(F)$ is the indicator of an appropriate "outlier rejection" set.

(b) Calculate the breakdown point for the above "outlier rejection plus mean" rule.

9. Another test for exponentiality is the Shapiro and Wilk (1965) test, which rejects outliers if

$$T_n = \frac{n\overline{X}_n^2}{\sum_{i=1}^n (X_i - \overline{X}_n)^2} > k_n.$$

(a) Can the test statistic T_n be expressed in terms of a functional?

(b) What is the breakdown point of the test statistic?

10. Define $g(\varepsilon) = T(F_{x,\varepsilon})$, and assume that $g(\varepsilon)$ has a continuous derivative on $[0, \varepsilon_0)$ for some $\varepsilon_0 > 0$. Then the influence function (3.2.2) of T at F can be obtained via

$$\lim_{\varepsilon \downarrow 0} \frac{\partial}{\partial \varepsilon} T(F_{x,\varepsilon}). \qquad (*)$$

Assume that F has a continuous positive derivative at its qth quantile.

(a) Consider the qth quantile functional $T(F) = F^{-1}(q)$. Show that if $x > F^{-1}(q)$ is fixed, then for all sufficiently small ε

$$T(F_{x,\varepsilon}) = F^{-1}\left(\frac{q}{1-\varepsilon}\right).$$

Use $(*)$ to obtain $\mathrm{IF}_q(x) = q/f(F^{-1}(q))$, $x > F^{-1}(q)$, and complete the derivation of (3.2.3).

(b) As an alternative to part (a), derive (3.2.3) by differentiating the identity $F_{x,\varepsilon}[F_{x,\varepsilon}^{-1}(t)] = t$ with respect to ε at $\varepsilon = 0$.

(c) Find the limiting distribution of

$$\frac{1}{n^{1/2}} \sum_{i=1}^{n} \mathrm{IF}_q(X_i)$$

and compare the result with (3.1.5).

11. Continuing Problem 3 on the linear functionals,

(a) Derive the influence function in terms of k. Under what conditions on k and F does the influence function exist? Under what conditions is it bounded and continuous for fixed k and F?

(b) Under what conditions on k is the breakdown point of the linear functional zero? positive?

12.* (a) Verify the details of the derivation of the influence function for L-estimators (3.2.11).

(b) Obtain the influence function (3.2.4) of the β-trimmed mean as a special case.

(c) Show that the formula for the asymptotic variance in (3.1.6) can be derived from $\mathrm{E}[\mathrm{IF}_{\beta,F}(X)]^2$.

13.* (a) Show that the Kolmogorov distance of $F_{x,\varepsilon}$ from F satisfies

$$D(F, F_{x,\varepsilon}) \leq \varepsilon.$$

Does the same inequality hold for any mixture $(1-\varepsilon)F + \varepsilon H$?

(b) Show that $D(F,G) \leq \varepsilon$ implies $F^{-1}(t-\varepsilon) \leq G^{-1}(t) \leq F^{-1}(t+\varepsilon)$ for $\varepsilon < \min\{t, 1-t\}$.

(c) Show that the quantile functional $T(G) = G^{-1}(t)$ is continuous at F (with respect to D) if and only if $F^{-1}(s)$ as a function of s is continuous at t.

(d) Show that the quantile functional is not continuous on any open set.

Section 3.3

14. Show that if the measure of scale T is equivariant in the sense of (3.3.1)(i), then the associated sequence $\{T_n\}$ of estimators [with $T_n = T(F_n)$] are scale equivariant, as defined after (2.3.1).

15. For any distribution F concentrated on the positive axis and $p > 0$ define the pth *norm* by $T_p(F) = \mu_p^{1/p}$, where μ_p is the pth moment $\mu_p = \mu_p(F) = \int x^p \, dF(x)$.

(a) Show that $\mu_p(F_n) = (\sum_{i=1}^n X_i^p)/n = p$th sample moment. Is either of $\mu_p(F)$ or $T_p(F)$ a measure of scale?

(b) Show that the breakdown point of T_p is $\varepsilon^* = 0$ for all $p > 0$.

(c) Show that the influence function of T_p at F satisfies

$$\mathrm{IF}(x) = \frac{T_p(F)}{p}\left(\frac{x^p}{\mu_p} - 1\right).$$

(d) Prove that if $\mu_{2p} < \infty$, then

$$n^{1/2}[T_p(F_n) - T_p(F)] \xrightarrow{d} N(0, V_p)$$

where

$$V_p = \frac{(\mu_{2p} - \mu_p^2)}{p^2} \frac{\mu_p^{2/p}}{\mu_p^2}.$$

Under what conditions does $\mathrm{Var}[n^{1/2}T_p(F_n)] \to V_p$?

(e) Show that the standardized variance of the pth norm is $v_p(F) = (1+2p)^{-1}$ when F belongs to the uniform scale parameter family.

(f) Show that the standardized variance of the pth norm is

$$v_p = \left[\frac{\Gamma(2p+1)}{\Gamma^2(p+1)} - 1\right] \frac{1}{p^2} = \frac{1}{p^2} \left[\binom{2p}{p} - 1\right]$$

for the exponential scale model.

(g) In each of the last two models, find the relative efficiency of the mean to the second norm is $e(T_1, T_2; F)$. Note that Bickel and Lehmann (1976) have shown that if $\mathcal{F}$ is the class of F on the positive axis with finite qth moment $\mu_q < \infty$, then

$$\inf_{F \in \mathcal{F}} e(T_p, T_q; F) = \frac{p^2}{q^2}.$$

Section 3.4*

16. (a) Verify the claim made in Example 1 of Section 3.4.1 that

$$\sup_{F \in \mathcal{F}_\delta} e(F^{-1}(t), T_0) = +\infty$$

for the neighborhood $\mathcal{F}_\delta$ of the exponential model. That is, show that there exists a sequence of $F_M \in \mathcal{F}_\delta$ for which the efficiency of a quantile estimator to the mean is unbounded. [*Hint*: take $f_M(x) = (1-\delta)e^{-x} + \delta h_M(x)$, where $h_M(x) = h(x)/\int_0^M h(y)\,dy$, and h has finite mean and infinite variance.]

(b) Verify the claim in Example 2 that

$$\sup_{F \in \mathcal{F}_\delta} e(T_\beta, T_0) = +\infty$$

i.e., trimmed means can be infinitely more efficient than the untrimmed mean in any neighborhood of the exponential model.

17. For $0 < a \leq 1$, $b > 0$, define a family of densities

$$f(x) = \begin{cases} c, & 0 < x \leq a \\ b, & a < x \leq 1 + 1 - t/b \end{cases}$$

where c is determined by a, b, and t. Clearly $F^{-1}(t) = 1$.

(a) Fix a and let $b \to 0$ to show $e(F^{-1}(t), \mu_0; f_{a,b}) \to 0$; that is, quantiles can be infinitely *less* efficient than the sample mean for decreasing densities.

(b) A similar result holds for trimmed means: let $\beta = 1 - t$ in the same example. Then

$$e(\mu_\beta, \mu_0; f_{a,b}) \to 0 \qquad \text{as} \quad a, b \to 0 \qquad \text{if} \quad a = o(b).$$

(c) By modifying the tail in the above example show that quantiles and trimmed means can be infinitely *more* efficient than the mean for decreasing densities.

18.* For any cumulative distribution function F define

$$f(y) = \lim_{h \downarrow 0} \left[\frac{F(y+h) - F(y-h)}{2h} \right]$$

when the limit exists as a real number, and otherwise define $f(y) = +\infty$. Then $\Delta_x = I_{[x,\infty]}(y)$ has density $\delta_x(y) = 0$ for $y \neq x$ and $\delta_x(x) = +\infty$. Some insight into the behavior of density estimators of the form

$$f_n(y) = \frac{F_n(y + h_n) - F_n(y - h_n)}{2h_n}$$

can be gained by finding the influence function of the functional

$$F \to f(y)$$

where y is fixed. Let $F_{x,\varepsilon} = (1 - \varepsilon)F + \varepsilon \Delta_x$, and denote its derivative by $f_{x,\varepsilon}$.

(a) Show that

$$\lim_{\varepsilon \downarrow 0} \left[\frac{f_{x,\varepsilon}(y) - f(y)}{\varepsilon} \right] = \delta_x(y) - f(y).$$

(b) Fix t, $0 < 0 < 1$. Derive the influence function of the functional

$$F \to f(F^{-1}(t))$$

and interpret it [for the estimator $f_n(F_n^{-1}(t))$].

19.* Fix $0 < \beta < 1$, and consider the β-trimmed mean $\mu_\beta(F)$.

(a) Find the change of log-variance curve $\dot{w}_\beta(x)$ for μ_β.
(b) Estimate lower and upper bounds on the relative efficiency of μ_β to μ_0 for ε-neighborhoods of the exponential scale family.

Section 3.5

20. (a) Use the results of example 1 of Section 3.5.2 to find the exact bootstrap distribution of the sample mean based on the seven observations from the exponential distribution which are analyzed in Section 3.5.1. Calculate its standard deviation.
(b) Carry out a simple Monte Carlo study using (3.5.2) with $B = 100$ and obtain the bootstrap empirical distribution of the sample mean; compare this distribution, including its standard deviation, with the results from part (a). Repeat for $B = 400$.
(c) Repeat (a) and (b) for the median discussed in example 2 of Section 3.5.2.

21. Verify the details of the cell lifetime data given in Section 3.5.1.

22. Verify the details in the derivation of (3.5.4) and (3.5.5).

3.7 COMPLEMENTS

3.7.1 Sensitivity Curve

The sensitivity curve was introduced by Tukey (1977) and is a finite sample version of the influence curve. Given a sample $x_1,\dots,x_n$ and an estimate $T_n = T_n(x_1,\dots,x_n)$ it calculates for each x the standardized effect on the estimate of an additional observation at x:

$$\mathrm{SC}(x) = \frac{T_{n+1}(x_1,\dots,x_n,x) - T_n(x_1,\dots,x_n)}{1/(n+1)}.$$

When the estimate can be expressed in functional form $T_n = T(F_n)$ for some T the sensitivity curve can be obtained from the difference quotient of (3.2.2) by substituting F_n for F and $1/(n+1)$ for ε in $[T(F_{x,\varepsilon}) - T(F)]/\varepsilon$.

Further discussion of the sensitivity curve and other measures of influence can be found in Cook and Weisberg (1982), Chapter 3, and in Hampel et al. (1986), Chapter 2.

3.7.2 Resistant Estimates and Qualitative Robustness

An estimate $T_n = T_n(x_1,\dots,x_n)$ is called *resistant* if, roughly speaking, small changes in all of the data (such as round-off errors) or large changes in a small proportion of the data do not affect T_n very much. This idea was introduced by Tukey (1977) and made precise by Boente, Fraiman, and Yohai (1987) in their definitions of qualitative resistance. The authors also relate their concepts to Hampel's (1968) definition of qualitative robustness.

While qualitative robustness for a statistical functional was originally defined in terms of continuity with regard to the domain topology generated by the Prohorov metric [see Hampel (1968)], other authors [Fernholz (1983), Gill (1989)] have used the Kolmogorov metric which generates a stronger topology.

3.7.3 Standard and Nonstandard Errors

Curiously statisticians have spent much more time showing that a particular estimator is more asymptotically efficient than another than deriving reliable estimates of the errors for finite samples. Even in the case of estimating p based on n Bernoulli trials, recent research by Blyth (1986) shows that the usual normal confidence intervals of the form $\hat{p} \pm k\sqrt{\hat{p}(1-\hat{p})/n}$ are too short to give adequate coverage probabilities even for p close to .5 unless n is quite large. And in the same binomial context Blyth (1980) argues the case for a different measure of error (expected absolute error) which is surprisingly simple to estimate and much more intuitive than the standard error.

We cited references in Section 2.6 to bootstrap methods for estimating the standard error. Another resampling scheme which preceded it is the jackknife [Quenouille (1949), Tukey (1958)]. It can be used in a variety of situations to produce estimates of variance and hence standard error. An excellent review of jackknife methods can be found in Miller (1974). The relationship between the jackknife and the bootstrap is discussed in Efron (1982). Efron shows that in general the jackknife estimate of the variance of $\hat{\theta}$ is essentially a bootstrap estimate of the variance of a linear approximation to $\hat{\theta}$. Because of the markedly nonlinear nature of the sample median, it is not surprising that the jackknife produces an inconsistent estimate of its variance. A review of methods of estimating the variance of the sample median can be found in Sheather (1987).

CHAPTER FOUR

Location–Dispersion Estimation

Statisticians are often interested in the "location problem," testing for a possible shift of a symmetric distribution, because it arises naturally in the differences of matched pairs as exemplified in Example 1 of Section 4.1.2 below. We postpone such testing until the next chapter since powerful, robust tests are usually based on efficient, robust estimators of location. In this chapter we concentrate on such estimators for the location of a symmetric distribution, but also treat the asymmetric case where location itself needs to be defined.

It is rare for the spread (dispersion) of a distribution to be known when the location is unknown, so in practice there are at least two quantities of interest: location and dispersion. In the case of parametric models the location–scale parameter represents the unknowns of location and dispersion; numerous examples are given in Section 4.2.

Section 4.3 contains the main results concerning L- and M-estimators of location as well as a few properties of the Hodges-Lehmann estimator. In Section 4.3.1 the notion of location is defined for nonparametric families which may include asymmetric distributions. After discussion of a number of examples of linear combinations of order statistics, the class of L-estimators for location is defined and the influence function derived. We prove an important result (due to Bickel and Lehmann) which shows that the efficiency of trimmed means relative to the untrimmed mean is bounded below by a positive constant. Then we derive the influence function estimate of the standard error of the trimmed mean. Finally we show how to obtain an estimate of the standard error of the median using kernel density estimators.

Section 4.3.3 contains the main results on M-estimators. First we find the influence function and see that it is proportional to the ψ function which defines the M-equation (4.3.6); and we discuss how to exploit this fact to obtain both robustness and efficiency of a location estimator. Next we show (under strong conditions) that M-estimators are consistent and asymptotically normal, and we discuss Huber's (1964) asymptotic minimax results.

In Theorem 4.3 we illustrate Hampel's method of bounding the influence function and determining the most efficient among such bias-robust M-estimators. The discussion is concluded with examples of the most widely used one-step M-estimators and redescending M-estimators.

In Section 4.4 we turn our attention to the concept of dispersion for nonparametric families and present numerous examples. The performance of several competing estimators of the standard deviation of a normal distribution is given. In particular we study the mean deviation, the median absolute deviation, the interquartile range, and the trimmed standard deviation.

In Section 4.5 we look at joint estimation of location and scale and in particular present Huber's Proposal 2 with the algorithm for solving the M-equations.

In Section 4.6 confidence intervals for the median based on order statistics are derived. A refinement due to Hettmansperger and Sheather (1986) which uses nonlinear interpolation to obtain "exact" confidence from the binomially determined probabilities is introduced.

Throughout the chapter we provide estimators of location which are accompanied by estimates of their standard error. In Section 4.7 most of the techniques presented earlier are applied to a variety of real data sets to obtain confidence intervals for the location parameter.

4.1 INTRODUCTION AND EXAMPLES

Given a sample $X_1, \ldots, X_n$ from F, where F is unknown, and the desire to estimate the "location" of F, we must ask several questions before proceeding to assume some parametric family.

4.1.1 Some Initial Questions

1. Is F symmetrically distributed about some fixed point μ? That is, is $P_F\{X \geq \mu + x\} = P_F\{X \leq \mu - x\}$ for all x? In terms of the cumulative distribution function, this property may be expressed as $1 - F(\mu + x^-) = F(\mu - x)$ for all x. This situation arises in the paired difference problem (see Example 1 below) in which the null hypothesis is that each $X_i = Y_i - Z_i$ arises as the difference of a pair Y_i, Z_i which have joint symmetric distribution: $F_{Y,Z}(y, z) = F_{Y,Z}(z, y)$ for all (y, z).

Table 4.1 Additional Hours of Sleep Gained with Drugs I and II and Differences between Them for 10 Individuals

Patient	I (Dextro-)	II (Laevo-)	II–I
1	+0.7	+1.9	+1.2
2	−1.6	+0.8	+2.4
3	−0.2	+1.1	+1.3
4	−1.2	+0.1	+1.3
5	−0.1	−0.1	+0.0
6	+3.4	+4.4	+1.0
7	+3.7	+5.5	+1.8
8	+0.8	+1.6	+0.8
9	+0.0	+4.6	+4.6
10	+2.0	+3.4	+1.4
mean	.75	2.33	1.58
st. dev.	1.79	2.00	1.23

Note that Y_i, Z_i need not be independent. The alternative hypothesis is that there has been a shift by μ in the difference, perhaps because of a shift by μ in Y. If F is not symmetric about some fixed point μ, then we must ask and answer the question: What is an appropriate measure of location in this context?

2. May we assume that F belongs to a parametric location–scale family? Some well-known examples are given in Section 4.2.2. Alternatively, F may be restricted to lying within a neighborhood of some unknown member of a parametric family, or not restricted at all.
3. Is location of prime importance, and the unknown scale or dispersion only a nuisance? Or do we want to estimate simultaneously location and scale?
4. Is the assumption of *independent* X_i's warranted? If not, what type of dependence structure is possible, and what are its effects on proposed estimators?

4.1.2 Examples

Example 1: The Cushny and Peebles Data. In 1904 Cushny and Peebles published their experimental results on "The action of optimal isomers" in the *Journal of Physiology*. The data show the differing effects of optical isomers of hyoscyamine hydrobromide in producing sleep. The average number (over several nights) of hours sleep gained per night by the use of two versions of the drug is given in Table 4.1.

The answer to question 1 of Section 4.1.1 is yes under a null hypothesis of no difference between the drug effects, provided the order in which the drugs are administered to each patient is randomized, and patients are ignorant of which drug they are taking. If the drugs have a different effect, it is not clear that the difference will be symmetrically distributed. Results are independent (question 4) for different patients, provided they do not discuss the effects. The importance of this data set arises from the fact that many famous statisticians, starting with W. A. Gosset, have used it as an example of "normal" data. (See further discussion of this point in Chapter 5.) Finally, to question 3, location is of primary interest, but an estimate of the standard error of the estimate is required.

If one assumes normality of each difference on the basis that each observation is an average of other (unlisted) observations, one may use the traditional estimates

$$\overline{x} = 1.58 \qquad \text{and} \qquad \widehat{\mathrm{SE}}[\overline{x}] = \frac{s}{\sqrt{n}} \approx .39.$$

The outlying value 4.6 raises doubts about the validity of the normality assumption, and other estimates are perhaps more reliable. One possibility is to use the 20% two-sided trimmed mean which removes 10% of the observations from each end of the sample. It yields $\overline{x}_{.2} = \frac{1}{8}\sum_{i=2}^{9} x_{(i)} = 1.40$. The influence function estimate of its standard error is based on the results derived in Example 4 of Section 4.3.2; it is determined by the sample variance $s_{W_{.2}}^2$ of the $(2 \times 10\%)$ Winsorized sample. To obtain $s_{W_{.2}}^2$, replace each of the observations smaller than the lth smallest $X_{(l)}$ by $X_{(l)}$, where $l \approx n/10$. More precisely $l = [n/10] + 1$. Similarly replace each of the 10% largest observations by $X_{(m)}$, where $m = n - [n/10]$. Then calculate the ordinary variance of this Winsorized sample and call it $s_{W_{.2}}^2$. The resulting estimated standard error of the trimmed mean is, by (4.3.5), equal to

$$\widehat{\mathrm{SE}}[\overline{x}_{.2}] = \frac{1}{(1-.2)}\frac{s_{W_{.2}}}{\sqrt{n}} \approx .23. \quad \square$$

Example 2: Lifetimes of EMT6 Cells. Consider the data from Figure 2.2: 10.4, 10.9, 8.8, 7.8, 9.5, 10.4, 8.4, 9.0, 22.2, 8.5, 9.1, 8.9, 10.5, 8.7, 10.4, 9.8, 7.7, 8.2, 10.3, 9.1. In this case the data are highly skewed, and estimates of both location γ (the minimum lifetime) and scale θ are desired. The famous transition probability model employs the location–scale family given in Problem 12 of Chapter 2, namely, the displaced exponential model, and assumes independence of the observations. For this model the expected lifetime is $\gamma + \theta$.

There is a qualitative difference between the estimation of the two parameters, for as shown in Problem 8(c), the minimum of the observations converges to the location parameter γ at a faster rate than does the MLE to the scale parameter θ. Thus if there are enough observations to accurately estimate the latter, the former will be even more accurately estimated by the minimum observation. Thus the problem reduces to the scale parameter problem discussed in Chapters 2 and 3. This situation arises whenever a location parameter is introduced to displace a scale parameter family for positive random variables. Another large class of location–scale models arises when the generating model is not bounded below (or above) by a known constant, such as the normal model in Example 1. In such cases location and scale are equally difficult to estimate. Moreover, the term *scale* is a bit of a misnomer, for it is used to describe the *spread* or *dispersion* of the distribution (in both directions) from the unknown location.

Another more realistic model for cell lifetimes is the displaced gamma distribution. There is certainly no consensus on which model is appropriate, but it seems clear that in answering the initial questions posed above, the model is asymmetric, that location and scale are of interest, and that assuming independence of the observations is *not* warrranted. □

4.2 LOCATION–SCALE PARAMETER FAMILIES

4.2.1 Definitions and Properties

Let X have distribution F and for each real μ denote the distribution of $X + \mu$ by F_μ. Then $\{F_\mu;\ -\infty < \mu < +\infty\}$ is called the *location parameter family* generated by F, and μ is called a *location parameter.* Such families have been subjected to extensive investigation, especially when F_μ is symmetric about μ, since these families arise naturally in the paired difference problem as exemplified by the Cushny and Peebles data given above. Whether or not symmetry holds, there is usually present a nuisance scale parameter σ which must be accounted for, since its value determines the error in estimates of μ. Thus we will concern ourselves primarily with location–scale families. As above, let $X \sim F$ and denote the distribution of $\sigma X + \mu$ by $F_{\mu,\sigma}$. Then $\{F_{\mu,\sigma};\ -\infty < \mu < +\infty, \sigma > 0\}$ is called the *location–scale family* generated by F, and (μ,σ) is a *location–scale parameter* for the family. One may show:

1. $F_{\mu,\sigma}(x) = F((x-\mu)/\sigma)$ for all x.
2. If a, b are real constants with $b > 0$, then

$$F_{\mu,\sigma}(x) = F\left(\frac{x-\mu}{\sigma}\right) = F_{a,b}\left(b\left(\frac{x-\mu}{\sigma}\right) + a\right)$$

Table 4.2 Some Continuous Models Used to Generate Location–Scale Families

Name	Generating f	Mean	Variance
Double Exponential	$\frac{e^{-\|x\|}}{2}$	0	2
Exponential	$e^{-x}I\{x>0\}$	1	1
Gamma, shape α	$\frac{x^{\alpha-1}e^{-x}I\{x>0\}}{\Gamma(\alpha)}$	α	α
Normal	$\frac{e^{-x^2/2}}{\sqrt{2\pi}}$	0	1
Student's t (n df)	$\frac{\Gamma\left(\frac{n+1}{2}\right)(1+x^2/n)^{-(n+1)/2}}{\sqrt{n\pi}\Gamma(n/2)}$	0, for $n>1$	$\frac{n}{n-2}$, for $n>2$
Uniform $(-1,1)$	$\frac{I\{-1<x<1\}}{2}$	0	$\frac{1}{3}$

so that the distribution $F_{\mu,\sigma}$ can be obtained from $F_{a,b}$ for any a, b. Thus $F_{a,b}$ also generates the location–scale family generated by F.

3. For any fixed constants a, b with $b > 0$, the pair $(a + b\mu, b\sigma)$ is a location–scale parameter for the same family as (μ, σ).
4. If F has density f, then

$$f_{\mu,\sigma}(x) = \frac{1}{\sigma} f\left(\frac{x-\mu}{\sigma}\right).$$

5. We write $\mathrm{E}_{\mu,\sigma}[g(X)]$ for $\int g(x)\,dF_{\mu,\sigma}(x)$ and note that when the rth moments exist, $\mathrm{E}_{\mu,\sigma}[X-\mu]^r = \sigma^r \mathrm{E}_{0,1}[X^r]$, whether or not μ is the mean of X or σ is the standard deviation of X. Thus calculations of $\mathrm{E}_{\mu,\sigma}[g(X)]$ can be carried out for a convenient choice of (μ, σ).

4.2.2 Examples of Location–Scale Families

For further reference we list in Table 4.2 some standard distributions used to generate location–scale families, as described above. The mean and the variance of the generating distribution are included.

4.3 ESTIMATORS OF LOCATION

There is a wealth of material on the finite sample and asymptotic properties of estimators of location and dispersion; see, for example, the material in the books Andrews et al. (1972), Huber (1981), or Hampel et al. (1986) for details and references. Here we will restrict attention to the most important estimators, expressing them as functionals (applied to the empirical distribution), and use the breakdown points and influence functions as guides to well-behaved estimators. The most important application will be to estimate the location of a "nearly normal" distribution in the presence of an unknown scale parameter, and several proposals will be discussed in the next section.

4.3.1 Descriptive Measures of Location

Let F belong to a possibly nonparametric family of distributions on the real line. We write $\mu(X)$ for $\mu(F)$ if $X \sim F$. A *measure of location* for F is a descriptive measure $\mu(F)$ which satisfies the following four conditions:

$$\begin{array}{ll} \text{(i)} & \mu(X+b)=\mu(X)+b \quad \text{for all } b \\ \text{(ii)} & \mu(-X)=-\mu(X) \\ \text{(iii)} & X\geq 0 \quad \text{implies} \quad \mu(X)\geq 0 \\ \text{(iv)} & \mu(aX)=a\mu(X) \quad \text{for all} \quad a>0. \end{array} \tag{4.3.1}$$

Condition (i) is called *location equivariance*. It plus the simple requirement "$\mu(b) = b$" were mentioned by David and Johnson as early as (1948) as the "usual conditions for a measure of location." They are in fact implied by (i) and (ii), which also imply that if F is symmetric about some point θ, then $\mu(F) = \theta$. We also find (iii) compelling since it, together with (i) and (ii), guarantees that $\mu(X)$ will lie in the range of values of X. Condition (iv), *scale equivariance*, guarantees that the corresponding estimators of location, namely, $\{\mu(F_n)\}$, will give the same result, independent of the unit of measurement. In addition to conditions (i)–(iv), Bickel and Lehmann (1975) have required measures of location to be stochastic order preserving; they strengthen (iii) to

(v) If X is stochastically larger[†] than Y, then $\mu(X) \geq \mu(Y)$.

This condition rules out the mode or midpoints of modal intervals as measures of location; see Problem 12 for details.

It will be understood, if not always stated explicitly, that every descriptive measure induces a sequence of estimators; for example, the mean $\mu(F)$ in

[†] X is stochastically larger than Y if $F_X(x) \leq F_Y(x)$ for all x.

example 1 below induces the sequence $\mu(F_n) = \overline{X}_n$, $n = 1,2,\ldots$. Thus we will even sometimes refer to this sequence as "the mean" and use the functional notation $\mu(F)$. Of course the induced estimators inherit the symmetry properties (4.3.1), and are thus location and scale equivariant.

While the descriptive measure is often defined for a large nonparametric class, we may desire that it also be Fisher-consistent for some parametric family $\{F_\theta : \theta \in \Theta\}$; this means that $\mu(F_\theta) = \theta$ for all $\theta \in \Theta$. This property, plus continuity of the descriptive measure $\mu(F)$, guarantees that the induced sequence of estimators will be consistent for the parameter θ in the usual sense of convergence in probability (see Sections 3.1.2 and 3.2.4). Note that any location measure satisfying (4.3.1) will be Fisher-consistent for the center of symmetry of a family of symmetric distributions.

Example 1: The Mean. Let $\mu(F) = \int x\,dF(x)$ for all F for which the integral exists and is finite. This descriptive measure satisfies properties (4.3.1) (i)–(iv) and is hence a location measure. [It also satisfies (v).] It has breakdown point $\varepsilon^* = 0$ and influence function $\mathrm{IF}_{\mu,F}(x) = x - \mu(F)$, for all real x.

Example 2: The Median. Define for each distribution F the real number $\mu_{1/2}(F) = F^{-1}(\frac{1}{2}) = \inf\{y : F(y) \geq \frac{1}{2}\}$. This descriptive measure will not satisfy property (4.3.1) (ii) at any F, which has a flat spot at its median (such as F which has equal probability $\frac{1}{4}$ on each of the points $-2,-1,1,2$), since we have defined the inverse to be the left-hand endpoint of an interval of constant value for F. Thus we need to restrict $\mu_{1/2}(F)$ to those F that do not have such an interval of medians, or we may modify the definition to select the midpoint of this interval; see Problem 7 of Chapter 1. The influence function for $\mu_{1/2}(F) = F^{-1}(\frac{1}{2})$ at a smooth F is given by (3.2.3) and shown in Figure 4.1(a).

Example 3: Linear Combinations of Two Quantiles. For any real a, b, α, β with $0 < \alpha$, $\beta < 1$, define $\mu(G) = aG^{-1}(\alpha) + bG^{-1}(\beta)$. For the sake of simplicity of exposition, we assume that G has a continuous positive density at the α, β quantiles. The conditions (4.3.1) impose severe restrictions on $\mu(G)$. For example, condition (i) implies that $a + b = 1$. Restricting attention to such $\mu(G)$, condition (iii) implies $0 < a < 1$, so that $\mu(G)$ is a *convex* combination of two quantiles. Continuing, the additional restriction (ii) implies that $\alpha = 1 - \beta$, and $a = \frac{1}{2}$. Thus conditions (i)–(iii) together imply that $\mu(G)$ is a symmetric mixture of quantiles:

$$\mu(G) = \frac{G^{-1}(\alpha) + G^{-1}(1-\alpha)}{2}$$

where $0 < \alpha < \frac{1}{2}$. Property (iv) is obviously satisfied.

The breakdown point for this location measure is α. Its influence function at a G with continuous positive density at each of the $\alpha, 1-\alpha$ quantiles is shown in Figure 4.1(b), and is simply the average of the influence functions for the respective quantiles, which are given by (3.2.3). Outliers to the right have the same positive effect, while those on the left have a negative effect; the effect is inversely proportional to the concentration of probability at the two quantiles. Inlying observations have a smaller effect on the symmetric average because they increase one quantile and decrease the other.

The limiting distribution of $n^{1/2}[\mu(G_n) - \mu(G)]$ is normal with parameters $(0, \alpha/2g^2(G^{-1}(\alpha)))$ when G is symmetric. This result may be found by using the joint limiting distribution of two order statistics given in Appendix B, and the fact that $\mu(G_n)$ is the average of two such order statistics. The details are worked out in the problems. There the reader is also asked to show that if $G = N(\mu, \sigma^2)$ is the normal distribution, the optimal choice of $\alpha = .27$. This may be obtained by straightforward calculations using the formula for the asymptotic variance given above. The efficiency of this simple linear combination of two quantiles $(X_{(.27n)} + X_{(.73n)})/2$ relative to the most efficient estimator $\overline{X}_n$ is surprisingly high—more than 80%—and it is far more robust against outliers, having breakdown point $\varepsilon^* = .27$.

Example 4: Trimmed Means. Define the (α, β)-trimmed mean by

$$T_{\alpha,\beta}(F) = \int_{F^{-1}(\alpha)}^{F^{-1}(1-\beta)} \frac{x\,dF(x)}{1-\alpha-\beta}.$$

This descriptive measure is a measure of location, i.e., it satisfies (4.3.1), provided that $\alpha = \beta$, and then it is called the 2β-trimmed mean (or symmetrically trimmed mean). Its influence function is derived in Section 4.3.2 and sketched in Figure 4.1(c). It shows that the induced estimators will be insensitive to outliers, provided that no more than $100\beta\%$ are on either side of the sample. Further discussion of this important class of estimators follows in the next section.

4.3.2 *L*-Estimators

In this section we will make some general remarks about the class of L-estimators and then show how to find estimates of standard errors for two important examples, trimmed means and the median. All the examples in Section 4.3.1 may be generalized to this class of symmetric mixtures of quantiles, with corresponding estimators which are linear combinations of order statistics. Let K be any probability distribution on $(0,1)$ which is symmetrical about $\frac{1}{2}$:

$$K\left(\tfrac{1}{2}-t^{-}\right)+K\left(\tfrac{1}{2}+t\right)=1, \qquad \text{for} \quad 0<t<\tfrac{1}{2}.$$

Then K defines a measure of location by averaging the quantiles:

$$\mu(F)=\int_0^1 F^{-1}(t)\,dK(t). \tag{4.3.2}$$

Because of the symmetry of K, we may rewrite this integral over the range $(0,\frac{1}{2}]$ and show that μ is a probability mixture of symmetric combinations of two quantiles:

$$\mu(F)=\int_0^{1/2}\frac{[F^{-1}(t)+F^{-1}(-t)]}{2}\,dK(t).$$

Therefore μ is a probability mixture of measures of location and hence can be shown to satisy the conditions (4.3.1) of a measure of location itself. The corresponding estimator is called a *linear combination of order statistics* or *L-estimator*:

$$\mu(F_n)=\sum_{i=1}^{n}k_iX_{(i)}, \qquad k_i=K\left(\frac{i}{n}\right)-K\left(\frac{i-1}{n}\right).$$

Special cases are 2β-trimmed means [take K uniform on the subinterval $(\beta,1-\beta)$], and finite combinations of order statistics (take K to be a finite discrete distribution).

When F has a density f with respect to Lebesgue measure, the influence function for the L-estimator may be defined in terms of K as follows. Define

$$h(t)=\int_0^t\frac{dK(s)}{f(F^{-1}(s))}, \qquad 0<t<1.$$

Then

$$\mathrm{IF}_{\mu,F}(x)=h(F(x))-\int_0^1 h(s)\,ds \tag{4.3.3}$$

and under certain conditions, $n^{1/2}[\mu(F_n)-\mu(F)]$ converges to the normal distribution with mean zero and variance given by

$$\mathrm{E}_F[\mathrm{IF}_{\mu,F}(X)]^2=\mathrm{Var}_F[h(F(X))]=\mathrm{Var}[h(U)] \tag{4.3.4}$$

where U is uniformly distributed on the unit interval.

The following theorem gives a method for comparing the asymptotic variances of different L-estimators.

Theorem 4.1: Bickel and Lehmann (1975). Let L-estimators T_1 and T_2 be determined by K_1 and K_2 in (4.3.2), and suppose that the densities of K_1 and K_2 satisfy

$$0 \leq \frac{dK_2(s)}{dK_1(s)} \leq A, \qquad \text{where} \quad 1 < A.$$

Then $V(T_2,F) \leq A^2 V(T_1,F)$, so the efficiency of T_2 to T_1 is at least A^{-2}.

Proof. Let U_1, U_2 be i.i.d. $U(0,1)$. For each $i = 1,2$,

$$\begin{aligned} 2V(T_i,F) &= \text{Var}[h_i(U_1) - h_i(U_2)] \\ &= \text{E}[h_i(U_1) - h_i(U_2)]^2 \\ &= \text{E}\left[\int_{U_1}^{U_2} \frac{dK_i(s)}{f(F^{-1}(s))}\right]^2. \end{aligned}$$

The result now follows from the linearity of the integral. □

Corollary to Theorem 4.1. The asymptotic efficiency of the 2β-trimmed mean relative to the untrimmed mean is bounded below by $(1-2\beta)^2$.

Proof. The proof is straightforward. □

Example 4: Trimmed Means (Continued from Section 4.3.1). The symmetrically trimmed means deserve special attention for at least four reasons:

1. They are robust to outliers, up to $100\beta\%$ on each side.
2. They have a very strong nonparametric efficiency property; namely, their asymptotic efficiency relative to the untrimmed mean never drops below $(1-2\beta)^2$, as shown by the corollary to Theorem 4.1.
3. They are simple to calculate, and their standard errors may be easily estimated from the 2β-Winsorized sample, as discussed below.
4. When the data *are* normally distributed, there is strong evidence that the distribution of the standardized trimmed mean has an approximate Student's t distribution with known degrees of freedom; hence robust confidence intervals of known coverage probability may be constructed for a normal mean, even when n is small. For more details, see Tukey and McLaughlin (1963) and Sections 5.2 and 6.5.

Property 1 is evident from the definition of the the descriptive measure defined earlier in example 4 of the last section. That definition leads to the induced estimator $T_{2\beta}(F_n)$, the average of the observations remaining after trimming the $[\beta n]$ smallest and $[\beta n]$ largest. Provided that F is without flat spots at the trimming points,

$$n^{1/2}[T_{2\beta}(F_n) - T_{2\beta}(F)] \xrightarrow{d} N(0, V(\beta, F))$$

where $V(\beta, F) = \mathrm{E}[\mathrm{IF}^2_{\beta,F}(X)]$; see Stigler (1969). The influence function is derived exactly as in the one-sided case treated in Chapter 3, and is given (for symmetric F) by:

$$(1-2\beta)\mathrm{IF}_{T_{2\beta},F}(x) = \begin{cases} F^{-1}(\beta) - W_{2\beta}(F), & x < F^{-1}(\beta) \\ x - W_{2\beta}(F) & F^{-1}(\beta) \le x \le F^{-1}(1-\beta) \\ F^{-1}(1-\beta) - W_{2\beta}(F), & F^{-1}(1-\beta) < x \end{cases}$$

where $W_{2\beta}(F)$ is the 2β-Winsorized mean:

$$W_{2\beta}(F) = (1-2\beta)T_{2\beta}(F) + \beta F^{-1}(\beta) + \beta F^{-1}(1-\beta).$$

Using this result the reader may verify that the influence function estimate of the standard error (defined in Section 3.5.1) is given by

$$\widehat{\mathrm{SE}}[T_{2\beta}(F_n)] = \frac{1}{(1-2\beta)} \frac{s_{W_{2\beta}}}{\sqrt{n}} \tag{4.3.5}$$

where $s^2_{W_{2\beta}}$ is the usual sample variance of the 2β-Winsorized sample. That is, let $m = [n\beta]$, and define

$$Y_{(i)} = \begin{cases} X_{(m+1)}, & i \le m \\ X_{(i)}, & m < i \le n-m \\ X_{(n-m)}, & n-m < i. \end{cases}$$

Then define

$$s^2_{W_{2\beta}} = \frac{1}{n-1} \sum_{i=1}^{n} (Y_i - \overline{Y})^2.$$

It is thus a simple matter to find the Winsorized sample, to calculate its variance, and hence to obtain the influence function estimate of the standard error for the trimmed mean (4.3.5). See also Appendix D.2.1.

Example 2: The Median (Continued from Section 4.3.1). Estimates of the standard error of the median are not so easy to find. The influence function depends on F through its density f, so the simple method of substituting

F_n for F is not available. A density estimate $\hat{f}$ is required if the asymptotic variance formula is to be exploited.

Suppose that the density of X_i, $i = 1,2,\ldots,n$, is continuous and positive in a neighborhood of θ; then by (3.1.5)

$$\frac{\hat{\theta}-\theta}{1/(2\sqrt{n}f(\theta))} \xrightarrow{d} N(0,1) \qquad \text{as} \quad n \to \infty.$$

Thus the asymptotic standard error of θ is given by $\mathrm{SE}(\hat{\theta}) = 1/2\sqrt{n}f(\theta)$. Our approach is to estimate $\mathrm{SE}(\hat{\theta})$ by estimating $f(\theta)$ with a density estimator. Here we use a *kernel density estimator* defined by

$$\hat{f}_k(x) = \frac{1}{nh}\sum_{i=1}^{n} k\left(\frac{x-X_i}{h}\right)$$

where the kernel k is a density function symmetric about 0 and the window width $h \to 0$ as $n \to \infty$. If for example we take $k(y) = \frac{1}{2}I(-1 \le y \le 1)$, then

$$\begin{aligned}\hat{f}_k(x) &= \frac{1}{nh}\sum_{i=1}^{n}\frac{1}{2}I\left(-1 \le \frac{x-X_i}{h} \le 1\right)\\ &= \frac{1}{2nh}\sum_{i=1}^{n} I(x-h \le X_i \le x+h)\\ &= \frac{F_n(x+h)-F_n(x-h)}{2h}.\end{aligned}$$

This is called Rosenblatt's *shifted histogram estimator.* The smoothing parameter h is also called the window width, since it determines how many observations go into the estimate of each value of f. The choice of the window width h is crucial to the performance of $\hat{f}_k(x)$, whereas the choice of k is far less important [Silverman (1986)].

We will first obtain an expression for the asymptotic MSE of the kernel density estimator, and then choose the window width that minimizes this MSE. The result h_{opt} depends on both the unknown density and an unknown scale parameter, and will be estimated using the data and the assumption that the data are normal. (This last requirement only ensures that the window width is optimal if the data are normal; it can still give reasonable estimates of the optimal window width when the data are nonnormal.) Now to the details.

We first establish expressions for the mean and the variance of $\hat{f}_k(x)$,

$$\begin{aligned}
\mathrm{E}\{\hat{f}_k(x)\} &= \mathrm{E}\left\{\frac{1}{nh}\sum_{i=1}^{n} k\left(\frac{x-X_i}{h}\right)\right\} \\
&= \int_{-\infty}^{\infty} \frac{1}{h} k\left(\frac{x-y}{h}\right) f(y)\,dy \\
&= \int_{-\infty}^{\infty} k(u) f(x-hu)\,du \\
&= \int_{-\infty}^{\infty} k(u)\left\{f(x) - huf'(x) + \frac{1}{2}h^2 f''(x) + \cdots\right\}du \\
&= f(x) + \frac{1}{2}h^2 f''(x)\int_{-\infty}^{\infty} u^2 k(u)\,du + 0(h^4)
\end{aligned}$$

since k is a density function symmetrical about 0. Therefore

$$\mathrm{bias}\{\hat{f}_k(x)\} = \frac{1}{2}h^2 f''(x)\int_{-\infty}^{\infty} u^2 k(u)\,du + 0(h^4)$$

and

$$\begin{aligned}
\mathrm{Var}\{\hat{f}_k(x)\} &= \mathrm{Var}\left\{\frac{1}{nh}\sum_{i=1}^{n} k\left(\frac{x-X_i}{h}\right)\right\} \\
&= \frac{1}{n}\left[\mathrm{E}\left\{\frac{1}{h^2}k^2\left(\frac{x-X_1}{h}\right)\right\} - \left(\mathrm{E}\left\{\frac{1}{h}k\left(\frac{x-X_1}{h}\right)\right\}\right)^2\right].
\end{aligned}$$

Now

$$\begin{aligned}
\mathrm{E}\left\{\frac{1}{h^2}k^2\left(\frac{x-X_1}{h}\right)\right\} &= \int_{-\infty}^{\infty}\frac{1}{h^2}k^2\left(\frac{x-y}{h}\right) f(y)\,dy \\
&\ \ \vdots \\
&= \frac{1}{h} f(x)\int_{-\infty}^{\infty} k^2(u)\,du + 0(1).
\end{aligned}$$

Thus

$$\mathrm{Var}\{\hat{f}_k(x)\} = \frac{1}{nh} f(x)\int_{-\infty}^{\infty} k^2(u)\,du + 0\left(\frac{1}{n}\right).$$

If we measure the accuracy of the estimate by its mean squared error,

$$\begin{aligned}\text{MSE}\{\hat{f}_k(x)\} &= \text{bias}^2\{\hat{f}_k(x)\} + \text{Var}\{\hat{f}_k(x)\}\\ &= \frac{1}{4}h^4\{f''(x)\}^2\left[\int_{-\infty}^{\infty} u^2 k(u)\,du\right]^2\\ &\quad + \frac{1}{nh}f(x)\int_{-\infty}^{\infty} k^2(u)\,du + 0\left(\frac{1}{n} + h^6\right).\end{aligned}$$

For a given kernel k, the value of h that minimizes the leading terms in $\text{MSE}\{\hat{f}_k(x)\}$ is given by

$$h_{\text{opt}}(x) = \left[\frac{\int_{-\infty}^{\infty} k^2(u)\,du}{(\int_{-\infty}^{\infty} u^2 k(u)\,du)^2}\right]^{1/5} \times \left[\frac{f(x)}{(f''(x))^2}\right]^{1/5} n^{-1/5}.$$

Define the scale parameter λ by $f(x) = (1/\lambda)f_1(x/\lambda)$. Then

$$f'(x) = \frac{1}{\lambda^2}f_1'\left(\frac{x}{\lambda}\right) \quad \text{and} \quad f''(x) = \frac{1}{\lambda^3}f_1''\left(\frac{x}{\lambda}\right).$$

Thus

$$h_{\text{opt}}(x) = \left[\frac{\int_{-\infty}^{\infty} k^2(u)\,du}{(\int_{-\infty}^{\infty} u^2 k(u)\,du)^2}\right]^{1/5} \times \left[\frac{f_1(x/\lambda)}{(f_1''(x/\lambda))^2}\right]^{1/5} \lambda n^{-1/5}.$$

A popular choice of h in practice is that h is proportional to $\hat{\lambda}n^{-1/5}$ where $\hat{\lambda}$ is an estimate of the scale parameter λ and the constant of proportionality is calculated assuming the data is normal. For $k(y) = \frac{1}{2}I(-1 \le y \le 1)$, it can be shown that $\int_{-\infty}^{\infty} k^2(u)\,du = \frac{1}{2}$ and $\int_{-\infty}^{\infty} u^2 k(u)\,du = \frac{1}{3}$ so that

$$\frac{\int_{-\infty}^{\infty} k^2(u)\,du}{(\int_{-\infty}^{\infty} u^2 k(u)\,du)^2} = \frac{\frac{1}{2}}{(\frac{1}{3})^2} = 4.5.$$

Let λ be the interquartile range and $f = \varphi$ the standard normal density. Then taking f_1 to be the normal density with mean 0 and interquartile range 1, it is easily found that $f_1(0)/(f_1''(0))^2 = (1.349)^{-5}\sqrt{2\pi}$, so the optimal window width is estimated to be $h_{\text{opt}} = 1.2\hat{\lambda}n^{-1/5}$.

The above approach to choosing the window width was developed by Scott (1979) and Freedman and Diaconis (1981) for the case of histogram density estimators. For an adaptive choice of window width see Sheather (1986b).

The Standard Error of the Median for an Asymmetric Example

To illustrate the application of the above technique on a particular data set, consider again the cell lifetimes from Example 2 of Section 4.1. The median is $\hat{\theta} = (x_{(9)} + x_{(10)})/2 = 9.1$. We wish to estimate f at $x = \hat{\theta} = 9.1$ based on the kernel estimate with $k(y) = \frac{1}{2}I(-1 \le y \le 1)$. The interquartile range[†] $\hat{\lambda} = 10.4 - 8.55 = 1.85$. So we take $h = 1.2\hat{\lambda}n^{-1/5} = 1.2 \times 1.85 \times 20^{-1/5} = 1.2$.

Now

$$\hat{f}_k(x) = \frac{1}{nh}\sum_{i=1}^{n} k\left(\frac{x - X_i}{h}\right)$$

where $k(y) = \frac{1}{2}I(-1 \le y \le 1)$. Therefore

$$\begin{aligned}\hat{f}_k(9.1) &= \frac{\#\{X_i \le 9.1 + h\} - \#\{X_i < 9.1 - h\}}{2nh} \\ &= \frac{\#\{X_i \le 10.3\} - \#\{X_i < 7.9\}}{48} = \frac{14 - 2}{48}.\end{aligned}$$

Thus a density estimate of the standard error of the median is

$$\widehat{\mathrm{SE}}_{\mathrm{density}}(\hat{\theta}) = \frac{1}{2\sqrt{n}\hat{f}_k(\hat{\theta})} = \frac{1}{2 \times \sqrt{20} \times .25} = 0.45.$$

By way of comparison the exact bootstrap estimate of the standard error of $\hat{\theta}$ is .42.

4.3.3 M-Estimators

We will first elaborate on a few of the properties of this very important class of estimators, and then concentrate on one example which has good efficiency and robustness properties, the Huber estimator and its one-step version. We will also briefly discuss the redescending M-estimators, which maintain high efficiency while completely rejecting very large outliers.

Let F be any distribution function on the real line, and let ψ be an odd, nondecreasing function which is not identically zero. We want to define $\mu_\psi(F)$ implicitly as the solution $\mu = \mu_\psi(F)$ of

$$\int \psi(x - \mu)\,dF(x) = 0. \tag{4.3.6}$$

However, for a given ψ and F there may be no solution, a unique solution, or multiple solutions to (4.3.6). To see this, consider the example $\psi(x) = \mathrm{sgn}(x)$, for which (4.3.6) becomes $P\{X > \mu\} - P\{X < \mu\} = 0$ or $1 = F(\mu) + F(\mu^-)$, and try various discrete and continuous F. The difficulties encountered in defining a solution are the same as those for defining the median of an arbitrary distribution; see Chapter 1, Problem 7.

[†]The first quartile is defined by $q_1 = (1 - r)X_{(i)} + rX_{(i+1)}$, where $i = [(n + 1)/4]$ and $r = (n + 1)/4 - i$. The third quartile is similarly defined, and the interquartile range is $q_3 - q_1$.

Define $h(\mu) = \mathrm{E}_F[\psi(X - \mu)] = \int \psi(x - \mu)\,dF(x)$; then $h(\mu)$ is nonincreasing in μ and takes on positive and negative values as μ ranges from $-\infty$ to $+\infty$. If there is no solution to (4.3.6) define one to be the point where $h(\mu)$ crosses the axis. If there is more than one solution, there is a finite interval of solutions, and we may take the midpoint of the interval as *the* solution.

The value of the descriptive measure $\mu_\psi(F)$ is estimated by $\mu_\psi(F_n)$, the solution of

$$\frac{1}{n}\sum_{i=1}^{n}\psi(X_i - \mu) = \int \psi(x - \mu)\,dF_n(x) = 0. \tag{4.3.7}$$

It may be the case that ψ is generated by the density of a symmetric distribution through the process of finding the maximum likelihood estimator for location

$$\psi(x - \mu) = \frac{\partial \ln f(x - \mu)}{\partial \mu} = -\frac{f'(x - \mu)}{f(x - \mu)}.$$

Even when ψ is not so generated, the estimators defined by (4.3.7) will be called *M-estimators* for location. Huber (1964) introduced such M-estimators and found general conditions under which they are consistent and asymptotically normal.

In general $\mu_\psi(F)$ satisfies properties (4.3.1) (i)–(iii) and (v), but not (iv). The verification of these properties is somewhat trickier than for explicitly defined measures. First, consider location equivariance, property (i):

$$\begin{aligned} 0 &= \mathrm{E}_F[\psi(X - \mu_\psi(F))] \\ &= \mathrm{E}_F[\psi((X + b) - (b + \mu_\psi(F))] \\ &= \mathrm{E}_{F_b}[\psi(Y - (b + \mu_\psi(F))] \end{aligned}$$

where $Y \sim F_b$ and $F_b(x) = F(x - b)$ for all x. Now by definition of $\mu_\psi(F_b)$.

$$0 = \mathrm{E}_{F_b}[\psi(Y - \mu_\psi(F_b)].$$

Comparing this expression with the last line of the string of equalities, we see that $\mu_\psi(F_b) = b + \mu_\psi(F)$, which verifies (i).

The verification of (ii) and (iii) is left to the reader in the problems. Property (iv) can be shown to be false by example, so M-estimators are not in general scale equivariant. This obstacle can be overcome; see (4.3.10) below.

Estimators which are solutions to estimating equations have two desirable properties: first, when ψ is odd, bounded, and monotonically increasing, the breakdown point $\varepsilon^* = \frac{1}{2}$, which means that the estimators can withstand large proportions of very bad observations without breaking down completely. Second, the influence function of the solution $\mu_\psi(F)$ of the estimating equation is proportional to ψ itself:

$$\mathrm{IF}_{\psi,F}(x) = \frac{\psi(x-\mu_\psi)}{\mathrm{E}_F[\psi'(X-\mu_\psi)]}. \tag{4.3.8}$$

This relationship is derived from (4.3.6) by substituting $F_{x,\varepsilon}$ for F, and finding the derivative with respect to ε at $\varepsilon = 0$ (see below for details). Also note that we have for simplicity identified the estimator μ_ψ with ψ in writing $\mathrm{IF}_{\psi,F}$ for $\mathrm{IF}_{\mu_\psi,F}$.

The relationship (4.3.8) shows that an M-estimator can in principle be defined by choice of ψ function to have the desirable properties of efficiency and robustness. Efficiency at a particular F is achieved in the location problem by taking ψ proportional to the derivative of the log–likelihood defined by the density of $F: \psi(x) = -(f'/f)(x)$. Robustness is achieved by choosing ψ that is smooth and bounded to reduce the influence of a small proportion of observations. Both efficiency and robustness are possible when a ψ function suggested by the MLE theory (which is typically not bounded) is modified to one that is bounded and continuous.

For example, when the F in question is the normal distribution, then the ψ function suggested by the MLE is just the identity function, which is unbounded. This ψ function may be modified to

$$\psi_k(x) = \begin{cases} -k, & x < -k \\ x, & -k \leq x \leq +k \\ +k, & +k < x \end{cases} \tag{4.3.9}$$

which for large k leads to very efficient estimation. The function ψ_k is called *Huber's ψ function* because he showed [Huber (1964)] that it had desirable properties. See below after Theorem 4.2 for some discussion of results from his paper, which is fundamental to robust statistics.

The reader may have noticed that the trimmed means have influence curves which are of the same form as (4.3.9), and thus—for proper choice of α relative to k—exactly the same asymptotic efficiency as the M-estimator. The trimmed mean will have a smaller breakdown point than the M-estimator, but it is easier to calculate because it is an explicit function of the observations.

Derivation of the Influence Function for M-Estimators

We will not give general conditions for which the influence function exists, but rather show that the influence function has a certain form when it does exist. We assume therefore the existence of the derivative at zero of

$\mu_\psi(F_{x,\varepsilon})$:

$$\mathrm{IF}_{\mu_\psi,F}(x) = \frac{\partial}{\partial \varepsilon}[\mu_\psi(F_{x,\varepsilon})]_{\varepsilon=0}.$$

By definition,

$$\begin{aligned} 0 &= \mathrm{E}_{F_{x,\varepsilon}}[\psi(X - \mu_\psi(F_{x,\varepsilon}))] \\ &= \varepsilon k(x,\varepsilon) + \mathrm{E}_F[\psi(X - \mu_\psi(F_{x,\varepsilon}))] \end{aligned}$$

where

$$k(x,\varepsilon) = \psi(x - \mu_\psi(F_{x,\varepsilon})) - \mathrm{E}_F[\psi(X - \mu_\psi(F_{x,\varepsilon}))].$$

Then provided ψ' exists and is integrable, we may differentiate with respect to ε at zero:

$$0 = k(x,0) + \mathrm{E}_F[\psi'(X - \mu_\psi(F))]\frac{\partial}{\partial \varepsilon}[\mu_\psi(F_{x,\varepsilon})]_{\varepsilon=0}.$$

Rearranging terms, we have

$$\mathrm{IF}_{\mu_\psi}(x) = \frac{\psi(x - \mu_\psi(F))}{\mathrm{E}_F[\psi'(X - \mu_\psi(F))]}.$$

The asymptotic theory of M-estimators, as with the traditional maximum likelihood estimators, can be quite involved. The following theorem only gives the main ideas on consistency and asymptotic normality, under assumptions for which the proof is short.

Theorem 4.2: Consistency, Asymptotic Normality of M-Estimators

(i) Assume that F is symmetric about $\mu_0 = \mu_\psi(F)$ and that ψ is odd and nondecreasing. Also assume that $h_F(\mu) = \mathrm{E}_F[\psi(X - \mu)]$ exists (as a real number) and that $h_F(\mu)$ is strictly decreasing in a neighborhood of μ_0. Let $\hat{\mu}_n = \mu(F_n)$ be the solution to (4.3.7). Then $\hat{\mu}_n \xrightarrow{p} \mu_0$.

(ii) Further assume that $h'_F(\mu)$ exists, equals $-\mathrm{E}_F[\psi'(X - \mu)] < 0$, and that $h''_F(\mu)$ exists and is bounded for all μ in a neighborhood of μ_0. Then

$$n^{1/2}[\hat{\mu}_n - \mu_0] \xrightarrow{d} N(0, V_{\psi,F})$$

where $V_{\psi,F} = \mathrm{E}_F[\mathrm{IF}_{\psi,F}(X)]^2$, the expected squared influence function given by (4.3.8).

Proof of (i). We need to show that for each $\epsilon > 0$ there exists an $N(\epsilon)$ such that for all $n \geq N(\epsilon)$

$$P\{|\hat{\mu}_n - \mu_0| < \epsilon\} \geq 1 - \epsilon.$$

If this is not the case, then for some $\epsilon_0 > 0$ and every integer k there exists a larger integer n_k such that the above inequality fails. In other words, for some $\epsilon_0 > 0$ there exists a subsequence $\{n_k\}$ of the positive integers with

$$P\{|\hat{\mu}_{n_k} - \mu_0| \geq \epsilon_0\} \geq \epsilon_0.$$

Without loss of generality we assume $\mu_{n_k} \leq \mu_0 - \epsilon_0$; for if not, there is a further subsequence lying entirely to the left (or the right) of the ϵ_0 neighborhood of μ_0. Then applying the law of large numbers to $h_{F_{n_k}}(\mu_0 - \epsilon_0)$, and using the strict monotonicity of h at μ_0, we obtain a contradiction:

$$0 = h_{F_{n_k}}(\mu_{n_k}) \geq h_{F_{n_k}}(\mu_0 - \epsilon_0) \xrightarrow{P} h_F(\mu_0 - \epsilon_0) > h_F(\mu_0) = 0.$$

It follows that the sequence of solutions converges: $\hat{\mu}_n \xrightarrow{P} \mu_0$.

Proof of (ii). Expand $h_{F_n}(\hat{\mu}_n)$ in a Taylor series about μ_0:

$$h_{F_n}(\hat{\mu}_n) = h_{F_n}(\mu_0) + (\hat{\mu}_n - \mu_0)h'_{F_n}(\mu_0) + \frac{(\hat{\mu}_n - \mu_0)^2 h''_{F_n}(\mu_n^*)}{2}$$

where $|\mu_n^* - \mu_0| \leq |\hat{\mu}_n - \mu_0|$. Then rearrange terms to obtain:

$$n^{-1/2}[\hat{\mu}_n - \mu_0] = \frac{n^{1/2}[h_{F_n}(\hat{\mu}_n) - h_{F_n}(\mu_0)]}{h'_{F_n}(\mu_0) + \dfrac{[\hat{\mu}_n - \mu_0]h''_{F_n}(\mu_n^*)}{2}}$$

By definition, $h_{F_n}(\hat{\mu}_n) = 0$, so the numerator on the right is a normalized sum of i.i.d. random variables and hence by the central limit theorem is asymptotically normal with parameters $(0, \mathrm{E}_F[\psi(X - \mu_0)]^2)$.

The first term in the denominator $h'_{F_n}(\mu_0) \xrightarrow{P} h'_F(\mu_0)$ by the law of large numbers. The second term is a product of two factors, one of which is bounded in probability, and one which converges to zero in probability by part (i). Hence by Slutsky's lemma, (Theorem B.4 of Appendix B.1), part (ii) of Theorem 4.2 follows. □

The above results can be proved under more general conditions, for example with discontinuous ψ, but they then require more involved proofs; see Huber (1964), for example. In that paper Huber also considered the choice of M-estimator as a two-person game in which the statistician would choose a ψ function, Nature would choose a distribution F, and the pay-off to Nature was the asymptotic variance $V(\psi, F)$. The statistician's strategy, knowing that Nature was free to choose any F of the form $F = (1-\varepsilon)\Phi + \varepsilon H$, where H was an arbitrary symmetric contaminating distribution, was to choose ψ so as to minimize the maximum pay-off to Nature. It turns out that the game has a solution in which Nature maximizes her minimum pay-off by choosing H so that the resulting F is normal in the central portion of the distribution, but has exponential tails. The statistician's minimax strategy which deals best with this least favorable distribution is an M-estimator with ψ function of the form (4.3.9). The surprising part of this result is that the tails of the least favorable distribution are not heavier. Unfortunately the asymptotic minimax approach is mathematically difficult and has not led to appealing solutions in many other statistical contexts.

Another approach to robust estimation using M-estimators places a bound on the influence function and subject to this constraint tries to maximize the efficiency by choice of M-estimator at some particular model F. The following result illustrates the basic idea of this approach, which has been so successfully exploited by Hampel et al. (1986), in the general location problem and in many other contexts, including the scale problem and regression.

Theorem 4.3: Optimal Bounded Influence M-Estimation Assume that F is the normal distribution with known variance and unknown mean μ. It is desired to choose an odd, nondecreasing ψ function and hence an M-estimator of μ, which minimizes the asymptotic variance $V(\psi,\Phi)$ among all estimators satisfying $|\mathrm{IF}_{\psi,\Phi}| \leq c < \infty$, for some preassigned $c \geq \sqrt{\pi/2}$. The solution is the Huber ψ function of (4.3.9), namely, $\psi_k(x) = \max\{-k, \min\{x,k\}\}$, where $c = k/(2\Phi(k)-1)$.

Proof. Since $V(\psi,\Phi)$ and $\mathrm{E}[\psi']$ do not depend on μ, without loss of generality we may assume that $\mu = 0$. The problem is to choose ψ so as to minimize $V(\psi,\Phi) = \mathrm{E}[\psi^2]/(\mathrm{E}[\psi'])^2$ subject to $|\mathrm{IF}_{\psi,\Phi}| = |\psi/\mathrm{E}[\psi']| \leq c$. The ψ that solves this problem, if any, is clearly determined only up to a multiplicative constant. Hence we may choose this constant so that $\mathrm{E}[\psi'] = d = 2\Phi(k) - 1$. The problem may be restated: choose ψ so as to minimize $\mathrm{E}[\psi^2]$ subject to $|\psi| \leq cd$. Now

$$\mathrm{E}[\psi(X) - X]^2 = \mathrm{E}[\psi(X)]^2 - 2\mathrm{E}[X\psi(X)] + \mathrm{E}[X^2].$$

Integration by parts shows that the second term on the right equals the constant $-2d$, so that minimizing $\mathrm{E}[\psi(X)]^2$ is the same as minimizing the left-hand side of the equality. This is accomplished, subject to $|\psi| \leq cd$, by the Huber ψ_k function, with $k = cd$. □

We summarize below some of the desirable properties of M-estimators:

1. They are robust to large proportions of outliers: when ψ is nondecreasing, the breakdown point is equal to .5. [This is proved in Huber (1981).]
2. The influence function is proportional to the ψ function which defines the estimating equation. Hence this function may be chosen to bound

the influence of outliers and achieve high efficiency for a particular model, as illustrated above by Huberizing in the normal model.

3. Because of property 2, for every asymptotically normal estimator T there is an M-estimator (defined by a ψ function equal to the influence function of T), which has exactly the same asymptotic behaviour. Thus from the point of view of asymptotic normality, only M-estimators need by studied. In addition to this property of generality, ideas developed for one-dimensional parameters carry over to higher dimensional parameters. This is not the case for L-estimators or R-estimators, because there is no natural ordering or ranking of higher dimensional observataions.
4. Finally M-estimators can be chosen to completely reject outliers, while maintaining a large breakdown point and high efficiency at the model; see example 6 below.

We also list some of the main drawbacks of M-estimators. We shall see that nearly all of them can be overcome.

1. The main disadvantage of M-estimators is the one which maximum likelihood estimators have long suffered; they are in general only implicitly defined and must be found by an iterative search. Given the general increase in computer awareness and familiarity with algorithms, this should present less of a problem to today's students; although some conceptual difficulties remain in dealing with implicitly defined functions. In practice, an iterative search for a location parameter converges very rapidly, given a reasonable initial estimate. In some cases, the first iteration leads to an estimate which is almost as reliable as the final estimate; such one-step estimates are discussed below in example 5.
2. As mentioned earlier, M-estimators as defined above will not be scale equivariant, and modifications are necessary which make them so. Given a ψ function for which (4.3.1) (i)–(iii) are satisfied, and a measure of dispersion $\tau(F)$ satisfying (4.4.1), define $\mu_\psi(F)$ as the solution μ of

$$\int \psi\left(\frac{x-\mu}{\tau(F)}\right) dF(x) = 0. \tag{4.3.10}$$

Then $\mu_\psi(F)$ satisfies all the properties (4.3.1) and is hence a measure of location.

The M-estimator induced by (4.3.10) is $\mu(F_n)$, the solution μ of

$$\sum_{i=1}^{n} \psi\left(\frac{X_i-\mu}{\tau(F_n)}\right) = 0. \tag{4.3.11}$$

This equation is often solved iteratively using Newton's method as described below for one step, but repeated until the results of successive iterations differ by a negligible amount. The first estimate of μ is usually taken to be the median, because of its high breakdown value. A good estimate of $\tau(F)$ is the median absolute deviation, described below in Section 4.4, example 3. This estimator also has breakdown point .5 and need not be very efficient, if we are only interested in estimating location, because the estimators of location and scale are (aysmptotically) independent when F is symmetric. When both location and scale parameters are of interest, methods of Section 4.5 are appropriate.

Example 5: One-Step M-Estimators. Denoting initial estimates of dispersion and location by $\hat{\tau}_0 = \tau(F_n)$ and $\hat{\mu}_0$ we may find another approximation $\hat{\mu}_1$ to the solution of (4.3.11) by Newton's method. Let $h(\mu) = \sum_{i=1}^{n} \psi((X_i - \mu)/\hat{\tau}_0)$, and set $h'(\mu)$ at $\mu = \hat{\mu}_0$ equal to $(0 - h(\hat{\mu}_0))/(\hat{\mu}_1 - \hat{\mu}_0)$. Thus $\hat{\mu}_1$ is the point where the tangent line to the graph of $h(\mu)$ at $\mu = \hat{\mu}_0$ crosses the axis. Then the *one-step M-estimator* is defined by

$$\hat{\mu}_1 = \hat{\mu}_0 - \frac{h(\hat{\mu}_0)}{h'(\hat{\mu}_0)} = \hat{\mu}_0 + \hat{\tau}_0 \frac{\sum_{i=1}^{n} \psi((X_i - \hat{\mu}_0)/\hat{\tau}_0)}{\sum_{i=1}^{n} \psi'((X_i - \hat{\mu}_0)/\hat{\tau}_0)}. \tag{4.3.12}$$

One-step estimators maintain most of the equivariance and asymptotic properties of the estimators which they approximate, and they are surprisingly efficient for finite sample sizes.

In particular, the one-step version of Huber's M-estimator defined above is, after some algebra,

$$\hat{\mu}_1 = \frac{k\hat{\tau}_0(i_2 - i_1) + \sum_{i=i_1+1}^{n-i_2} X_{(i)}}{n - i_1 - i_2}$$

where i_1 is the number of observations X_i for which $(X_i - \hat{\mu}_0)/\hat{\tau}_0 < -k$ and i_2 is the number for which $(X_i - \hat{\mu}_0)/\hat{\tau}_0 > k$. When $\hat{\mu}_0$ is taken to be the median, and $\hat{\tau}_0$ is taken to be a suitable multiple of the median absolute deviation (see example 3 of Section 4.4), this estimator is efficient and robust for mixture models centered at the normal distribution. [See Andrews et al. (1972) for comparative simulation studies, in which the above estimator is called P15.] See also Appendix D.2.3.

As an application, consider again the Cushny and Peebles data for the difference in effects of two drugs, which we assume has an approximately normal distribution. Then the initial estimates are the median and the median absolute deviation: $\hat{\mu}_0 = \text{med}_i\{X_{(i)}\} = 1.3$ and $\hat{\tau}_0 = \text{MAD}/\Phi^{-1}(\frac{3}{4}) = .4/.6745 = .563$. If we take $k = 1.28 = \Phi^{-1}(.9)$, then the influence function of Huber's M-estimate will match that of the 2β-trimmed mean with $\beta = .1$.

The one-step estimate is then $\hat{\mu}_1 = 1.36$, and the influence function estimate of its standard error is .22; these results should be compared with the earlier estimates given in Section 4.1.1, Example 1.

Example 6: Redescending M-Estimators. Redescending M-estimators have ψ functions which are nondecreasing near the origin but then decrease toward the axis far from the origin. They usually satisfy $\psi(x) = 0$ for all x with $|x| \geq r$, where r is a finite number which may be considered the minimum rejection point. They have the same desired effect on very large outliers that outlier rejection techniques are supposed to have, namely, they completely reject them. However, they have several advantages. First, they do not suffer from the masking effect that outlier rejection techniques often do, and which causes them to have very low breakdown points. In fact when the solution to a redescending M-estimator equation is found by iteration, beginning with an initial estimator which has breakdown point .5, the resulting estimator also has a high breakdown point; see Huber (1984).

Another advantage is that the ψ function can be chosen to redescend smoothly to 0, which means that the information in moderately large outliers is not ignored completely. This greatly improves the efficiency of the redescending M-estimator. Finally, the redescending M-estimators do not suffer from the arbitrariness involved in outlier rejection techniques, which arises when one starts to visually decide whether there is one outlier on the right, or two on the left, etc. One note of caution: redescending estimators may not be well defined in the sense that there may be multiple solutions to the M-estimating equations (4.3.7).

We will define and briefly discuss two examples which were shown to be efficient in the Princeton robustness study [Andrews et al. (1972)]. First, Hampel's *three-part redescending M*-estimators have ψ functions which are odd functions [$\psi(-x) = -\psi(x)$ for all x] and are defined for positive x by

$$\psi_{a,b,r}(x) = \begin{cases} x, & 0 \leq x < a \\ a, & a \leq x < b \\ a(r-x)/(r-b), & b \leq x < r \\ 0, & r \leq x. \end{cases} \tag{4.3.13}$$

This function is plotted in Figure 4.1(d).

Second, Tukey's *biweight* is defined by

$$\psi_{\mathrm{bi}(r)} = x(r^2 - x^2)^2 I_{[-r,r]}(x) \tag{4.3.14}$$

and is plotted in Figure 4.1(e). To implement these M-estimators, begin with the sample median and the median absolute deviation, and iterate to a solution as described near (4.3.11).

A thorough discussion of these estimators is beyond the scope of this text. We will only select a few results from Hampel et al. (1986), Chapter 2.6; in particular, Table 3. They chose the biweight with $r = 4$, and the three-part redescender with $a = 1.31$, $b = 2.04$, and $r = 4$. For the sake of comparison with the monotone ψ functions, they also consider the Huber estimator (4.3.9) with $k = 1.4088$. With these choices of constants all three estimators have the same maximal bias at the normal model as measured by the gross-error sensitivity $\gamma^* = \sup_x |\mathrm{IF}_{\psi,\Phi}(x)| = 1.6749$. At the normal model the Huber estimator has ARE (relative to the mean) of .9563, while the Hampel redescender has ARE = .9119, and the Tukey biweight has ARE = .9100. The redescenders are slightly more efficient than the Huber estimator for several symmetric, wider tailed distributions, but much more efficient (about 20% more) than the Huber estimator for the Cauchy distribution. This is because they completely reject gross outliers, while the Huber estimator effectively treats them the same as moderate outliers.

4.3.4 *R*-Estimators

By definition R-estimators of location are those obtained by inverting a test for location which is based only on the ranks of the absolute values of the observations [see Lehmann (1975) or Hettmansperger (1984a)]. While they can be expressed in the functional form, the formalities do not seem to give much insight for the effort involved. Thus we will be content here to study the analog of the Wilcoxon test, the Hodges-Lehmann estimator, because it has strong robustness and efficiency properties.

Let X, Y be i.i.d. F and fix $\beta > 0$. Define $\mu(F)$ as the median of the distribution of $\beta X + (1 - \beta)Y$. It may be shown that $\mu(F)$ satisfies (4.3.1). The corresponding class of estimators

$$\mu(F_n) = \mathrm{med}_{i,j}\{\beta X_i + (1 - \beta)X_j\}$$

includes the Hodges-Lehmann estimator ($\beta = \frac{1}{2}$). Some rather unintuitive choices ($\beta = .9$, $\beta = 2$) lead to estimators with good efficiency and robustness properties for medium to wide tailed distributions; the small sample and asymptotic properties of the class are explored in Maritz, Wu, and Staudte (1977). The breakdown point is

$$\varepsilon^* = 1 - \frac{1}{\sqrt{2}} \approx .29$$

for all choices of β. In addition the influence function is bounded and as smooth as the underlying F.

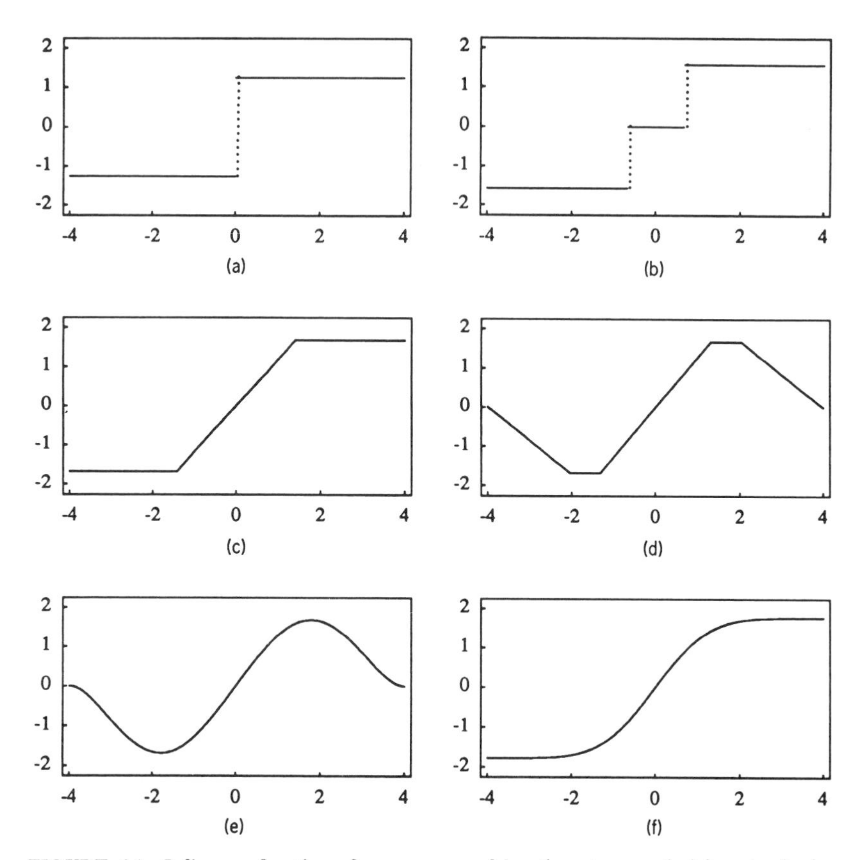

FIGURE 4.1 Influence functions for measures of location at an underlying standard normal distribution. (a) Median, (b) linear combination of two quantiles, (c) trimmed mean and Huber minimax solution, (d) Hampel three-part redescender, (e) Tukey biweight, and (f) Hodges-Lehmann estimator. If the underlying distribution is not normal, but is still symmetric and continuous, then the influence functions would be similar to those above.

Example 7: The Hodges-Lehmann Estimator. The Hodges-Lehmann estimator is the median of the set of all averages of pairs of observations. Perhaps because it combines the robust properties of the median with the efficiency properties of averaging, it performs well for a variety of distributions. It has asymptotic efficiency $\geq .86$ relative to the sample mean for estimating the location of *any* symmetric distribution, and efficiency at the normal distribution of $3/\pi \approx .95$. It can be infinitely more efficient than the sample mean. These facts can be gleaned through study of the asymptotic variance, given below.

Moreover, the Hodges-Lehmann estimator is robust with influence function (for F symmetric)

$$\mathrm{IF}_{\mathrm{HL}}(x) = \frac{2F(x-\mu)-1}{2\int f^2(y)\,dy}. \tag{4.3.15}$$

This influence function is sketched in Figure 4.1(f). It shows that outliers at x have a bounded influence and that the sensitivity of the estimator depends on the smoothness of F. The reader may check that the variance of the asymptotic distribution is

$$V(\mathrm{HL}, F) = \mathrm{E}_F[\mathrm{IF}_{\mathrm{HL}}(X)]^2 = \left\{\frac{1}{12\int f^2(y)\,dy}\right\}^2.$$

A review of methods for estimating this asymptotic variance is given in Draper (1988). An efficient method for calculating the Hodges-Lehmann estimator is found in Robinson and Sheather (1989).

4.4 ESTIMATORS OF DISPERSION

Let $X_1, \ldots, X_n$ be i.i.d. $N(\mu, \sigma^2)$. Two estimators are often suggested for estimating the standard deviation σ. The first is the usual sample standard deviation s_n, and the second is an alternative based on the difference of two order statistics. Remove the largest $2\frac{1}{2}\%$ of the observations and the smallest $2\frac{1}{2}\%$ of the observations and find the range of those remaining. The alternative estimate of σ is this 5% trimmed range, divided by 4. Since the smaller and larger observations are easy to spot by eye for moderate sample sizes, the latter rule is a quick method of estimating the standard deviation σ. It has the virtue of being robust to a small proportion of outliers, which the sample standard deviation is not. But how efficient is the trimmed-range method? Is $2\frac{1}{2}\%$ the best proportion to trim, given the desire for computational ease, efficiency, and robustness? In this section we will answer these questions and look at several other estimators of dispersion, including the median absolute deviation and the trimmed standard deviation.

4.4.1 Descriptive Measures of Dispersion

Turning to the problem of measuring the dispersion of an arbitrary distribution F, assume first that F is symmetric about a known point μ. If σ is a measure of scale [satisfying conditions (3.3.1)], then $\tau(X) = \sigma(|X-\mu|)$ has the usual properties of a measure of dispersion mentioned by David and Johnson (1948), namely, scale equivariance and location and sign invariance. Symbolically,

$$
\begin{array}{lll}
\text{(i)} & \tau(aX) = a\tau(X), & a > 0 \\
\text{(ii)} & \tau(X + b) = \tau(X), & \text{for all } b \\
\text{(iii)} & \tau(X) = \tau(-X). &
\end{array}
\tag{4.4.1}
$$

Any nonnegative functional $\tau(F)$ which has these properties will be called a *measure of dispersion*. To see how these conditions may restrict the choice of a descriptive measure, consider the following class.

Example 1: Linear Combination of Two Quantiles. Let $\tau(G) = cG^{-1}(\alpha) + dG^{-1}(\beta)$. Assume without loss of generality that $\alpha < \beta$. Then (i) of (4.4.1) is clearly satisfied; and (ii) implies $c = -d$. Continuing, (iii) is satisfied for a large class of G only if $\beta = 1 - \alpha$. Hence $\tau(G)$ satisfies

$$\tau(G) = d(G^{-1}(1-\beta) - G^{-1}(\beta)) \qquad \text{for some} \quad 0 < \beta < \tfrac{1}{2}.$$

The constant d in the above linear combination is usually determined by the desire for Fisher consistency for the unknown scale parameter of a particular F. For example, if we want $\tau(F_n) \xrightarrow{P} \sigma$ when the underlying distribution happens to be $N(\mu, \sigma^2)$, then d is determined by $1/d = G^{-1}(1-\beta) - G^{-1}(\beta)$, where $G = \Phi$, the standard normal distribution. The optimal choice of β, in terms of minimizing the standardized asymptotic variance, turns out to be $\beta = .07$, and the efficiency of this estimator relative to the sample standard deviation is only .65.

Remarks

1. If $\mu(X)$ is a measure of location [satisfying conditions (4.3.1)] and σ is a measure of scale satisfying conditions (3.3.1), then $\tau(X) = \sigma(|X - \mu(X)|)$ satisfies (4.4.1). Some common examples are given below.
2. Location and dispersion measures generalize the notion of location–scale parameters in the sense that for any fixed location–scale family $F_{\theta,\sigma}(x) = F((x-\theta)/\sigma)$, we have $\mu(F_{\theta,\sigma}) = a + b\theta$ and $\tau(F_{\theta,\sigma}) = b\sigma$, so that this pair forms a location–scale parameter for the same family.
3. In a location–scale parameter, the second component often refers to a population characteristic which is a measure of *dispersion*, but the terminology is now too entrenched to be modified.
4. In comparing estimators of dispersion, we will use the *standardized* asymptotic variance, for the same reasons given for scale estimators in Chapter 3.

Example 2: Mean Deviation. Take $\mu(X) = \mathrm{E}_F[X]$, $\sigma(Y) = \mathrm{E}_F[Y]$, and $\tau(X) = \sigma(|X - \mu(X)|)$. The natural estimator is

$$\tau(F_n) = \mathrm{E}_{F_n}|X - \mu(F_n)| = \frac{1}{n}\sum |X_i - \overline{X}|.$$

This estimator has breakdown point $\varepsilon^* = 0$, and an influence function that is sketched in Figure 4.2(a). Although it is nonrobust to outliers, its asymptotic efficiency relative to the sample standard deviation is remarkably superior for slightly contaminated normal mixtures. In particular, when

$$F(x) = (1 - \varepsilon)\Phi\left(\frac{x - \mu}{\sigma}\right) + \varepsilon\Phi\left(\frac{x - \mu}{3\sigma}\right)$$

the ARE is .876 for $\varepsilon = 0$ but greater then 1 for all $.002 < \varepsilon < .5$. For more details and information on the historical importance of this example see Tukey (1960) and Huber (1981).

Example 3: Median Absolute Deviation. The corresponding estimator is often referred to as MAD. Let $\mu(X) = \operatorname{med} X = F_X^{-1}(\frac{1}{2})$, $\sigma(Y) = \operatorname{med} Y$, and $\tau(X) = \sigma(|X - \mu(X)|)$. Then

$$\tau(F_n) = \operatorname{med}|X_i - \operatorname{med} X_j|.$$

It may be shown that the breakdown point $\varepsilon^* = .5$. The influence function is derived using the chain rule and is shown in Figure 4.2(b). This estimator is often used as an initial estimate of dispersion because of its high breakdown point. However, it has very low efficiency relative to the sample standard deviation for normal data of only .3674.

Example 4: Interquartile Range. A special case of example 1 in common use is obtained when $\beta = \frac{1}{4}$:

$$\tau(F) = F^{-1}(\tfrac{3}{4}) - F^{-1}(\tfrac{1}{4}).$$

Clearly $\varepsilon^* = .25$ and the influence function is the difference of the influence functions for the respective quantiles as determined by (3.2.3). Note that it has an influence function identical to the MAD for symmetric F and thus shares its asymptotic properties.

Example 5: The β-Trimmed Standard Deviation. Let $x_\beta = F^{-1}(\beta)$, $0 < \beta < \frac{1}{2}$. Define $\tau(X) = \sigma(|X - \mu(X)|)$, where $\mu(X)$ is the 2β-symmetrically trimmed mean:

$$\mu(X) = \frac{1}{1 - 2\beta}\int_{x_\beta}^{x_{1-\beta}} x\, dF(x)$$

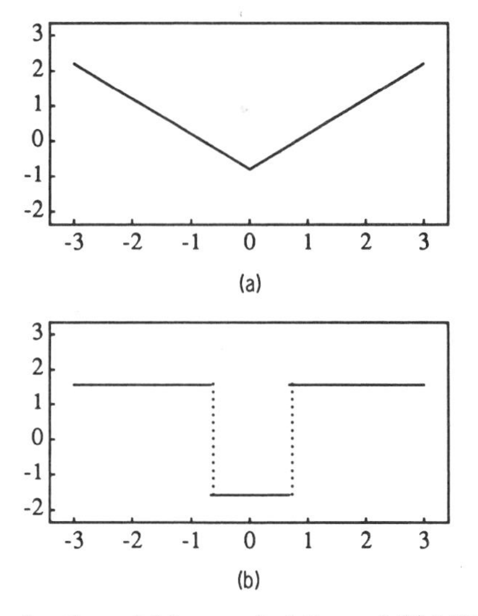

FIGURE 4.2 Influence functions of (a) mean deviation and (b) MAD at a symmetric distribution.

and σ_β is the β-trimmed measure of scale applied to the absolute deviation; that is,

$$\sigma_\beta^2(Y) = \frac{1}{1-\beta}\int_0^{y_{1-\beta}} y^2\, dG(y)$$

where G is the distribution of $Y = |X - \mu(X)|$. The induced estimator $\tau(F_n)$ of $\tau(F)$ finds a symmetrically trimmed mean and then from it the standard deviation of the $[n(1-\beta)]$ observations that are closest to it. It is simpler computationally to discard the $[\beta n/2]$ smallest and $[\beta n/2]$ largest observations initially and then compute the usual ($\beta = 0$) mean and standard deviation. The latter estimator of dispersion is defined by the functional τ^*, which computes the standard deviation of the conditional distribution of X given that X lies between $F^{-1}(\beta/2)$ and $F^{-1}(1-\beta/2)$. When F is *symmetric*, $\tau(F) = \tau^*(F)$; so we will use the latter functional and estimator in such cases.

4.4.2 Performance of Some Dispersion Estimators

In this section we compare the estimators of dispersion defined above with the sample standard deviation for estimating the standard deviation of a normal distribution (Table 4.3). Each estimator needs to be multiplied by a suitable constant to make it consistent for σ, and these constants are found

Table 4.3 Standardized Variances of Estimators of a Normal Standard Deviation for Various Contamination Models

Model	Standard Deviation	Mean Deviation	MAD	Interquartile Range	.1-trimmed Standardized Deviation
Φ	.529	.597	1.326	1.178	.647
	(.509)	(.578)	(1.226)		(.619)
$.9\Phi(x) + .1\Phi(x/3)$	1.626	1.051	1.455	1.242	1.018
$.95\Phi + .05C$	> 1000	689.4	1.397	1.216	1.174
$.95\Phi(x) + .05E(x/3)$	3.364	1.468	1.387	1.205	1.063
	(3.402)	(1.163)	(1.211)		(1.058)

in the problems. The estimators are compared on the basis of their estimated standardized variances for samples of size 20 from various models, which are mixtures of the normal distribution and 5–10% of contamination. The estimates are based on 4000 replications [see Staudte (1980) for further details].

Remarks

1. The contaminating distributions include C for Cauchy and $E(x/3)$ for the one-sided exponential with mean 3.
2. Most of the above results were found assuming location was unknown, so that "median deviation" means $\text{med}_j|x_j - \text{med}_i x_i|$. The values in parentheses were made assuming location was known, and then "median deviation" means $\text{med}_j|x_j|$. The assumption of known location makes little difference to the performance of these estimates of dispersion.
3. On the basis of these results, if there is any doubt about the normality assumption, one would rank these estimators as follows: .1-trimmed standard deviation, interquartile range, MAD, mean deviation, and standard deviation. Only the .1-trimmed standard deviation has reasonably high efficiency among the robust estimators at the normal model. Note also that the interquartile range is not the most efficient of the simple estimators defined in Example 1 of Section 4.1.
4. Finally, we should be concerned whether these $n = 20$ results bear any resemblance to asymptotic results. Some comparisons are possible and given in Table 4.4, using asymptotic results from Bickel and

Table 4.4 Comparison of Efficiencies of Competing Estimators of a Normal Standard Deviation Relative to the Sample Standard Deviation

Model	Mean Deviation	Interquartile Range	.1-trimmed Standardized Deviation
Φ	.88	.43	.82
	[.88]	[.37]	[.78]
$.9\Phi(x) + .1\Phi\left(\frac{x}{4}\right)$	2.00	2.02	1.71
	[2.21]	[2.05]	[3.48]

Lehmann (1976). Each entry in the table is the efficiency of an estimator relative to the standard deviation, as measured by the ratio of their variances. The first number of each pair is the sample size 20 efficiency, and the second in square brackets is the asymptotic efficiency. The agreement between asymptotic and $n = 20$ results is good except for the case of trimmed standard deviation in the second model. This large discrepancy could be caused by the difference in the two trimming procedures described in example 5 of Section 4.4.1.

4.5 JOINT ESTIMATION OF LOCATION AND DISPERSION

4.5.1 M-Estimators of Location and Dispersion

Let $F_{\mu,\sigma}(x) = F((x - \mu)/\sigma)$ denote a location–scale family generated by F, and assume that F has two derivatives f and f'. The classical maximum likelihood approach to estimation of the location–scale parameter may often be carried out by setting the partial derivatives with respect to μ and σ, respectively, of the log–likelihood equal to zero, and solving the resulting likelihood equations. Letting $\psi(x) = -(f'/f)(x)$, these equations may be expressed:

$$\sum_{i=1}^{n} \psi\left(\frac{x_i - \mu}{\sigma}\right) = 0$$
$$\sum_{i=1}^{n} \left[\left(\frac{x_i - \mu}{\sigma}\right)\psi\left(\frac{x_i - \mu}{\sigma}\right) - 1\right] = 0. \tag{4.5.1}$$

Alternatively, these estimating equations may be expressed as the evaluation at $G = F_n$ of

$$\begin{aligned} \mathrm{E}_G\left[\psi\left(\frac{X-\mu}{\sigma}\right)\right] &= 0 \\ \mathrm{E}_G\left[\left(\frac{X-\mu}{\sigma}\right)\psi\left(\frac{X-\mu}{\sigma}\right)-1\right] &= 0 \end{aligned} \tag{4.5.2}$$

where $X \sim G$. When F_n is based on a sample from $G = F_{\mu,\sigma}$, the MLEs so derived will (under regularity conditions) be consistent, asymptotically normal, and efficient. However, they may be nonrobust with small breakdown points or unbounded influence functions. In the most important case of normal data, the breakdown point is zero, and the influence functions are unbounded. The approach described below in Section 4.5.2 modifies the MLEs so as to maintain high efficiency while reducing the sensitivity to outliers.

The estimating equations in (4.5.2) motivate the following general definition of M-estimators of location and dispersion. Let ψ be an odd, increasing function and let χ be an even function. Then define $\mu(G)$ and $\sigma(G)$ as the simultaneous solutions μ and σ to

$$\begin{aligned} \mathrm{E}_G\left[\psi\left(\frac{X-\mu}{\sigma}\right)\right] &= 0 \\ \mathrm{E}_G\left[\chi\left(\frac{X-\mu}{\sigma}\right)\right] &= 0. \end{aligned} \tag{4.5.3}$$

Huber (1981) treats questions of existence and uniqueness of solutions to (4.5.3), as well as the derivation of the influence curves for $\mu(G)$ and $\sigma(G)$. In particular he shows that for symmetric G the location and dispersion estimates are asymptotically independent, and that the influence function of the location estimator arising from (4.5.3) is the same as that of the one-step M-estimator based on a consistent initial estimator of location. (See example 5 of Section 4.3.3.)

4.5.2 Huber's Proposal 2

If the ψ function in (4.5.2) was motivated by the maximum likelihood approach in the context of a normal model, then $\psi(x) = x$ is unbounded. Thus Huber (1964) proposed the choices $\psi(x) = \psi_k(x) = \max\{-k, \min\{x, k\}\}$ and $\chi(x) = \psi_k^2(x) - a(k)$ in (4.5.3), where the constant $a(k)$ is chosen to make the estimator of the normal standard deviation $\sigma(F)$ Fisher-consistent:

$$\mathrm{E}_\Phi[\psi_k^2(Z)] = a(k). \tag{4.5.4}$$

Table 4.5 Asymptotic Efficiency of Huber's M-Estimator Relative to the Mean at the Normal Distribution for Various Choices of k

k	$a(k)$	$b(k)$	$E_\Phi(\psi_k, \overline{X})$
.0	.0000	.0000	.6366
.5	.1851	.3829	.7921
1.0	.5161	.6827	.9031
1.5	.7785	.8664	.9642
2.0	.9205	.9549	.9897

Huber [(1981), Section 6.7] suggests different methods for solving his Proposal 2 system of equations:

$$\sum_{i=1}^{n} \psi\left(\frac{x_i - \mu}{\sigma}\right) = 0$$
$$\sum_{i=1}^{n} \psi^2\left(\frac{x_i - \mu}{\sigma}\right) = (n-1)a(k). \tag{4.5.5}$$

One method starts with μ_0 and σ_0, the very robust median and the median absolute deviation, and then proceeds iteratively with

$$[\sigma_{m+1}]^2 = \frac{1}{(n-1)a(k)} \sum_{i=1}^{n} \psi^2(y_{m,i})[\sigma_m]^2$$
$$\mu_{m+1} = \mu_m + \frac{\sum_{i=1}^{n} \psi(y_{m,i})\sigma_m}{\sum_{i=1}^{n} \psi'(y_{m,i})} \tag{4.5.6}$$

where $y_{m,i} = (x_i - \mu_m)/\sigma_m$.

The choice of the tuning constant $a(k)$ in (4.5.4) depends on the conflicting merits of high asymptotic efficiency of the location estimator and a low bound on the influence function. The asymptotic relative efficiency of Huber's M-estimator ψ_k to the sample mean at the normal distribution is given by the ratio of the asymptotic variances, which by Theorem 4.2 and (4.3.8) are

$$E_\Phi(\psi_k, \overline{X}) = \frac{V(\psi_\infty)}{V(\psi_k)} = \frac{\mathrm{E}_\Phi^2[\psi'(X)]}{\mathrm{E}_\Phi[\psi_k^2(X)]}. \tag{4.5.7}$$

Thus the asymptotic relative efficiency of Huber's M-estimator to $\overline{X}$ at the normal distribution is $E_\Phi(\psi_k, \overline{X}) = b(k)/a(k)$, where $b(k) = 2\Phi(k) - 1$ and $a(k) = 2\Phi(k) - 1 - 2k\varphi(k) + 2k^2(1 - \Phi(k))$. Some selected values are given in Table 4.5. More details are given in Huber (1964, 1981), Section 6.6, and Hampel et al. (1986), Section 2.5d.

4.6 CONFIDENCE INTERVALS FOR THE MEDIAN

In Section 4.3 a large number of location estimators were shown to be asymptotically normal, and estimates of their standard errors were given. Thus large-sample approximate $1-\alpha$ confidence intervals for location may be found; we describe them in Section 4.6.2 below. But first we discuss two methods for obtaining exact confidence intervals.

4.6.1 Distribution-Free Confidence Intervals for the Median

In this section we first discuss a method which gives exact $1-\alpha$ confidence intervals for the median of a possibly asymmetric distribution. All that is required is that the underlying distribution be continuous.

The median is defined by the functional $\theta = \theta(F) = F^{-1}(\frac{1}{2})$ in Example 2 of Section 4.3. Let $X_{(1)} \leq X_{(2)} \leq \cdots \leq X_{(n)}$ denote the order statistics of a sample from F, and let k be an integer satisfying $1 \leq k < n/2$. Then $[X_{(k)}, X_{(n-k+1)}]$ is a confidence interval for θ with confidence coefficient determined by the binomial $(n, .5)$ distribution. To see this, introduce the sign statistic

$$T_n = \sum_{i=1}^{n} I(X_i > \theta) = \#\{X_i > \theta\} \sim B(n, .5).$$

Then it is not hard to see that

$$P(k \leq T_n \leq n-k) = P(X_{(k)} < \theta < X_{(n-k+1)}),$$

and hence the confidence coefficient may be easily determined. However, because the binomial distribution is discrete, it is not possible in general to find a confidence interval based on the sign statistic with confidence coefficient exactly equal to the desired one. For example, the closest available confidence coefficients to .95 are .891 and .978 when $n = 10$, .885 and .959 when $n = 20$, and .943 and .964 when $n = 100$.

We will interpolate intervals to achieve .95 confidence. Begin with $(X_{(k+1)}, X_{(n-k)})$, a γ_{k+1} confidence interval for θ, and $(X_{(k)}, X_{(n-k+1)})$, a γ_k confidence interval for θ. Define the interpolated interval (X_L, X_U) by

$$X_L = \lambda X_{(k+1)} + (1-\lambda)X_{(k)} \qquad \text{and} \qquad X_U = \lambda X_{(n-k)} + (1-\lambda)X_{(n-k+1)}.$$

Denote the confidence coefficient of (X_L, X_U) by γ; that is, let $\gamma = P(X_L < \theta < X_U))$. For a given γ with $\gamma_{k+1} < \gamma < \gamma_k$ we want to choose the appropriate λ. First we need to define the interpolation factor I by

$$I(\gamma) = \frac{\gamma_k - \gamma}{\gamma_k - \gamma_{k+1}}.$$

We seek a formula for λ in terms of $I = I(\gamma)$. Note that linear interpolation takes $\lambda = I$. However linear interpolation gives poor results. Hettmansperger and Sheather (1986) showed that $I = k/n$ when $\lambda = \frac{1}{2}$ for all densities symmetric about θ. Using a normal approximation (with continuity correction) to the binomial distribution gives $k \approx (n/2) - z\sqrt{n/4} + \frac{1}{2}$, so $k/n \ll \frac{1}{2} = \lambda$. Thus linear interpolation is inappropriate.

Hettmansperger and Sheather (1986) calculated I as a function of λ for a number of underlying distributions. They found very little difference between the different underlying distributions and recommended the formula

$$\lambda = \frac{(n-k)I}{k + (n-2k)I}. \tag{4.6.1}$$

This formula is exact if the data are from a double exponential distribution. Two recent studies have provided further empirical evidence of the accuracy of (4.6.1) for a wide range of distributions [see Sheather and McKean (1987), and Hall and Sheather (1988)]. Formula (4.6.1) is used in version 6 of Minitab and is referred to as **NLI** in the output of the sign confidence interval command **sint**. To illustrate this command, we analyze the cell lifetime data of Example 2 from Section 4.1. Below is Minitab output describing the data:

```
MTB > desc c1
-     N      MEAN     MEDIAN     TRMEAN     STDEV     SEMEAN
C1   20     9.930      9.100      9.372     3.042      0.680
-
-      MIN         MAX         Q1          Q3
C1   7.700      22.200      8.550      10.400
MTB > sint c1.
SIGN CONFIDENCE INTERVAL FOR MEDIAN
-                       ACHIEVED
-     N  MEDIAN     CONFIDENCE    CONFIDENCE INTERVAL     POSITION
C1   20    9.10         0.8847          (8.80, 10.30)            7
-                       0.9500          (8.72, 10.38)          NLI
-                       0.9586          (8.70, 10.40)            6
MTB > boxplot c1;
SUBC > notch 95.

          ---(+ )-----                                0
   +---------+---------+---------+---------+---------+------C1
  6.0       9.0      12.0      15.0      18.0      21.0
```

The upper and lower dashed lines have length approximately equal to the interquartile range $r = x_{.75} - x_{.25} = Q3 - Q1$ and mark the location of the middle 50% of the data. The plus sign + indicates the location of the median. The "whiskers" emanating from the sides of the box extend outward to the furthest observation within distance $1.5r$ of each quartile. Observations between $1.5r$ and $3r$ from the quartiles are deemed probable outliers and their position is marked by an asterisk (there are none for these data). Observations further than $3r$ from the quartiles are marked with an O.

It is common practice to mark sign confidence intervals for the median as so-called "notches" on the box plot. In the above plot this has been done for the interpolated 95% confidence interval [8.72, 10.38].

Finally to end this section we note that the *length* of a 95% confidence interval of the form $[X_{(k)}, X_{(n-k+1)}]$ will have finite sample breakdown point $\varepsilon_n^* \approx \frac{1}{2} - 1/\sqrt{n}$. Thus for $n = 9$, $\varepsilon_9^* \approx \frac{1}{6}$; for $n = 25$, $\varepsilon_{25}^* \approx \frac{3}{10}$.

Confidence Intervals Based on Pairwise Averages

In this subsection we assume that the distribution is symmetric about an unknown median θ. For each $1 \leq i \leq j \leq n$ define the pairwise average of X_i and X_j by $A_{ij} = (X_i + X_j)/2$. These averages are sometimes called Walsh averages. Denote the $m = \binom{n}{2}$ ordered pairwise averages by $A_{(1)}, A_{(2)}, \ldots$. Then it may be shown (see for example Lehmann, (1975)) that for all θ

$$P_\theta\{A_{(i)} \leq \theta < A_{(i+1)}\} = P_0\{T = i\}$$

where T has the null distribution of the one-sample Wilcoxon signed-rank statistic defined in (5.2.15). This distribution is given in most books on nonparametric statistics or books of statistical tables. The confidence interval $[L, U]$ with $L = A_{(i)}$ and $U = A_{(m-i+1)}$ has exact confidence coefficient $2P\{T \leq i\} - 1$. For large n the confidence coefficient is computed using the fact that T is approximately normal with mean $\mathrm{E}[T] = n(n+1)/4$ and variance $\mathrm{Var}[T] = n(n+1)(2n+1)/24$.

4.6.2 Large Sample Approximate Confidence Intervals

All the confidence intervals in this section are of the form

$$[L, U] = \left[\hat{\theta} - c\widehat{\mathrm{SE}}[\hat{\theta}],\ \hat{\theta} + c\widehat{\mathrm{SE}}[\hat{\theta}]\right]$$

for an appropriate choice of c. For example the classical t-intervals are given by $\hat{\theta} = \overline{x}$, $\widehat{\mathrm{SE}}[\hat{\theta}] = s/\sqrt{n}$ and $c = t_{n-1,1-\alpha/2}$.

For the 2β-trimmed mean c is the $1 - \alpha/2$ quantile of the Student t-distribution with $n - [2\beta n] - 1$ degrees of freedom (see the results in Sections 4.3 and 5.2.) Also see Appendix D.2.1.

To find the Huber M-estimate we use the $\psi = \psi_k$ function defined in (4.3.9) and solve the equation (4.3.11) iteratively. The standard choice of robust estimate of scale is $\tau_n = \tau(F_n) = 1.483 \times \text{MAD}$. The bending constant k chosen so that the resulting estimator has the desired asymptotic properties. For example if k is chosen to be $1.28 = \Phi^{-1}(.9)$, then the influence function of Huber's estimator will be the same as that of the $2 \times .1$-trimmed mean. Its asymptotic efficiency relative to the sample mean for $k = 1.28$ is approximately .93 for normal data by the results in Table 4.5. Its standard error may be estimated using the influence function method. It is the square root of the estimated asymptotic variance, namely

$$\frac{\sum_i \psi^2((x_i - \hat{\theta})/\tau_n)}{(\sum_i \psi'(x_i - \hat{\theta})/\tau_n))^2}. \tag{4.6.2}$$

These estimates are readily calculated using Minitab Macros D.2.3–D.2.5 as illustrated in the examples to follow.

4.7 EXAMPLES

The exact distributions of robust estimators are often difficult to derive, so asymptotic methods and simulation studies are commonly used to assess their performance. Alternatively one may apply a variety of robust estimators to real data sets and use some other means of assessing their performance. This has been done by Stigler (1977) and Rocke, Downes, and Rocke (1982). We select three examples from their collections of carefully chosen data sets, plus one from a recently compiled source by Andrews and Herzberg (1985). Examination of such data sets brings the statistician closer to the situation faced by experimental scientists, but of course it is no substitute for involvement with the experiment itself.

In an article which asked: Do robust estimators work with *real* data? Stigler (1977) chose data sets from physical science experiments of the eighteenth and nineteenth centuries for which the unknown parameter to be estimated is now known. On the basis of his analysis of 11 estimators on some 20 data sets he concludes that a lightly trimmed mean performs best and that the untrimmed mean works better than some robust proposals. As Stigler acknowledges, one problem with comparing estimates with "true values" is that the experimental results may include systematic biases which affect all the estimators and make it difficult to detect real differences between them. Rocke, Downes, and Rocke (1982) emphasize this point and use the estimated variances of the estimates to make their comparisons, both on Stigler's data sets and on some more recently obtained analytical

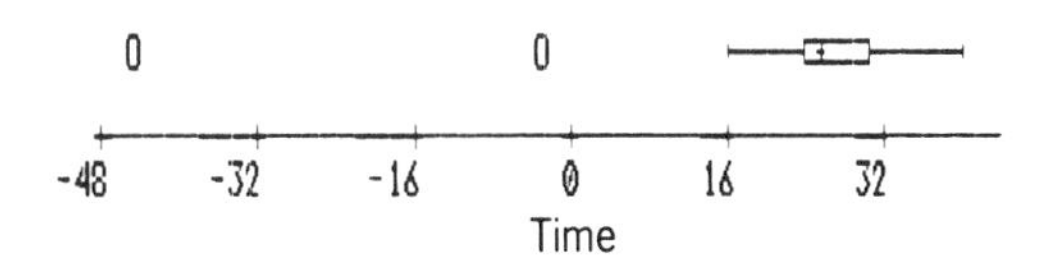

FIGURE 4.3 Estimates of the time required by light to travel a known distance.

chemistry data. Their article is entitled: "Are robust estimators really necessary?" Their answer is affirmative, but their recommendations are different from Stigler's. They conclude that heavily trimmed means or M-estimators such as Huber's bounded influence estimator or Tukey's biweight are most efficient, and recommend the latter type over the trimmed means because of its superior efficiency for normal data, and because the smooth down-weighting of outliers seems more reasonable than outright trimming.

Some of the points raised in these articles will be raised in the following study of a few specific examples.

Example 1: Newcomb's Measurements of the Velocity of Light. In 1882 Newcomb estimated the velocity of light by repeatedly measuring the time in millionths of a second which it took light to cover a known distance and then converting the average time to a velocity. A boxplot of the data (which are given in Appendix C.3) is shown in Figure 4.3. If there is no a priori reason for assuming that the measurements will be normally or even symmetrically distributed, then only the sign statistic method of obtaining a confidence interval for the median seems appropriate. However, Newcomb may have had enough prior experience with the method to make assumptions about the distributional shape of such measurements. We may safely assume the observations are independent. We find the confidence intervals $[L, U]$ based on the Student's t-statistic, the sign statistic (with nonlinear interpolation) discussed in Section 4.6, and the Wilcoxon type confidence intervals based on Walsh averages. These intervals are effortlessly computed on Minitab with the respective commands TINT, SINT, and WINT.

Minitab also calculates the point estimates corresponding to the sign and Wilcoxon statistics, namely the median and Hodges-Lehmann estimators, respectively. For the sake of comparison we have included these estimates in the tables below, even though we do not have estimates of their standard errors. Such estimates of standard errors require a density estimate, for example using the method given in Example 2 of Section 4.3.2.

In addition, we use the Minitab macros TMEAN and HUBER1-3 given in Appendix D to find the confidence intervals based on the $2 \times 5\%$ and

Table 4.6 Point Estimates of Location, Estimated Standard Errors, and 95% Confidence Intervals for Location

Method	$\hat{\theta}$	$\widehat{\text{SE}}[\hat{\theta}]$	L	U
Student's t	26.21	1.32	23.57	28.85
Sign	27.00	—	26.00	28.40
Wilcoxon	27.50	—	26.00	28.50
$2 \times 5\%$ trmean	27.40	.70	26.00	28.80
$2 \times 10\%$ trmean	27.43	.68	26.06	28.79
Huber $\psi_{1.28}$	27.38	.62	26.16	28.60

$2 \times 10\%$ trimmed means and the Huber M-estimator with $k = 1.28$. To calculate a $2 \times 5\%$ percent symmetrically trimmed mean we remove approximately $\beta n = .05 \times 66 = 3.3$ observations from each end of the sample. We trimmed exactly 3 observations from each end. For the $2 \times 10\%$ symmetrically trimmed mean we removed 6 observations from each end before averaging. These results are listed in Table 4.6.

From the boxplot in Figure 4.3 it is easy to see that there are two gross outliers, and these points are what cause the Student's t-interval to be so much longer than the others. The "true value" of time being estimated in the experiment has subsequently been shown to be 33.02, so it is clear that experimental *bias* in all the results was more important than the differences between the estimators. This fact should not obscure the main point that the traditional Student's t-procedures are badly affected by the outliers.

Finally we remark that it may be tempting to simply reject the outliers and take the mean of those remaining. However, there is no way to measure the standard errors or the confidence level of intervals based on such ad hoc methods. □

Example 2: Percentage of Shrimp in Shrimp Cocktail. This example and the next are two of those analyzed by Rocke, Downes, and Rocke (1982). In both cases measurements are made in different laboratories, so they are independent. They also appear to be roughly symmetric, with about 10% of the observations casting doubt on an assumption of normality. These data are originally from King and Ryan (1976), wherein a method of determining the amount of shrimp in shrimp cocktail was studied to determine compliance with proposed labeling regulations. The amount of shrimp is measured as a percentage of the declared total weight of the contents of the commercial container. Differences in measurements can arise because of differing amounts of shrimp, but also because of absorption of cocktail sauce and difficulties in distinguishing small pieces of shrimp from horseradish. It is desired to have a reliable measure of the standard error inherent in a given

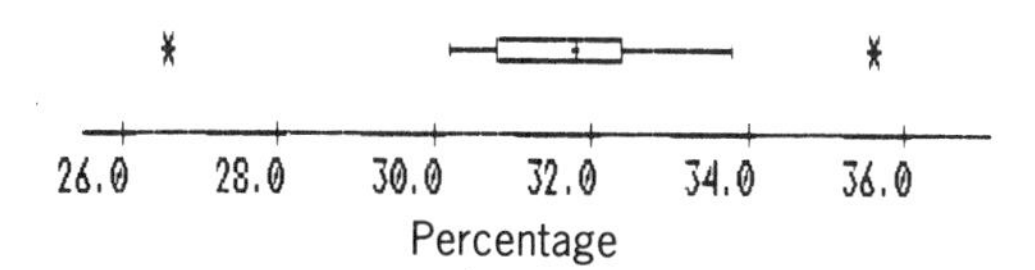

FIGURE 4.4 Percentage of shrimp in shrimp cocktail.

Table 4.7 Point Estimates of the Unknown Percentage, Their Standard Errors, and 95% Confidence Intervals for the Percentage

Method	$\hat{\theta}$	$\widehat{\mathrm{SE}}[\hat{\theta}]$	L	U
Student's t	31.79	.43	30.88	32.71
Sign	31.95	—	31.01	32.45
Wilcoxon	31.85	—	31.20	32.50
2 × 5% trmean	31.86	.30	31.24	32.51
2 × 10% trmean	31.85	.29	31.21	32.49
Huber $\psi_{1.28}$	31.86	.30	31.27	32.45

method of measurement. The results of 18 different collaborators applying one method to the same size container of shrimp cocktail yielded the following results:

$$32.2,\ 33.0,\ 30.8,\ 33.8,\ 32.2,\ 33.3,\ 31.7,\ 35.7,\ 32.4,$$
$$31.2,\ 26.6,\ 30.7,\ 32.5,\ 30.7,\ 31.2,\ 30.3,\ 32.3,\ 31.7.$$

In the boxplot of these data in Figure 4.4 there are two observations which are outlying, and the t-procedures will be much more affected by them than the other (robust) methods. The results of the various methods are listed in Table 4.7. □

Example 3: Detecting the Pesticide DDT in Kale. Another data set analyzed by Rocke, Downes, and Rocke (1982) was originally published by Finsterwalder (1976), who compared methods for detecting different pesticides in eggs and kale. The following 15 observations are the ppm (parts per million) of p,p'-DDT detected in kale by 15 different laboratories using the multiple pesticide residue method (Figure 4.5):

$$2.79,\ 2.93,\ 3.22,\ 3.78,\ 3.22,\ 3.38,\ 3.18,$$
$$3.33,\ 3.34,\ 3.06,\ 3.07,\ 3.56,\ 3.08,\ 4.62,\ 3.34.$$

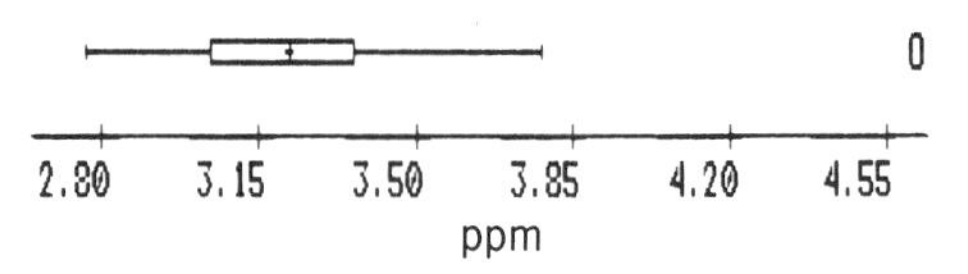

FIGURE 4.5 The ppm of DDT in kale as calculated in fifteen laboratories.

Table 4.8 Point Estimates of Location, Their Standard Errors, and 95% Confidence Intervals for Location

Method	$\hat{\theta}$	$\widehat{\text{SE}}[\hat{\theta}]$	L	U
Student's t	3.33	.11	3.09	3.57
Sign	3.22	—	3.07	3.37
Wilcoxon	3.26	—	3.12	3.47
2 × 5% trmean	3.27	.08	3.10	3.44
2 × 10% trmean	3.25	.07	3.10	3.40
Huber $\psi_{1.28}$	3.25	.07	3.11	3.40

The mean and the standard deviation of these observations are 3.327 and .433. In the article by Finsterwalder the observation 4.62 was "rejected as outlier, not used in statistical evaluation." It was rejected on the basis of Dixon's test (1953). The standard deviation of the remaining 14 observations is by our computations $s = .2529$, slightly different from the .262 reported in the article, but much smaller than the .433 of the original observations.

If the standard error is estimated by $s/\sqrt{n} = .2529/\sqrt{14} = .07$ using the 14 remaining observations, then this estimate is smaller than that for the robust estimators applied to all 15 observations. However, it is conditional on the observation being rejected. No one knows what the unconditional standard error of the outlier rejection method plus mean technique is because it is subjective, depending first on whether or not an outlier rejection technique is used (after looking at the data), second on what significance level is used, and third on whether a one- or two-sided test for outliers is used. The results of various methods are listed in Table 4.8.

Example 4: Darwin's Data on Heights of Plants. In 1878 Darwin carried out experiments with the plant *Zea mays* in which he compared the heights of cross-fertilized plants with self-fertilized plants grown in the same pots. The data are listed in Appendix C, and a thorough description is given

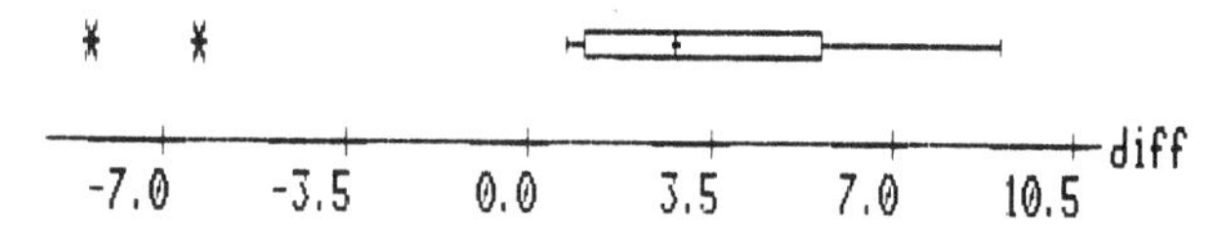

FIGURE 4.6 Difference in heights of cross-fertilized and self-fertilized *Zea mays*.

Table 4.9 Point Estimates of the Mean Difference, Their Standard Errors, and 95% Confidence Intervals for the Mean Difference

Method	$\hat{\theta}$	$\widehat{\mathrm{SE}}[\hat{\theta}]$	L	U
Student's t	2.62	1.22	.00	5.23
Sign	3.00	—	1.28	5.75
Wilcoxon	3.125	—	.50	5.19
2 x 5% trmean	2.94	1.24	.24	5.65
2 x 10% trmean	3.34	.85	1.45	5.23
Huber $\psi_{1.28}$	3.34	.94	1.50	5.18

by Andrews and Herzberg (1985). We may show that the pairs (crossed, self) have a product moment correlation coefficient of $-.335$, although the significance of this result may be questionable given the presence of some outlying observations. In any case only four pots were used to grow the 15 pairs of plants, so if there is an effect due to pots, the observation pairs are not really independent. Table 4.9 lists the results. It is similar to those presented earlier in that it does assume independence of the differences of the 15 pairs. They are, in inches,

$$6.125,\ -8.375,\ 1,\ 2,\ .75,\ 2.875,\ 3.5,\ 5.125,$$

$$1.75,\ 3.625,\ 7,\ 3,\ 9.375,\ 7.5,\ -6.$$

A boxplot of these results is shown in Figure 4.6. Note that there are two possible outliers on the left, although outlier rejection methods may not identify them as such due to masking. In any case the results in Table 4.9 show that there is a marked difference between the results. In particular the t-test is the only one that would not find the data significant at level .05. This does not *prove* that the t-test is leading to the wrong decision, but it does indicate the great effect that two moderate outliers can have on it. □

4.8 PROBLEMS

Section 4.1

1. Verify the details of Example 1 in Section 4.1.2 by finding the 20%-trimmed mean, the 20%-Winsorized sample, and the estimate of the standard error of the trimmed mean.

2. In Example 2 of Section 4.1.2 find the maximum likelihood estimators of γ and θ; and evaluate them for the given data.

3. Let (Y,Z) have the bivariate normal distribution with correlation $\rho_{Y,Z}$ and respective means (μ_Y, μ_Z) and variances (σ_Y^2, σ_Z^2).
   (a) Find the distribution of the difference $X = Y - Z$.
   (b) Under what restrictions on the parameters does the joint distribution of (Y,Z) satisfy $F_{Y,Z}(y,z) = F_{Y,Z}(z,y)$ for all (y,z)?
   (c) If (Y,Z) satisfies the symmetry condition in (b), what is the distribution of the difference $X = Y - Z$?

4. Assume that F has a density f with respect to Lebesgue measure, and that the distribution is symmetric about μ as defined in Section 4.1.1. Express the condition of symmetry in terms of the density.

5. Let Y,Z have a joint density which is symmetric in its two arguments: $f_{Y,Z}(y,z) = f_{Y,Z}(z,y)$ for all (y,z). Show that the difference of Y,Z has a symmetric density, whether or not Y,Z are independent.

Section 4.2

6. Verify the five properties of location–scale families given in Section 4.2.1.

7. Derive the mean and variance for each of the densities shown in Table 4.2. Explain how to use these results to obtain the mean and variance for any member of a location–scale family generated by each density.

8. The MLE (maximum likelihood estimates) of (μ, σ) for a location–scale family with differentiable density may often be found by setting the partial derivatives (with respect to μ and σ) of the log–likelihood

$$\ln[L(\mu,\sigma)] = \sum_{i=1}^{n} \ln\left[\frac{1}{\sigma} f\left(\frac{x_i - \mu}{\sigma}\right)\right]$$

equal to zero and solving the equations simultaneously.

(a) Find the estimating equations described above in terms of f, f' and (μ, σ).

(b) Solve the general equations found in part (a) for as many of the densities in Table 4.1 as possible.

(c) It is claimed in Section 4.1.2 that for the displaced exponential model, $X_{(1)}$ converges to γ at a faster rate than the MLE does to θ. Compare the mean squared error of $X_{(1)} - \gamma$ with that of $\hat{\theta} - \theta$ as $n \to \infty$.

Section 4.3

9. **(a)** Show that if a distribution F is symmetric about θ, then (4.3.1)(i) and (ii) imply that $\mu(F) = \theta$.

(b) Show that if $a \leq X \leq b$ with probability 1, then conditions (4.3.1) (i)–(iii) together imply that $a \leq X \leq b$.

10. In example 3 of Section 4.3.1 assume that the underlying distribution is normal with unknown parameters, and find the choice of α that yields the most efficient estimator of the mean which is a linear combination of two order statistics. That is, find α which minimizes the asymptotic variance of $n^{1/2}[\mu(F_n) - \mu(F)]$. Find the ARE of this estimator relative to that of the sample mean.

11. **(a)** Show that probability averages of measures of location are measures of location; i.e., if μ_s satisfies (i)–(iv) for an index set of values of s, and K is a probability distribution on these values, then $\int \mu_s \, dK(s)$ also satisfies (i)–(iv).

(b) In particular show that symmetric mixtures of quantiles

$$\mu(F) = \frac{F^{-1}(t) + F^{-1}(1-t)}{2}$$

satisfy (i)–(iv).

12. **(a)** Show that condition (v) of Section 4.3.1 implies condition (iii).

(b) Define $T(X)$ to be the midpoint of the modal interval of length $2c$, where $c > 0$ is fixed; then $T(X)$ is a number m which maximizes $P_F\{|X - m| \leq c\}$. Restrict the domain of T to those distributions F which have unique modal intervals of finite length. Show that T satisfies (i)–(iv) but not (v) of location measures by construction of an appropiate F.

13. Let $\mu(F)$ be a location measure, and let

$$F_{\theta,\tau}(x) = F\left[\frac{x-\theta}{\tau}\right]$$

denote a location–scale family. Then the influence function $\mu(F)$ at $F_{\theta,\tau}$ satisfies

$$\mathrm{IF}_{\theta,\tau}(x) = \tau \mathrm{IF}_{0,1}\left[\frac{x-\theta}{\tau}\right].$$

Thus it suffices to calculate the influence function for one (convenient) member of the family. *Hint*: sampling from $(1-\varepsilon)F_{\theta,\tau} + \varepsilon\Delta_x$ may be considered a two-step procedure—first select a distribution from either $F_{\theta,\tau}$ or Δ_x and then obtain an observation from it.

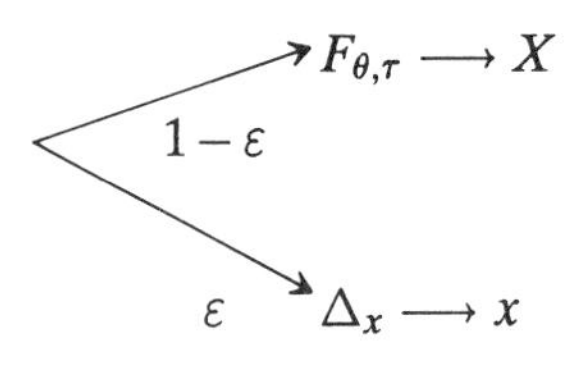

This is equivalent to taking a linear function of the outcome of sampling from a mixture of $F_{0,1}$ and $\Delta_{x-\theta/\tau}$:

$1-\varepsilon$: $F_{0,1} \longrightarrow Y \rightarrow \theta + \tau Y$

ε: $\Delta_{(x-\theta)/\tau} \longrightarrow \dfrac{x-\theta}{\tau} \rightarrow \theta + \tau\left(\dfrac{x-\theta}{\tau}\right)$

14. **(a)** Verify the details of finding the breakdown points in examples 1–4 of Section 4.3.1.

(b) Find the influence functions for the same examples assuming F is symmetric, and compare your results with those in the text and Figure 4.1.

15. **(a)*** Show that the breakdown point for an L-estimator is the minimum of α, β, where $[\alpha, 1-\beta]$ is the smallest closed interval which contains the support of K.

(b) Derive the formula for the influence function of an L-estimator given in Section 4.3.2, and use it to find that of the 2β-trimmed mean.

16. Prove the corollary to Theorem 4.1, and use it to calculate the asymptotic efficiency of the .2-trimmed mean relative to the mean.

17. In Section 4.3.3 assume that ψ has a positive derivative.

(a) Show that $\mu_\psi(F)$ satisfies (4.3.1)(ii), (iii), and (v).

(b)*Derive the breakdown point of $\mu_\psi(F)$.

(c) Show that $\mu_\psi(F)$ has an influence function which is proportional to ψ, and find the constant of proportionality.

(d) Show by example that $\mu_\psi(F)$ does not satisfy condition (iv).

(e) Let $\tau(F)$ be a measure of dispersion. Show that the solution $\mu = \mu(F)$ of

$$\int \psi\left(\frac{x-\mu}{\tau(F)}\right) dF(x) = 0$$

satisfies the conditions (4.3.1) for a measure of dispersion.

18. For the one-step M-estimator of example 5 of Section 4.3.3, determine whether the estimator inherits the desirable properties of M-estimators.

(a) Does it satisfy conditions (4.3.1)?

(b) Does it have breakdown point .5?

(c) Is its influence function proportional to the defining ψ function?

(d) What is its asymptotic efficiency relative to that of the M-estimator which it approximates?

19. Derive the influence function and breakdown point for the Hodges-Lehmann estimator (example 7). Find its efficiency relative to the mean for normal data.

20. Consider the following random sample of size 11 from a double exponential distribution with unknown location and dispersion:

13.1 10.4 10.0 9.6 14.3 9.1 10.3 10.0 8.9 10.2 8.8.

(a) Verify the density estimate of the standard error for the median of the above 11 observations. [*Answer*: The median and the interquartile range are $\hat{\theta} = x_{(6)} = 10$ and $\hat{\lambda} = 10.4 - 9.1 = 1.3$. Therefore we take $h_{\text{opt}} = 1.4 \times 1.2 \times 11^{-1/5} = 1.0$. Then $\hat{f}_k(10) = .3182$ and the $\text{SE}[\hat{\theta}] \approx .47$.]

(b) Using results from Section 3.5.2, find the exact bootstrap distribution and use it to calculate $\text{SE}_{\text{boot}}\hat{\theta}$. *Hint*:

$$\text{SE}^2_{\text{boot}} = \sum_{k=1}^{11} p_k x^2_{(k)} - \left[\sum_{k=1}^{11} p_k x_{(k)}\right]^2$$

where

k	1	2	3	4	5	6
p_k	.00017	.00703	.04404	.12150	.20588	.24274

The bootstrap estimate of standard error is

$$\text{SE}_{\text{boot}} = .40.$$

(c) If the bootstrap distribution of the median could not be found precisely as in (b), a Monte Carlo experiment could yield approximately the same result. Carry out such an experiment for $B = 100$ bootstrap samples and compare your results with those found in (b).

21. In this problem we carry out the steps of finding the estimated standard error of the median using the kernel density estimator. Recall the seven differences .3, .4, .5, .5, .6, .9, 1.7 of cell lifetimes from Example 2 of Section 2.1 (see also Problem 12, Section 2.5). The median is clearly $\hat{\theta} = x_{(4)} = .5$.

(a) Estimate f at $x = \hat{\theta} = .5$ by the kernel estimate with $k(y) = \frac{1}{2}I\{-1 \leq y \leq 1\}$. [*Solution*: $\hat{\lambda} = .5$ and $\hat{f}_k(.5) = 1.07$.]

(b) Show that $\widehat{\text{SE}}_{\text{density}}(\hat{\theta}) = .18$.

Section 4.4

22. Verify the details of Example 1.

23. Derive the breakdown points for the measures of dispersion given in Examples 1–4. Compare the ordering which they suggest with that of the corresponding variances in Table 4.3 against gross contamination by the Cauchy model.

24. In this problem we find the most efficient location and dispersion estimators based on a few order statistics for the double exponential distribution.

(a) Define $\mu(F)$ as in Example 3 of Section 4.3.1, and show that the limiting distribution of $n^{1/2}[\mu(F_n) - \mu(F)]$ is $N(0, V_\mu(F))$, where

$$V_\mu(F) = \frac{\beta}{2f^2(F^{-1}(\beta))}.$$

Show that this expression is minimized by the choice $\beta = .5$ when F is the double exponential distribution, so that the median is more efficient than any convex combination of two order statistics.

(b) Define $\tau(F)$ as is Example 1 of Section 4.4.1, and show that the limiting distribution of $n^{1/2}[\tau(F_n)-\tau(F)]$ is $N(0, V_\tau(F))$, where

$$V_\tau(F) = \frac{2\beta d^2(1-2\beta)}{f^2(F^{-1}(\beta))}.$$

Choose d so that $\tau(F)$ is Fisher-consistent for the standard deviation of the double exponential distribution. Show that $V_\tau(F)$ is minimized by choosing $\beta \approx .1$.

(c) Suppose that both location and scale must be estimated with only two order statistics, and that it is desired that the asymptotic variances be equal: $V_\mu(F) = V_\tau(F)$. Show that this is achieved for $\beta \approx .25$. Also show that the ARE of the resulting location estimator relative to the median is only .5, and the ARE of the resulting dispersion estimator relative to that obtained in (b) is only .75; so that one can estimate location and dispersion for this distribution far more efficiently with three order statistics than with only two.

25. In this problem we answer some of the questions raised about the trimmed range estimate of a normal standard deviation which are raised in the opening to Section 4.4.

(a) What is the "rationale" for this alternative trimmed range estimator? That is, why should we expect it to be close to σ?

(b) Define for any $0 < \beta < .5$ and distribution G the *2β-trimmed range* functional

$$T_\beta(G) = \frac{G^{-1}(1-\beta) - G^{-1}(\beta)}{k_\beta}$$

where k_β is a constant to be specified. Define the sample 2β-trimmed range by $T_\beta(F_n)$, the functional applied to the usual empirical distribution function of the X_i's. Show that $k_\beta = 2z_{1-\beta} = 2\Phi^{-1}(1-\beta)$ makes $T_\beta(G)$ Fisher-consistent for σ at the normal model $G = N(\mu, \sigma^2)$. *Hint*: first show that $x_\beta = G^{-1}(\beta) = \sigma z_\beta + \mu$.

(c) Explain why a study of the class of estimators defined in part (b) also includes a study of the 5% trimmed range defined earlier.

(d) The influence function of the qth-quantile functional $G^{-1}(q)$ is given by (3.2.3). Use it to effortlessly find and sketch the influence function $\mathrm{IF}_{T_\beta,G}$ of the 2β-trimmed range, assuming that g is symmetric about μ. Interpret the influence function.

(e) Show that the mean and the variance of the influence function are, respectively, $\mathrm{E}_G[\mathrm{IF}_{T_\beta,G}(X)] = 0$ and

$$\mathrm{Var}_G[\mathrm{IF}_{T_\beta,G}(X)] = \frac{\beta(1-2\beta)}{2g^2(x_\beta)}.$$

(f) Explain why $T_\beta(F_n)$ is asymptotically normal and give an expression (in terms of β and Φ) for the asymptotic relative efficiency of the 2β-trimmed range relative to the sample standard deviation for estimating σ when the data are normal.

(g) What are the relative pros and cons of using the 5%-trimmed range instead of the sample standard deviation?

26. In this problem we examine two definitions of outlier, one based on moments and the other based on quantiles. According to the first definition, an outlier is any observation x_i which lies more than k standard deviations from the mean, that is, outside the interval

$$[\overline{x}_n - ks_n, \overline{x}_n + ks_n]$$

where k is a specified constant. In the second definition, an outlier is any observation lying outside the interval determined by two symmetrically chosen order statistics:

$$[x_{(n\alpha)}, x_{(n(1-\alpha))}]$$

where $x_{(n\alpha)} = F_n^{-1}(\alpha)$ is the sample α quantile, for some α which is specified in advance. (For the sake of simplicity, we assume throughout that $n\alpha$ is an integer.)

(a) Let $x_1, \ldots, x_n$ be the realizations of n independent observations on the double exponential distribution, which has density

$$\frac{1}{\sigma} f\left(\frac{x-\mu}{\sigma}\right)$$

where $f(x) = e^{-|x|}/2\sqrt{2}$, and μ, σ are the unknown mean and standard deviation. Show that for $k = \ln(10)$ and $\alpha = .05$ approximately 10% of the observations will be classified as outliers by the respective definitions, at least for large n.

(b) Show that if L_{ln} is the random length of the sample size n moment interval, then $n^{1/2}[L_{\text{ln}} - a_{\text{ln}}] \xrightarrow{d} N(0, V_1)$ for some choice of constants $\{a_{\text{ln}}\}$. Similarly show that the length of the quantile interval is asymptotically normal. Since the two sequences of intervals have the same length in the limit, we prefer the definition which leads to the intervals with less variability, since they more reliably identify the 10% outlying observations. Compare the random *lengths* of the two intervals specified in (a) in terms of their asymptotic variances.

(c) Compare the two intervals on the basis of their breakdown points (finite or asymptotic).

(d) Assume now that f is an arbitrary positive continuous density. What can be said about the relative proportion of observations which will be declared outliers by the respective definitions?

Section 4.5

27. Derive the influence functions for the location and dispersion measures defined by (4.5.3).

28. Explain why the value of $a(k)$ given by (4.5.4) will make Huber's Proposal 2 estimator consistent for the standard deviation σ of a normal distribution. Verify the entries in Table 4.5 for $k = 1$.

Section 4.6

29. Verify the 95% confidence interval given for the median cell lifetime, using the nonlinear interpolation formula.

30. Find a 95% confidence interval for the median difference in the number of hours of extra sleep induced by the drugs in Example 1 of Section 4.1. Use the interpolation formula to obtain the correct coefficient, and compare your results with those obtained with the classical t-interval. The solutions are

```
- tinterval 95 c1
-        N      MEAN     STDEV    SE MEAN     95.0 PERCENT C.I.
2-1     10     1.580     1.230      0.389       ( 0.700, 2.460)
- sinterval 95 c1
SIGN CONFIDENCE INTERVAL FOR MEDIAN
-                             ACHIEVED
-        N      MEDIAN    CONFIDENCE    CONFIDENCE INTERVAL    POSITION
2-1     10       1.300        0.8906        ( 1.000, 1.800)           3
-                             0.9500        ( 0.932, 2.005)         NLI
-                             0.9785        ( 0.800, 2.400)           2
```

Use your interval to carry out a level .05 test of no difference between the drugs.

31. Derive a general formula for the finite sample breakdown point of the length of a level $1-\alpha$ sign statistic confidence interval. (The solution is $\varepsilon^*{}_n \approx \frac{1}{2} + z_{\alpha/2}/2\sqrt{n}$.)

Section 4.7

32. **(a)** The student can and should effortlessly confirm the results in Tables 4.6–4.9 using basic Minitab commands plus the macros in Appendix D.

(b) Choose one of the four examples in Secton 4.7 and verify the confidence interval based on the $2 \times 10\%$ trimmed mean using only the text and a hand calculator.

(c) Choose one of the four examples in Secton 4.7 and verify the confidence interval based on the Huber estimator using only the text and a hand calculator.

33. Consider the descriptive measure T defined for each distribution F for which two moments exist by

$$T(F) = E_F[X^2] - E_F^2[X].$$

(a) List the properties of a measureof dispersion and determine whether T is such a descriptive measure.

(b) Find $T(F_n)$ where F_n is the usual empirical distribution based on an independent, identically distributed sample of size n from F.

(c) Is T Fisher consistent for the variance of a normal distribution? Is $T(F_n)$ weakly consistent for $T(F)$ whenever F has two moments?

(d) What are the finite sample and asymptotic breakdown points of T at an F having two moments?

(e) Derive, plot, and interpret the influence curve $\mathrm{IF}_{T,F}(x)$ for T at F.

(f) Give sufficient conditions for which

$$A_n = n^{-1/2} \sum_{i=1}^{n} \mathrm{IF}_{T,F}(X_i) \xrightarrow{d} N(0, V)$$

and find V.

4.9 COMPLEMENTS

4.9.1 Location Estimators

Because there has been so much research into the location of a symmetric distribution we have only been able to discuss a representative subset of the available methods. In particular we have omitted Andrews' (1974) proposal for a redescending M-estimator [$\psi(x) = c\sin(x)$]; it has performance comparable to Tukey's biweight [see Andrews et al. (1972)].

For asymmetric distributions there are approaches to estimating location other than the distribution-free approach discussed in Section 4.3. For example, Hogg (1974) proposed adaptive trimmed means where the amount of trimming depends on the estimated shape.

4.9.2 Scale Estimators

Rousseeuw and Leroy (1987) introduced a new scale (in the sense of dispersion) estimator which is proportional to the length of the shortest half of the sample. The shortest half of the sample is defined to be the minimum of $x_{(h)} - x_{(1)}, \ldots, x_{(n)} - x_{(n-h+1)}$ where $h = [n/2] + 1$. It has been dubbed the "shorth." It has breakdown point $[n/2]/n$ and influence function equal to that of the interquartile range and median absolute deviation when the underlying distribution is symmetric.

While the median absolute deviation is the most frequently used of the high breakdown estimators of dispersion, it may be worthwhile to give up some security for a more efficient estimator. This can be obtained for normal and moderately large-tailed distributions by taking a larger quantile than the median of the absolute deviations. To this end define the class of dispersion measures which are a multiple of $T_q(X) = \sigma_q(|X - \text{med}(X)|)$, where $\sigma_q(Y)$ is the qth quantile of the distribution of Y. When $q = .5$ this estimate of dispersion $T_{.5}$ reduces to the median absolute deviation.

If X has the double exponential distribution, then $Y = |X - \text{med}(X)|$ has the exponential distribution, and we saw in Section 3.3 that the choice $q \approx .8$ gave the most efficient estimator of scale. Thus $T_{.8}$ would be more efficient than $T_{.5}$ for estimating dispersion of the double exponential distribution. Similarly for normal data, we could calculate the choice of q which minimizes the standardized variance of the scale estimate of the folded normal distribution. According to Rey (1983), Table 8 in Chapter 7, the optimal $q \approx .86$. On the basis of these and other results for various t-distributions Rey recommends $q = .7$ as a relatively efficient alternative to the median absolute deviation.

CHAPTER FIVE

Testing for a Location Parameter

The property of robustness I believe to be even more important in practice than that the test should have maximum power and that the statistics employed should be fully efficient.

G. E. P. Box (1953)

We continue the location parameter context of the last chapter, but now our efforts are directed to tests of hypotheses concerning the location of a symmetric distribution. First we give a brief historical review of the t-test and alternative tests. Then we formally define the one-sample location problem, the power function for tests of the null hypothesis versus one-sided alternatives, and the asymptotic power function for sequences of alternatives which approach the null hypothesis at the right rate.

We then calculate the asymptotic power functions for a number of tests, namely, the sign test, the t-test, the 2β-trimmed t-test, and the Wilcoxon test. The asymptotic power functions of the sign test and of the trimmed t-test are compared with finite sample power functions and are seen to be good approximations for moderate sample sizes.

Along the way we will define the Pitman efficiency of one test to another, and express this asymptotic relative efficiency of tests in terms of their efficacies. These asymptotic results are rounded off by some practical considerations, such as finding the sample size required to achieve power against a specific alternative for a fixed level.

In Section 5.3 we motivate a model for "undetectable outliers," which allows us to monitor the effects of outliers on the tests through the asymptotic power function. We will see that only the sign test, among those studied, has robustness of level for this kind of contamination. There are great differences among the other tests, with the Student's t-test faring worst. The asymptotic power function is also found to depend on the influence function of the estimator of location corresponding to the test statistic.

Finally, in Section 5.4 we derive the asymptotic power function of the t-test in the presence of autoregressive dependence, and find that positive dependence causes the test to overstate the significance of the test, while simultaneously lowering the power for large alternatives. For negative dependence the results are reversed.

5.1 WOULD W. S. GOSSET USE THE STUDENT t-TEST?

In 1908 W. S. Gosset, writing under the name Student, published in *Biometrika* his derivation of the t-distribution and included cumulative tables for sample sizes ranging from 4 to 10. He also carried out *by hand* a Monte Carlo study of data from normal sources and showed that the empirical distribution "fitted" his derived law. Fourteen years later he wrote to R. A. Fisher "I am sending you a copy of Student's Tables as you are the only man that's ever likely to use them!" This lesson in humility is reported in Cochran's contribution to the collection *On the History of Statistics and Probability*, edited by Owen (1976). In divising new methods and carrying out simulations to verify experimentally his theory W. S. Gosset was way ahead of his time.

Over the intervening years Wilcoxon (1945) published his signed rank test, which Pitman showed (1948) to be nearly as efficient as the t-test for location of a normal mean. Hodges and Lehmann (1956) showed that the efficiency of Wilcoxon's test relative to the t-test never falls below .86 for symmetric distributions but is unbounded above. These asymptotic efficiencies have been shown to be surprisingly informative in terms of predicting behavior for small samples [see Lehmann (1975), p. 174]. Moreover, the Wilcoxon test almost maintains its nominal level in the presence of outliers, while the t-test is nonrobust in this regard, as shown in the Section 5.3.

The Wilcoxon test is monotone, while the t-test is not. To illustrate this lack of monotonicity, we consider the Cushny and Peebles data set (Example 1, Section 4.1.2) which Gosset used to illustrate his t-test. The sample mean and the standard deviation of the difference in sleep between drug I and drug II are 1.58 and 1.23, respectively. If θ denotes the mean difference in hours of sleep between drugs I and II, the hypothesis $H : \theta = 0$ is rejected in favor of $K : \theta > 0$ when $\sqrt{n}\bar{x}_n/s_n > t_{n-1,1-\alpha}$; so for a level $\alpha = .05$ test we compare $\sqrt{10} \times 1.58/1.23 = 4.06$ with $t_{9,.95} = 1.833$ and reject H. Now suppose that the largest difference $x_{(10)} = 4.6$ had been 16 hours instead of 4.6 hours. The t-test no longer rejects H at level .05! One might argue that 16 is such an obvious outlier (the drug having made the patient unconscious), that anyone would reject this observation, and quite likely the drug. However, it is disquieting that more evidence in favor of an alternative can lead to a change of decision from rejecting H to not rejecting it.

Both tests are nonrobust to dependence; see Box (1979). And both tests are readily available on computer packages at the touch of a button. Which one would Gosset choose?

5.2 THE ASYMPTOTIC POWER FUNCTION

The asymptotic power function reveals much about the large sample behavior of a statistical test, and it can give surprisingly good approximations to small sample power functions, as we shall see below (Figure 5.1, Table 5.1, and Figures 6.1 and 6.2). Not only does it give the numerical power of the test against "reasonable" alternatives to the null hypothesis, but it also provides a basis for comparing the relative efficiency of the two tests. Furthermore, it will reveal the effects on level and power when the assumptions of underlying distribution, independence, and symmetry are violated.

5.2.1 The One-Sample Location Problem

Throughout this section we assume that:

$$\begin{array}{ll} \text{(i)} & F \text{ has density } f \text{ which is symmetrical about } 0. \\ \text{(ii)} & X_1,\ldots,X_n \text{ are i.i.d. with distribution function } F(x-\theta). \end{array} \tag{5.2.1}$$

The hypothesis $H : \theta = 0$ is rejected in favor of $K : \theta > 0$ when a test statistic T_n exceeds a critical point:

$$T_n = T_n(X_1,\ldots,X_n) \geq c_n.$$

We write $P_\theta\{\ \}$ to indicate that the probability in parentheses is calculated assuming (5.2.1) (ii). The critical point c_n is chosen so that the test has *size* not exceeding a preassigned significance level $\alpha : P_0\{T_n \geq c_n\} \leq \alpha$. The *power* of detecting an alternative $\theta > 0$ with this sample size n test is then defined by $\Pi_n(\theta) = P_\theta\{T_n \geq c_n\}$.

The *power function* $\Pi_n(\theta)$, $\theta > 0$, can be extended to include $\theta = 0$, and then it contains the information required to assess the performance of the sequence of test statistics $\{T_n : n = 1,2,\ldots\}$. Ideally $\Pi_n(\theta)$ is small ($\leq \alpha$) at $\theta = 0$ and large (≈ 1) for all $\theta > 0$.

The dependence of $\Pi_n(\theta)$ on the sample size n is an unnecessary complication which can be removed by letting θ approach zero at the right rate as n approaches infinity. For often it is the case that for $\theta_n = \theta/\sqrt{n}$,

$$\frac{T_n - \mathrm{E}_{\theta_n}[T_n]}{\sqrt{\mathrm{Var}_{\theta_n}[T_n]}} \xrightarrow{d} N(0,1)$$

and this may be used to show that the following limit exists:

$$\Pi(\theta) = \lim_{n\to\infty} \Pi_n(\theta_n).$$

This limit defines the *asymptotic power function* of the test statistics $\{T_n\}$ for the alternatives $\{\theta_n\}$. To illustrate this method, we consider a series of examples.

5.2.2 Examples of Power Functions

Example 1: The Sign Test. Let

$$\mathrm{sgn}(a) = \begin{cases} 1, & \text{if } a > 0 \\ 0, & \text{if } a = 0 \\ -1, & \text{if } a < 0. \end{cases}$$

The sign test rejects H in favor of K when $\sum_{i=1}^{n} \mathrm{sgn}(X_i)$ is too large. This test is equivalent to one based on the test statistic

$$T_n = \sum_{i=1}^{n} I\{X_i > 0\} = \frac{1}{2}\left[\sum_{i=1}^{n} \mathrm{sgn}(X_i) + n\right]. \tag{5.2.2}$$

Since $P_\theta\{X_i > 0\} = 1 - F(-\theta) = F(\theta)$ it follows that $T_n \sim \mathrm{Bin}(n, F(\theta))$. Hence, to determine the critical point c_n for a level α test, we need to solve for the largest c_n for which

$$\alpha \geq P_0\{T_n \geq c_n\} = \sum_{k=c_n}^{n} \binom{n}{k} \frac{1}{2^n}.$$

As n increases, the distribution $\mathrm{Bin}(n, \frac{1}{2})$ may be approximated by the normal distribution $N(n/2, n/4)$, which yields, with $\Phi(z_\alpha) = \alpha$, the critical point $c_n = (\sqrt{n} z_{1-\alpha} + n)/2$. The justification for this result is

$$\begin{aligned} \Pi_n(0) = P_0\{T_n \geq c_n\} &= P_0\left\{\frac{T_n - \mathrm{E}_0[T_n]}{\sqrt{\mathrm{Var}_0[T_n]}} \geq \frac{c_n - n/2}{\sqrt{n/4}}\right\} \\ &= P_0\left\{Z_n \geq \frac{2c_n - n}{\sqrt{n}}\right\} \to 1 - \Phi(z_{1-\alpha}) = \alpha \end{aligned} \tag{5.2.3}$$

if $Z_n \xrightarrow{d} Z \sim N(0,1)$, and c_n is chosen so that $(2c_n - n)/\sqrt{n} \to z_{1-\alpha}$ as $n \to \infty$.[†] When the sequence of critical points $\{c_n\}$ is chosen so that $\Pi_n(0) \to \alpha$ as $n \to \infty$ as in (5.2.3), we say the test has *asymptotic level* α.

[†] There are two limiting operations which are combined here, $Z_n \xrightarrow{d} Z$ and $(2c_n - n)/\sqrt{n} \to z_{1-\alpha}$; justification for the combined result is found in Problem 7.

To compute the limiting power, standardize the test statistic in the expression for the power of the test:

$$\Pi_n(\theta) = P_\theta \left\{ \frac{T_n - \mathrm{E}_\theta[T_n]}{\sqrt{\mathrm{Var}_\theta[T_n]}} \geq \frac{c_n - nF(\theta)}{\sqrt{nF(\theta)(1-F(\theta))}} \right\}. \tag{5.2.4}$$

Then by the central limit theorem applied to T_n,

$$\Pi_n(\theta) \to 1 - \Phi(L) \qquad \text{as} \quad n \to \infty,$$

where

$$L = \frac{z_{1-\alpha}}{2\sqrt{F(\theta)(1-F(\theta))}} + \lim_{n\to\infty} \frac{\sqrt{n}(1-2F(\theta))}{2\sqrt{F(\theta)(1-F(\theta))}}. \tag{5.2.5}$$

From (5.2.5) it is clear that for all $\theta > 0$, $L = -\infty$, so $\Pi_n(\theta) \to 1$ as $n \to \infty$.

This property of having asymptotic power 1 against all alternatives at a fixed level α is called *consistency* of the test. Consistency is the minimum we would expect of a test, since with sample sizes increasing without bound the test should be certain to detect alternatives to H. By considering alternatives to $\theta = 0$ of the form

$$\theta_n = \frac{\theta}{\sqrt{n}}, \qquad \theta > 0 \tag{5.2.6}$$

for sample size n, we find that the limit L in (5.2.5) is finite, the power function is strictly between 0 and 1, and hence more information is yielded by the limiting power function. More precisely, if f is continuous at 0,

$$\sqrt{n}[1-2F(\theta_n)] = -2\theta \frac{[F(\theta/\sqrt{n}) - F(0)]}{\theta/\sqrt{n}} \to -2\theta f(0) \qquad \text{as} \quad n \to \infty. \tag{5.2.7}$$

Substituting $\theta_n = \theta/\sqrt{n}$ for θ in (5.2.4) and (5.2.5) yields $L = z_{1-\alpha} - 2\theta f(0)$, and

$$\Pi(\theta) = \lim_{n\to\infty} \Pi_n(\theta_n) = \Phi(2\theta f(0) - z_{1-\alpha}). \tag{5.2.8}$$

The limiting power $\Pi(\theta)$ against the sequence of alternatives $\{\theta/\sqrt{n}\}$, considered as a function of $\theta \geq 0$, is called the *asymptotic power function* of the test. In this sign test example the power for large samples depends on the value of the density at the center of symmetry.

The asymptotic power function is also remarkably accurate even for small sample sizes, as shown by a comparison of $\Pi(\theta)$ and $\Pi_n(\theta/\sqrt{n})$, which are shown in Figure 5.1. For normal data,

$$\Pi(\theta) = \Phi\left(\sqrt{\frac{2}{\pi}}\frac{\theta}{\sigma} - z_{1-\alpha}\right).$$

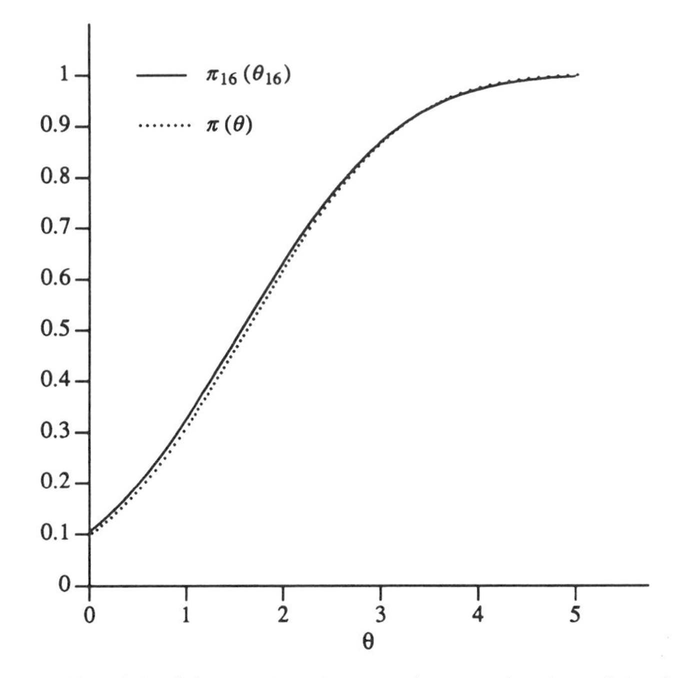

FIGURE 5.1 Plot of the finite sample and asymptotic power functions of the sign test for normal data.

The other curve in Figure 5.1 is found by computing $\Pi_n(\theta_n) = P_{\theta_n}\{T_n > c_n\}$, using the fact that $T_n \sim \text{Bin}(n, F(\theta_n))$. For example, when $n = 16$, $\alpha = .1$, we require $c_{16} = 11$. (Actually this gives a size $\alpha = .11$ test.) Then the finite sample power against $\theta_{16} = 1/\sqrt{16} = .25$ is $P\{T_{16} \geq 11\} \approx .33$, since $T_{16} \sim \text{Bin}(16, \Phi(.25))$. The asymptotic power function yields

$$\Pi(1) = \Phi\left(\sqrt{\frac{2}{\pi}} - z_{.9}\right) = \Phi(-.48) \approx .32. \quad \square$$

Example 2:* *The t-Test. Now we compare the sign test with the t-test using the asymptotic power functions. Thanks to W. A. Gosset, we may choose the t-test critical points to give *exact* level α for every n, and hence also an asymptotic level α:

$$c_n = t_{n-1,1-\alpha} \to z_{1-\alpha} \qquad \text{as} \quad n \to \infty.$$

The asymptotic power function of the t-test is then the limit, as n increases without bound, of

$$\Pi_n(\theta_n) = P_{\theta_n}\left\{\frac{\sqrt{n}\overline{X}_n}{s_n} > c_n\right\}$$

$$= P_{\theta_n}\left\{\frac{\overline{X}_n - \mathrm{E}_{\theta_n}[\overline{X}_n]}{\sqrt{\mathrm{Var}_{\theta_n}[\overline{X}_n]}} > \frac{c_n(s_n/\sqrt{n}) - \theta_n}{\sigma/\sqrt{n}}\right\}$$

$$= \Phi\left(\frac{\theta}{\sigma} - \frac{s_n}{\sigma}c_n\right) \tag{5.2.9}$$

assuming that the observations are normally distributed. Even if the observations only have a finite second moment, the central limit theorem guarantees that with increasing n the difference between the last two expressions becomes negligible. Hence quite generally the asymptotic power function of the t-test is

$$\Pi(\theta) = \lim_{n\to\infty} \Pi_n(\theta_n) = \Phi\left(\frac{\theta}{\sigma} - z_{1-\alpha}\right) \tag{5.2.10}$$

using the fact that $s_n/\sigma \xrightarrow{P} 1$ and Slutsky's lemma.

To compare the performance of the t-test with the sign test under the conditions (5.2.1) we will use the asymptotic power functions (5.2.8) and (5.2.10), denoted by $\Pi_S(\theta)$ and $\Pi_t(\theta)$, respectively. If the t-test is based on m observations, it has approximate power against the alternative $\eta = \theta/\sqrt{m}$ of $\Pi_t(\sqrt{m}\eta)$. Similarly, the sign test based on n observations has approximate power $\Pi_S(\sqrt{n}\eta)$ against the alternative $\eta = \theta/\sqrt{n}$. These two approximating values are equal when $\Pi_S(\sqrt{n}\eta) = \Pi_t(\sqrt{m}\eta)$ or $2f(0)\sqrt{n}\eta = \sqrt{m}\eta/\sigma$. Hence the two tests will have approximately the same power at the same alternative η when $m/n \approx 4\sigma^2 f^2(0)$. This ratio of m to n, which gives the same power against the same alternative at the same level, may approach a limit for m, $n \to \infty$ and then it is called the *asymptotic relative efficiency* or *Pitman efficiency* of the sign test to the t-test; it is denoted by $\mathrm{E}_{S,t}(F)$ and equals

$$\mathrm{E}_{S,t}(F) = 4f^2(0)\sigma^2. \tag{5.2.11}$$

In the special case where $F \sim N(0, \sigma^2)$, $f(0) = 1/\sqrt{2\pi}\sigma$ and $\mathrm{E}_{s,t}(\Phi) = 2/\pi \approx .64$. The sign test is much less efficient than the t-test for normal data; it requires $n = 100$ observations to have the same power of detecting an alternative as the t-test based on only $m = 64$ observations.

Of course when F is not normally distributed, the result may change radically. For example, when $F = \mathrm{DE}$, the double exponential distribution, the density with standard deviation $\sigma = \sqrt{\mathrm{Var}_F[X]}$ is

$$f(x) = \frac{e^{-\sqrt{2}x/\sigma}}{\sqrt{2}\sigma}$$

so $f(0) = 1/\sqrt{2}\sigma$ and $\mathrm{E}_{S,t}(\mathrm{DE}) = 2$; the sign test is twice as efficient as the t-test for detecting a shift in the double exponential distribution when the sample sizes are large. □

Example 3: The 2β-Trimmed t-Test. This test is an alternative to the t-test which is apparently robust to $100\beta\%$ of outliers on either side of the sample (but see Section 5.3). It shares with the t-test the property of being asymptotically distribution free in the sense that its level for large sample sizes is the desired α, whether or not the observations are normally distributed; unlike the t-test it does not require finite moments for this property to hold. The disadvantage is that it is not as efficient as the t-test for normally distributed data. To show this, we will find the asymptotic power function. The influence function estimate of the standard error of $\overline{X}_{n,2\beta}$ is found in Example 4 of Section 4.3.2 to be $\mathrm{SE}[\overline{X}_{n,2\beta}] \approx s_{W_{2\beta}}/(1-2\beta)\sqrt{n}$. Define the *$2\beta$-trimmed t-test* as that which rejects $H : \theta = 0$ in favor of $K : \theta > 0$ when

$$\frac{(1-2\beta)\sqrt{n}\,\overline{X}_{n,2\beta}}{s_{W_{2\beta}}} \geq t_{k-1,1-\alpha} \tag{5.2.12}$$

where $k = n - [2\beta n]$. As $n \to \infty$, the left-hand side approaches the standard normal distribution, so the sequence of tests has asymptotic level α.

The choice of t-distribution critical points has not been justified analytically, although a number of independent Monte Carlo studies have provided evidence that they are accurate enough for all practical purposes. Table 5.1 shows some of the estimated critical points and t-critical points for various sample sizes. For each combination of sample size n and trimming proportion β there are four entries. The values in parentheses are the .95 and .975 critical points from the t-distribution with $n - [2\beta n] - 1$ d.f. The values to their left are the estimated critical points given by Padmanabhan (1985) for the standardized 2β-trimmed t-distribution, assuming that the underlying distribution F is normal. All values have been rounded

Table 5.1 Monte Carlo and ($t_{n-[2\beta n]-1,1-\alpha}$) Critical Points for the 2β-Trimmed t-Test Applied to Normal Data

β	$1-\alpha$	$n=10$	$n=20$	$n=30$
.15	.95	2.00 (1.94)	1.79 (1.77)	1.69 (1.72)
	.975	2.51 (2.45)	2.28 (2.16)	2.06 (2.09)
.20	.95	2.11 (2.01)	1.80 (1.80)	1.74 (1.74)
	.975	2.71 (2.57)	2.21 (2.20)	2.11 (2.11)
.25	.95	2.31 (2.35)	1.87 (1.83)	1.88 (1.76)
	.975	3.12 (3.18)	2.35 (2.26)	2.30 (2.14)

Table 5.2 ARE of the 2β-Trimmed t-Test to the t-Test for Various Trimming Proportions

β	$x_{1-\beta}$	σ_β	$E(\overline{X}_{2\beta}, \overline{X})$	$x_{1-\beta}V_\beta^{-1/2}(1-2\beta)^{-1}$
.01	2.326	.9346	.996	2.28
.05	1.645	.7893	.974	1.80
.10	1.282	.6616	.943	1.56
.15	1.036	.5564	.909	1.41
.20	.8416	.4632	.874	1.31
.25	.6745	.3777	.836	1.23

to two decimal places. Padmanabhan also shows that these critical points are remarkably stable as F varies over a selection of short- and long-tailed distributions.

The asymptotic power function is derived in the same way as that of the t-test. It is

$$\Pi_{2\beta}(\theta) = \Phi(V_\beta^{-1/2}\theta - z_{1-\alpha}) \tag{5.2.13}$$

where

$$\begin{aligned} x_\beta &= F^{-1}(\beta) \\ \sigma_\beta^2 &= \frac{1}{1-2\beta}\int_{x_\beta}^{x_{1-\beta}} x^2\, dF(x) \\ V_\beta &= \frac{\sigma_\beta^2}{1-2\beta} + \frac{2\beta x_\beta^2}{(1-2\beta)^2}. \end{aligned} \tag{5.2.14}$$

The reader may show that this power function includes that of the sign test (let $\beta \to \frac{1}{2}$) and the untrimmed t-test (let $\beta \to 0$) as limiting cases. The efficiency (ARE) of $\overline{X}_{n,2\beta}$ to $\overline{X}_n$ is clearly $V_0/V_\beta = \mathrm{E}(\overline{X}_{2\beta}, \overline{X})$. Some values for the normal distribution are tabled in Table 5.2. The last column of figures will be referred to later in Section 5.3.2.

Some finite sample comparisons with efficiencies are possible with results given by Johnson and Kotz (1970), p. 60. For $n = 10$ and $\beta = .2$ the efficiency of the trimmed to untrimmed mean is 89%, compared with the asymptotic value of 87.4% shown above. For $n = 15$ and four observations trimmed off each side (which corresponds to $\beta \approx .27$) the efficiency is 83%, compared to the asymptotic value (for $\beta = .25$) of 83.6%. □

Example 4: The Wilcoxon Signed Rank Test. Let $R_1^+, \ldots, R_n^+$ be the ranks of the absolute values of the observations $|X_1|, \ldots, |X_n|$, and define $W_n^+ = \sum_{i=1}^n R_i^+ \operatorname{sgn}(X_i)$, the signed rank statistic. Wilcoxon (1945) proposed the test which rejects $H : \theta = 0$ in favor of $K : \theta > 0$ when W_n^+ exceeds a critical value. This test may be shown to be equivalent to one based on the statistic

$$T_n = \sum_{i \neq j} I\{X_i + X_j > 0\} \tag{5.2.15}$$

which is in a form more amenable to mathematical analysis.

The asymptotic power function is derived in exactly the same way as in the previous examples, and is

$$\Pi_W(\theta) = \Phi(c\theta + z_\alpha) \tag{5.2.16}$$

where $c = \sqrt{12} \int f^2(y)dy$. Armed with this result, the reader may derive the asymptotic relative efficiency of the Wilcoxon test relative to the previous tests. It cannot be emphasized enough that this test competes very favorably with the t-test over a wide range of distributions, in terms of power. Moreover, it is *distribution-free* in the sense that its finite sample level is free of F, assuming only that F is symmetric and continuous. The t-test is only *asymptotically* distribution-free in that its *asymptotic* level is free of the distribution F (assuming F has a second moment). In Section 5.3 we will see that the Wilcoxon test is far more robust to outliers than the t-test. □

5.2.3 The General Derivation of Asymptotic Power and Efficiency

The reader who has derived the asymptotic power functions for the above examples will have noticed an underlying structure, found by Pitman and extended by Noether (1955). It is discussed in more detail in Lehmann (1975), but presented here because of its importance:

The test statistics T_n are asymptotically normal when H is true. Thus there exists $\mu_n(0)$ and $\sigma_n(0)$ such that $[T_n - \mu_n(0)]/\sigma_n(0)$ converges to the standard normal disribution under $H : \theta = 0$. Then, with $c_n = \mu_n(0)+$

$z_{1-\alpha}\sigma_n(0)$, the test has asymptotic level α:

$$\Pi(0) = \lim_{n\to\infty} \Pi_n(0) = \lim_{n\to\infty} P\{T_n \geq c_n\} = \alpha. \tag{5.2.17}$$

With this choice of critical points $\{c_n\}$ and alternatives $\theta_n = \theta/\sqrt{n}$, $\theta > 0$,

$$\begin{aligned}\Pi_n(\theta_n) &= P_{\theta_n}\{T_n \geq c_n\}\\ &= P_{\theta_n}\left\{\frac{T_n - \mu_n(\theta_n)}{\sigma_n(\theta_n)} \geq \frac{\sigma_n(0)}{\sigma_n(\theta_n)}\left[z_{1-\alpha} - \frac{\mu_n(\theta_n) - \mu_n(0)}{\sigma_n(0)}\right]\right\} \qquad (5.2.18)\\ &\to \Phi(e_T\theta + z_\alpha).\end{aligned}$$

where e_T is a constant defined below. The argument leading to the limit in (5.2.18) requires that the following three results hold.

1. Under the alternatives $\{\theta_n\}$, T_n is asymptotically normal with parameters $\mu_n(\theta_n), \sigma_n(\theta_n)$. It is often the case, but not necessarily so, that $\mu_n(\theta) = \mathrm{E}_\theta[T_n]$ and $\sigma_n^2(\theta) = \mathrm{Var}_\theta[T_n]$. Moreover, often T_n can be multiplied by a suitable constant depending on n so that $\mu_n(\theta) = \mu(\theta)$ is free of n. For example the sign test statistic in (5.2.2) can be multiplied by n^{-1}, and the Wilcoxon signed rank statistic in (5.2.15) by $\binom{n}{2}^{-1}$. When this is done, it is also often the case that $n\sigma_n^2(\theta_n)$ converges to a positive constant.
2. The ratio of $\sigma_n(0)$ to $\sigma_n(\theta_n)$ must approach a positive limit as n increases to infinity. As anticipated by (5.2.18), this limit is often 1.
3. Finally, $[\mu_n(\theta_n) - \mu_n(0)]/\sigma_n(0) \to e_T\theta$. This will be the case whenever $\sqrt{n}\sigma_n(0) \to \sigma(0) > 0$. Under the special condition described in 1. above, the constant $e_T = \mu'(0)/\sigma(0)$.

The constant e_T, which appears in the asymptotic power function, is called the *efficacy* of the test; it is the limiting rate of change of the mean of the test statistic, divided by its standard deviation, in the neighborhood of the null hypothesis. If the efficacy is large, the power function will rise rapidly and alternatives will likely be detected. The relative *efficiency* of two tests is often defined in terms of the efficacy, with justification given by an argument presented in Example 2, Section 5.2.2. Letting e_S, e_T be the respective efficacies of test statistics S_n, T_n satisfying conditions 1–4, the Pitman efficiency or *asymptotic relative efficiency* of S_n to T_n is defined by

$$E(S_n, T_n) = \left(\frac{e_S}{e_T}\right)^2. \tag{5.2.19}$$

5.2.4 Some Practical Considerations

In addition to the theoretical insights that the asymptotic power function gives us regarding the tests, it also has the immediate practical use of assisting us to choose the required sample size to achieve a desired power at a given alternative $\theta/\sqrt{n}$. To achieve power $1-\beta$ of detecting an alternative θ_1 say, simply set

$$1-\beta = \Pi_n(\theta_1) \approx \Pi(\theta_1\sqrt{n}) = \Phi(\theta_1\sqrt{n}e_T + z_\alpha)$$

and solve for

$$n \approx \left[\frac{z_\alpha + z_\beta}{e_T\theta_1}\right]^2. \tag{5.2.20}$$

Of course this argument tacitly assumes that the normal approximation will be reasonably accurate for the sample size that is being sought.

Having chosen a sample size, and carried out an experiment, the test statistic may be compared with the exact critical values for small sample sizes in order to make a decision regarding the null hypothesis at the stated significance level. Critical values are readily available for all of the tests described in Section 5.2. The P-value of the test statistic may be calculated, either exactly using the same sources, or approximately using the normal approximation to T_n. For example, if the observed value of $T_n = k$, the p-value for the one-sided location problem is approximated using the asymptotic mean and standard deviation of (5.2.17):

$$P_0\{T_n \geq k\} = P\left\{\frac{T_n - \mu_n(0)}{\sigma_n(0)} \geq \frac{k - \mu_n(0)}{\sigma_n(0)}\right\}$$
$$\approx \Phi\left(\frac{\mu_n(0) - k}{\sigma_n(0)}\right).$$

The approximation can often be improved with an appropriate continuity correction when T_n takes on discrete values.

Throughout this section we have assumed that the null hypothesis value of θ was zero, but if it were otherwise (say θ_0) then the observations $X_1, \ldots, X_n$ could be replaced by $X_1 - \theta_0, \ldots, X_n - \theta_0$, and testing could proceed as described above. In choosing which of the test statistics to use, statisticians have traditionally been guided primarily by the efficiency of the tests for a particular model: the sign test is suggested for double exponential data, the t-test for normal data, the signed rank test for logistic data, etc.

When uncertain of the model, statisticians are more likely to employ one whose significance level does not depend on knowledge of the distribution. For example, the sign test may be used to test whether the median of the distribution is shifted or not; even symmetry of F is not required.

The Wilcoxon signed rank test requires only that the observations be from a symmetric, continuous F for its stated significance levels to hold.

Statisticians may have been concerned with ease of calculation of the test statistic and P-value as well, although this is no longer a problem with the available hand calculators and computer statistical packages. Conceptual clarity of the procedure is also important, as well as an understanding of the performance of the test under various departures from the assumptions, a subject to which we now turn our attention.

5.3 ROBUSTNESS AGAINST OUTLIERS

We continue our discussion of the one-sample tests for location, but now study the effects of undetected outliers. It is obvious that if the observations are mostly from the assumed distribution (5.2.1) but there are also a fixed proportion of outliers to the right of the null hypothesis value of $\theta = 0$, then the statistical tests described above will have a greater chance of rejecting H, whether or not it is true. However, not all tests respond to outliers to the same degree, and the asymptotic power function will reveal the differences.

5.3.1 A Model for Undetected Outliers

We will suppose that the statistician acts as though conditions (5.2.1) hold for a known F and employs one of the statistical tests for location described in Section 5.2. We want to know how these tests will perform if the data actually satisfy

$$X_1,\ldots,X_n \quad \text{are i.i.d.} \quad \left(1-\frac{\varepsilon}{\sqrt{n}}\right)F(\cdot-\theta)+\frac{\varepsilon}{\sqrt{n}}\Delta_x(\cdot) \tag{5.3.1}$$

where F is symmetric about 0, x is a fixed representative "outlier" greater than θ, and $0 < \varepsilon \leq 1$ is a constant. The outliers at x would be obvious if ε or x were too large, relative to the square root of n. To see this, suppose for the moment that x is three standard deviations to the right of the center of symmetry. Then the choice $\varepsilon = 1$ leads to the untenable view that there will be proportion $1/\sqrt{n}$ of outliers that are undetectable in a sample of size n: if $n = 25$, then 5 observations are assumed to be undetectable outliers, while if $n = 100$, 10 observations are considered so. A more realistic choice is $\varepsilon = \frac{1}{5}$ or $\frac{1}{10}$.

But are these outliers really undetectable? Let us see whether Grubb's (1969) outlier rejection rule based on the maximum studentized residual would be of help. [According to Hampel et al. (1986), this rule "is probably one of the most popular rejection rules."] We selected samples of size $n =$

25 from the model (5.3.1) with $F = \Phi$, the standard normal distribution, $\varepsilon = .2$, and outliers at $x = 4$. Using the level .05 one-sided test which rejects an outlier on the right only if $(x_{(n)} - \overline{x})/s_x \geq 2.66$ [see Table 1, Grubbs (1969)], the number of rejections in 1000 samples was only 411, for an empirical power of .411. Note that in using this one-sided test we are assuming that the experimenter would only reject outliers on the right; if both outliers on the left and right are considered worthy of rejection, the critical point would be 2.82 and the power of detection of an outlier would be even less than .411.

The situation for $x = 3$ is much worse; the level .05 one-sided outlier rejection rule only has power .160 of detecting an outlier.

Under the assumption (5.3.1), but using the critical levels based on the assumption $\varepsilon = 0$, the asymptotic power function may be derived as in Section 5.2. This gives the actual power under (5.3.1) when using the test derived assuming (5.2.1).

5.3.2 The Effect of Outliers on Level and Power

We begin again with the sign test. If X_i has distribution (5.3.1), then

$$P_\theta\{X_i > 0\} = F(\theta) + \frac{\varepsilon}{\sqrt{n}} F(-\theta) \tag{5.3.2}$$

and the asymptotic power function of the sign test is (with $\theta_n = \theta/\sqrt{n}$)

$$\Pi_S(\theta) = \lim_{n \to \infty} \Pi_n(\theta_n) = \Phi(2\theta f(0) + z_\alpha + \varepsilon). \tag{5.3.3}$$

Thus the power function is increased for all $\theta > 0$ (which is good) but *also* increased for $\theta = 0$, which means the test is biased: it has asymptotic level $\Phi(\varepsilon + z_\alpha) > \alpha$, instead of the nominal level α. Table 5.3 gives a few values of the actual (asymptotic) size of the sign test for a proportion $\varepsilon/\sqrt{n}$ of outliers at x. (The size of x is irrelevant.) Note that a proportion $\varepsilon/\sqrt{n}$ of undetected outliers with $\varepsilon = .4$ will lead to a doubling of the traditional levels over the nominal values of .05 and .01. Of course $.4/\sqrt{n}$ outliers is an extreme situation, and such outliers would probably be noticed. More realistically, in practice $.1/\sqrt{n}$ of unnoticed outliers would be present. If ε is only .1, then the level is only increased slightly. For better approximations to the change in level for the sign test which are based on "small sample asymptotics," see Field and Ronchetti (1985).

When the level of a test changes only slightly in the presence of outliers or another departure from the assumed distribution, we say that the test has robustness of validity (or level). It is difficult to make this vague

Table 5.3 Selected Values of $\Phi(\varepsilon + z_\alpha)$ for Finding the Asymptotic Level and Power of Various Tests

	ε							
α	.1	.2	.3	.4	.5	.6	.7	.8
.05	.061	.074	.089	.106	.126	.148	.172	.199
.1	.119	.140	.163	.189	.218	.248	.281	.316

definition precise. However, because of the presumed importance of not making the type I error, a test whose nominal size is inflated by 20% due to undetectable outliers would just be acceptable. Thus a nominal .05 test which was inflated to .06 would be okay, but .07 (a 40% inflation) would be unacceptable to us. The sign test barely qualifies on this account if "undetectable" is interpreted as $.1/\sqrt{n}$ outliers. Those statisticians who want the level as stated will be disappointed to find that all the other tests proposed in Section 5.2 are even less robust to this kind of asymmetric contamination, as we shall see.

The expression (5.3.3) shows that the asymptotic power of the sign test is increased by the presence of outliers on the right, which is not surprising.

It is clear that if the outliers are to the left of the null hypothesis value, so that if in (5.3.1) we take $x < 0$, then the actual level would be somewhat smaller than the stated level, and the power would be reduced for all $\theta > 0$. In practice, outliers could of course arise from mistakes on either side of the sample, and partially cancel out each other's effects. However, here we are more concerned with the more uncomfortable situation of asymmetric contamination.

The asymptotic power function of the t-test under the model (5.3.1) is easily found to be

$$\Pi_t(\theta) = \Phi\left(\frac{\theta + x\varepsilon}{\sigma} + z_\alpha\right) \tag{5.3.4}$$

which shows that the t-test is not level robust to positive outliers, and in fact is much more sensitive to $\varepsilon/\sqrt{n}$ outliers at x than the sign test, since typically $x/\sigma \geq 3$ in practical situations where x represents an outlier. The reader may consult Table 5.3 (with $\varepsilon x/\sigma$ substituted for ε) to determine the devastating effect on the level of the t-test of contamination. To be specific, with proportion $1/5\sqrt{n}$ of outliers at $x = 4\sigma$, the stated level of .05 is actually closer to .2, and a stated level of .1 is in fact closer to .3.

The trimmed t-test works quite well by comparison, for the asymptotic power function is given by

$$\Pi_{2\beta}(\theta) = \begin{cases} \Phi\left(V_\beta^{-1/2}\left(\theta + \frac{\varepsilon x}{1-2\beta}\right) + z_\alpha\right) & x \le x_{1-\beta} \\ \Phi\left(V_\beta^{-1/2}\left(\theta + \frac{\varepsilon x_{1-\beta}}{1-2\beta}\right) + z_\alpha\right) & x > x_{1-\beta}. \end{cases} \tag{5.3.5}$$

The derivation of (5.3.5) is straightforward but lengthy; an outline is given in the problems. No matter how large the outliers at x are, the multiple of ε in (5.3.5) cannot exceed $x_{1-\beta}V_\beta^{-1/2}(1-2\beta)^{-1}$. Some values are listed for normal data in Table 5.2. For example, the trimmed mean which removes 10% from each end of the sample will have maximum level $\Phi(1.56\varepsilon + z_\alpha)$. When the proportion of outliers is $1/5\sqrt{n}$, the stated level of .05 is *at most* .1, and the stated level of .1 is at most .17. This is a great improvement on the corresponding breakdown of levels for the t-test. Morover, this trimmed t-test has asymptotic efficiency relative to the t-test of .943; see Table 5.2.

We postpone the derivation and interpretation of the asymptotic power function of the Wilcoxon signed rank test to the problems; it is

$$\Pi_W(\theta) = \Phi(e_W\theta + \sqrt{3}[2F(x) - 1]\varepsilon + z_\alpha) \tag{5.3.6}$$

where the efficacy $e_W = \sqrt{12}\int f^2(y)\,dy$. Perhaps it is worth noting that the maximum coefficient of ε is $\sqrt{3}$ independent of the underlying distribution. Thus when the proportion of outliers is $1/5\sqrt{n}$, the changes in level are *at most* double their nominal levels.

5.3.3 Connection to Influence Function of Estimators

All of the above tests are related via the test statistic to an estimator of θ; thus it is not surprising that the power function, in the presence of a small contamination at x, depends on the influence function of the estimator. For example, the 2β-trimmed t-test has asymptotic power function from (5.3.5) of

$$\Pi_{2\beta}(\theta) = \Phi(e_\beta(\theta + \varepsilon \mathrm{IF}_{\overline{X}_{2\beta}}(x)) + z_\alpha) \tag{5.3.7}$$

where $e_\beta = V_\beta^{-1/2}$.

Similarly, the signed rank test, which is derived from the Hodges-Lehmann estimator, has the asymptotic power function

$$\Pi_W(\theta) = \Phi(e_W(\theta + \varepsilon \mathrm{IF}_{\mathrm{HL}}(x)) + z_\alpha) \tag{5.3.8}$$

where e_W is the efficacy of the signed rank test.

Suppose now that F generates a location family. The above examples lead us to surmise that under suitable regularity conditions, for any location measure $T(F)$ with induced estimators $T_n = T(F_n)$, the asymptotic power function of the one-sided test of location based on T_n in the presence of contamination (5.3.1) is given by

$$\Pi_T(\theta) = \Phi(e_T\theta + e_T\varepsilon \mathrm{IF}_{T,F}(x) + z_\alpha) \tag{5.3.9}$$

where e_T is the efficacy of the test statistic.

There is a substantial literature on influence functions for tests which was initiated by Ronchetti (1979) and Rousseeuw and Ronchetti (1981). See also Chapter 3 of the book [Hampel et al. (1986)] for a formal introduction and further references. In particular, Ronchetti and Rouseeuuw define an influence function for the level of a test, another influence function for the power of a test, and use them to make extrapolations of level and power for all distributions in a neighborhood of the underlying F.

5.4* ROBUSTNESS AGAINST DEPENDENCE

A careful study of a set of data may reveal local dependencies of the sort described below, but since they are unsuspected, they are not often looked for. What is the effect of such dependencies on tests for changes of location? We will consider the effects of first-order autoregressive dependence on the one-sided t-test for location as described in Section 5.2. We will see that if the usual t-test is carried out in the presence of positive autoregressive dependence, then the actual level is larger than the stated level. Also the power against small positive values of shift alternatives is enhanced, while the power against large positive alternatives is decreased. These results are reversed for negatively correlated variables.

Similar behavior has been observed for other tests for location, but we will not go into such results here because of their complexity. See Gastwirth and Rubin (1975a), (1975b) Cressie (1980), and Hettmansperger (1984a) for further discussion.

5.4.1 Possible Sources of Local Dependence

While it is mathematically convenient to assume that observations are independent random variables, it may be that unsuspected dependence occurs. Such dependence may be inherent to the data, or introduced via machine or human error. As an example of the latter from our experience, consider a treadmill exercise test in which the patient is tested for blood pressure elevation during exercise. The patient walks continuously on a moving treadmill for up to 20 minutes while blood pressure measurements are taken and recorded at 2-minute intervals. One medical researcher observes the patient's systolic/diastolic pressures and relays the results verbally to an assistant, who records the data. The noise of the treadmill is considerable, and even if the correct readings are made, they may not be heard correctly by the assistant. The assistant is largely influenced by the most recently recorded values in cases where noise has blocked out the spoken value. Since the researchers giving the test are also occupied with talking to the patient and monitoring stress levels, it is easy for mistakes to take place. (Of course the entire session could be recorded to help avoid some of the errors, but this is not often done.) Another possibility is to use machines to measure and record blood pressure, but then similar recording errors can result from the device "sticking" in position from time to time.

What is the effect of such dependencies on tests for changes in blood pressure as a result of drug treatment, for example? We may gain some insight by finding and interpreting the asymptotic power function of the t-test, derived under the assumption of first-order autoregressive dependence. This model renders observations that have correlation of the form ρ^k, if the observations are taken k units of time apart. Thus consecutive observations have correlation ρ, while the correlation between other observations diminishes rapidly with time. This seems to be an appropriate model for the type of "local" dependence described above.

5.4.2 The Effects of Autoregressive Dependence on the t-Test

For each n we assume that $\epsilon_1, \ldots, \epsilon_n$ are i.i.d. with distribution

$$N\left((1-\rho)\frac{\theta}{\sqrt{n}},\ \sigma^2(1-\rho^2)\right) \tag{5.4.1}$$

where $-1 < \rho < 1$, $\theta \geq 0$, and $\sigma \geq 0$. Define $X_0 = \epsilon_0$, and

$$X_i = \rho X_{i-1} + \epsilon_i, \qquad i = 1, \ldots, n. \tag{5.4.2}$$

Note that when $\rho = 0$ in (5.4.1), the observations $X_1, \ldots, X_n$ are i.i.d. $N(\theta/\sqrt{n}, \sigma^2)$, and thus this model is of the form considered earlier in (5.2.1) with alternatives (5.2.6). The statistician who believes independence to be the case could use the standard level α t-test which rejects $H : \theta = 0$ in favor of $K : \theta > 0$ when

$$t = \sqrt{n}\frac{\overline{X}_n}{S_n} \geq t_{n-1}(1-\alpha).$$

Under the assumptions (5.4.1) and (5.4.2) it may be shown that

$$X_m = \sum_{i=0}^{m} \rho^{m-i} \epsilon_i, \qquad m = 1, \ldots, n.$$

Hence each X_m is normally distributed with parameters

$$\begin{aligned} \mathrm{E}[X_m] &= \frac{\theta}{\sqrt{n}}(1 - \rho^{m+1}) \\ \mathrm{Var}[X_m] &= \sigma^2(1 - \rho^{2m+2}) \end{aligned} \tag{5.4.3}$$

for $m = 1, \ldots, n$. Furthermore,

$$\mathrm{Cov}[X_m, X_{m+k}] = \sigma^2 \rho^k (1 - \rho^{2m+2}).$$

It follows that as $n \to \infty$, the sample mean and the standard deviation satisfy

$$\begin{aligned} n^{1/2}\left(\mathrm{E}[\overline{X}_n] - \frac{\theta}{\sqrt{n}}\right) &\to 0 \\ n\,\mathrm{Var}[\overline{X}_n] &\to \sigma^2\left(\frac{1+\rho}{1-\rho}\right) \\ s_n &\xrightarrow{P} \sigma. \end{aligned} \tag{5.4.4}$$

With the above results in hand, we may readily derive the asymptotic power function of the t-test which is carried out in the presence of autoregressive dependence. It is

$$P_\theta\left\{\sqrt{n}\frac{\overline{x}_n}{s_n} \geq t_{n-1}(1-\alpha)\right\} \to \Phi\left(\frac{b\theta}{\sigma} + bz_\alpha\right) \tag{5.4.5}$$

where Φ is the standard normal cdf and $\Phi(z_\alpha) = \alpha$. The constant b satisfies

$$b = \sqrt{\frac{1-\rho}{1+\rho}}. \tag{5.4.6}$$

With this result we may explain the effects of positive and of negative ρ on the test asymptotic level and power.

First take $\theta = 0$ in the asymptotic power function (5.4.5) to assess the effects on level. When ρ is positive, the constant b will be less than 1, so the actual level is greater than the stated level; on the other hand, for negative ρ, the actual level is smaller than stated. Thus the t-test is not level robust to positive dependence, but it is so to negative dependence.

To assess the effect on power, it is easiest to contemplate the graph of the power function; see Figure 5.2. Broadly speaking, the effect of positive

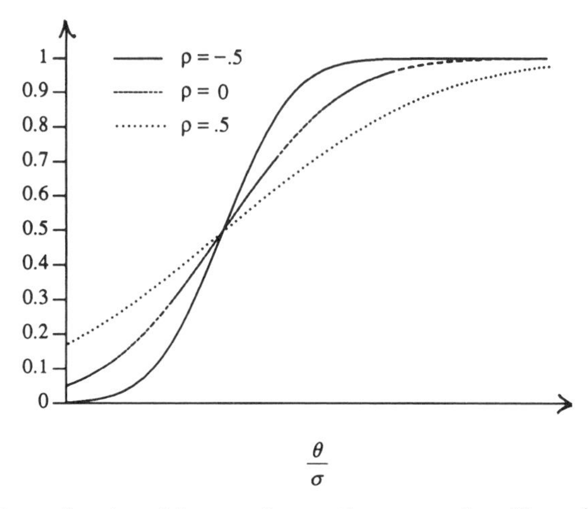

FIGURE 5.2 Power function of the t-test for negative, zero, and positive serial correlation ρ.

dependence is to flatten the power function, and for negative dependence, to sharpen its rise. Thus generally speaking, we can "live with" negative dependence, but should not use the t-test in the presence of positive dependence. Other standard tests exhibit similar behavior. No test which does not specifically measure dependence seems to be robust against both negative and positive autoregressive dependence.

5.5 TEST BIAS AND POWER FACTORS

We have shown that under various modifications of the basic assumptions (5.2.1), the asymptotic power function of a test T of location has the general form

$$\Pi_T(\theta) = \Phi(ce_T\theta + a + bz_\alpha). \tag{5.5.1}$$

See (5.2.10), (5.3.4) and (5.4.5) for special cases. Under the classical assumptions (5.2.1), $a = 0$, $b = c = 1$, and e_T is the efficacy of the test.

When $\theta = 0$, the expression (5.5.1) yields the asymptotic level. The constants a and b will affect this level and are hence called the *test bias term* and *test bias factor*, respectively. The test is not of level α (asymptotically) if $a + bz_\alpha > z_\alpha$. Thus we want to avoid situations where, say, $a = 0$ and $b > 1$, or where $a > 0$ and $b = 1$; and in general where $a > z_\alpha(1 - b)$.

The constant c can only affect the power, and is called the *test power factor*. The power function is increased for all alternatives $\theta > 0$ if $c > 1$, so such changes are desirable. Of course, if $a = 0$, $b = 1$, and $c < 1$, then the test has less power than under the assumptions of the model.

The terms test bias factor and test power factor were introduced by Brown (1982), in the context of the two-sample problem, which is discussed in the next chapter.

5.6 PROBLEMS

Section 5.1

1. The Cushny and Peebles data may be used to test the hypothesis that the mean difference between the two drugs is zero against the alternative that drug II leads to a significant increase in hours of sleep over drug I. Carry out the following one-sided tests at level .05 assuming that the data are normally distributed. Determine the P–value of the data in each case, or state why it cannot be calculated.
   (a) The sign test.
   (b) The t-test.
   (c) The t-test, assuming the observation $4.6 = x_{(10)}$ has been already rejected by an outlier rejection technique.

2. For each of the tests in Problem 1, give the least stringent assumptions concerning F which are required to carry out the test at the stated level.

Section 5.2

3. **(a)** Verify some of the details of Figure 5.1 in which the sign test asymptotic power function is compared with the finite sample power functions. In particular calculate and compare $\Pi_{16}(\frac{1}{4}) = \Pi_{16}(1/\sqrt{16})$ with $\Pi(1)$ at level $\alpha = .1$, and $\Pi_{16}(\frac{1}{8}) = \Pi_{16}(1/2\sqrt{16})$ with $\Pi(\frac{1}{2})$.
   (b) Use the asymptotic power function to determine what sample size is required with the sign test in this context to simultaneously achieve:
      (i) Asymptotic level $\alpha = .05$.
      (ii) Power .9 against alternative $\theta = \sigma$.

4. **(a)** Verify the details in Example 2 which claims that the sign test is twice as efficient as the t-test for detecting a shift in double exponential data.

(b) Compare the actual power functions of the two tests for $n = 16$ to see how they compare with the asymptotic predictions of part (a). You may need to carry out a simulation experiment. Assume $\alpha = .1$.

5. **(a)** Derive the asymptotic power function of the t-test (5.2.10).

(b) Find the actual power function of the t-test assuming sample size $n = 16$ and normal data using noncentral t-tables, and compare your results with part (a).

6. In this problem we study the asymptotic power function of the 2β-trimmed t-test.

(a) Derive the asymptotic power function (5.2.13) using results from example 4 of Section 4.3.2.

(b) Verify the details of Table 5.2 and compare the efficiency with the lower bound of the corollary of Theorem 4.1.

(c) Calculate the ARE of the 2β-trimmed t-test ($\beta = .1$) relative to the sign test for normal data.

(d) Repeat (b) for double exponential data.

7. In order to understand the comment after (5.2.3) we will first consider an example where it is *not* true that

$$\left.\begin{array}{l} Y_n \xrightarrow{d} Y \\ c_n \rightarrow c \end{array}\right\} \Rightarrow P\{Y_n \leq c_n\} \rightarrow P\{Y \leq c\}. \qquad (*)$$

Then we will show that if the distribution of Y is continuous at c, the desired result holds.

(a) Let $Y_n \sim U(0, 1/n)$, $n = 1$, and let $Y \sim \Delta_0$. Show that $Y_n \xrightarrow{d} Y$. Let $c_n = -1/n$, $n = 1, 2, \ldots$. Then clearly $c_n \rightarrow 0$ but $\lim P\{Y_n \leq c_n\} \neq P\{Y \leq 0\}$.

(b) Use Polya's theorem (see Appendix B) to show that if Y has a continuous distribution at c, then the result $(*)$ holds.

8. In this problem we study the asymptotic power function of the Wilcoxon signed rank test.

(a) Show that the test based on W_n^+ is equivalent to one based on T_n of (5.2.15).

(b) Derive (5.2.16).

9. Continue the project begun in Problems 1 and 2 by repeating them for

(a) The 2β-trimmed t-test, $\beta = .1$
(b) The Wilcoxon signed rank test.

10. Calculate the *efficacies* for each of the four tests in Section 5.2, assuming that the data are from a double exponential distribution, and use them to calculate the Pitman efficiencies of each pair of tests.

11. Write out the details of the derivation of (5.2.19) and use it to find the sample size required in Example 1 of Section 4.1.2 to attain power .9 at level .05 of detecting a difference of 2 hours of sleep. Assume normal data and find n for each of the four tests described in this section.

12. Show that $x_{1-\beta}/(1-2\beta) \to 1/2f(0)$ as $\beta \to \frac{1}{2}$ and hence that $\lim_{\beta\to 1/2} \Pi_{2\beta}(\theta) = \Pi_S(\theta)$. Why should the sign test be asymptotically equivalent to the 2β-trimmed t-test as $\beta \to \frac{1}{2}$?

Section 5.3

13. The model for asymmetric contamination is reasonable only if a statistician is unlikely to detect a proportion $\varepsilon/\sqrt{n}$ of outliers at x; otherwise the statistician would presumably try to do something about it.
(a) Choose one of the outlier rejection techniques and ascertain its power for finite n or asymptotically of detecting the outliers arising under the assumption (5.3.1).
(b) Let x in (5.3.1) be the outlier boundary (moment or quantile definition) given in Problem 24 of Chapter 4. Would the statistician be able to detect x?
(c) We have argued informally that $\epsilon = .2$ or $.1$ is appropriate in (5.3.1). What value of ϵ is appropiate in your opinion, and why?

14. Carry out the details of deriving the power function of the sign test in the presence of contamination (5.3.1), and assess the consequences of such contamination.
(a) Verify (5.3.2) and (5.3.3).
(b) Verify the entries in Table 5.3.

15. (a) Derive (5.3.4), the asymptotic power function of the t-test.
(b) Draw the graph of the power function found in (a) on the same axes with that of the sign test, for the sake of comparison. What conclusions can you draw?

(c) In Section 4.1 it was noted that the t-test was not monotone, and in fact loses power as one observation is carried off to infinity. How is this fact reconciled with the fact that the asymptotic power function (5.3.4) approaches 1 as $n \to \infty$?

16.* In this problem we will derive the asymptotic power function of the 2β-trimmed t-test.

(a) Let G be the mixture distribution of (5.3.1). Note that G depends on F, θ, ε, n, and x, although we suppress this dependence notationally. Assuming $\theta < x < x_{1-\beta} = F^{-1}(1-\beta)$, show that as $n \to \infty$,

$$G^{-1}(\beta) \sim \frac{\theta}{\sqrt{n}} + F^{-1}\left(\beta + \frac{\beta\epsilon}{\sqrt{n}}\right)$$

and

$$G^{-1}(1-\beta) \sim \frac{\theta}{\sqrt{n}} + F^{-1}\left(1 - \beta - \frac{\beta\epsilon}{\sqrt{n}}\right).$$

(b) Show that

$$\begin{aligned}\overline{X}_{n,2\beta} &= \frac{1}{n}(1-2\beta)\sum_{i=1}^{n} X_i I\{X_{(n\beta)} < X_i \le X_{(n-n\beta)}\} \\ &= \frac{1}{n}(1-2\beta)\sum_{i=1}^{n} X_i I\{G^{-1}(\beta) < X_i \le G^{-1}(1-\beta)\} + r_n\end{aligned}$$

where $\mathrm{E}[r_n^2] \to 0$.

(c) Let $Y = XI\{G^{-1}(\beta) < X \le G^{-1}(1-\beta)\}$, where $X \sim G$. Then

$$\mathrm{E}_G[\overline{X}_{n,2\beta}] \sim \frac{\mathrm{E}_G[Y]}{1-2\beta} \sim \frac{1}{\sqrt{n}}\left(\theta + f\left(\frac{\epsilon x}{1-2\beta}\right)\right).$$

(d) Complete the derivation of (5.3.5) and sketch it on the same axes as that of the sign and t-tests. Use $\beta = .1$, and assume that $F = \Phi$.

17. Derive the asymptotic power function (5.3.6) of the signed rank test in the presence of contamination, and sketch its graph with those of the other tests, again assuming normality as in 16(d).

Section 5.4

18. (a) Verify the details of the derivation of (5.4.3) to (5.4.5) outlined in Section 5.4.2.

(b) Which of the results in (a) can be carried out without the assumption of *normally* distributed variables?

19. Show that the asymptotic power function of the t-test has the graph as drawn in Figure 5.2 for negative, zero, and positive ρ.

Section 5.5

20. Verify the claims following (5.5.1).

5.7 COMPLEMENTS

A systematic study of the effects of heavy tails, correlation, and skewness on the behavior of the t-test has been carried out by Cressie (1980). His approach is to find traditional measures of skewness and kurtosis for a statistic equivalent to the t-statistic in terms of the moments of the underlying marginal distribution or possible correlation of the observations. These results allow him to conclude that compared to the normal distribution, heavy-tailed data render a light-tailed t-statistic, with the opposite effect for light-tailed data. Also, positively correlated data yield a heavy-tailed t-statistic, while negatively correlated data lead to a light-tailed one. Finally, positively skewed data render a negatively skewed t-statistic, while negatively skewed data lead to a positively skewed t-statistic. These general results enable to predict consequences for the level (or P-values) of the t-test.

Modifications to the degrees of freedom which account for such departures from the assumption of i.i.d. observations are suggested in Cressie, Sheffield, and Whitford (1984). See also Hall (1983). The effects of autoregressive dependence on one-sample statistics are studied in some detail in Gastwirth and Rubin (1975b).

CHAPTER SIX

The Two-Sample Problem

We begin by looking at the classical pooled t-test, and the Welch t-test which estimates the variances for each sample separately. The insights gained by study of the asymptotic power function for these tests in Section 6.2 are found to hold for quite small sample sizes, as confirmed in Section 6.3. General results for tests based on M-estimators were derived by Brown (1982) and some of them are summarized here in Section 6.4. In Section 6.5 the asymptotic power function of the two-sample trimmed t-test is derived. It supports the view that there are tests other than the classical ones which are far more robust to both outliers and unequal variances. Methods for easily implementing this test, and for finding confidence intervals for the difference between populations are illustrated. Finally, in Section 6.6 another robust method of testing for or estimating the difference in two populations is presented; this method is based on one-sample sign statistic intervals.

6.1 INTRODUCTION AND EXAMPLES

The two-sample problem arises in situations where independent samples are selected from two populations thought to differ not at all (the null hypothesis) or to differ only by a shift of location (the alternative hypothesis). The problem is how best to use the observations so as to detect a shift when it exists, while keeping the type 1 error probability bounded by some desired low level α.

The model which is perhaps most commonly employed is $X_1,\ldots,X_m$ i.i.d. $N(\mu_X,\sigma_X^2)$, independent of $Y_1,\ldots,Y_n$ i.i.d. $N(\mu_X+\theta,\sigma_X^2)$. It is desired to test

$$
\begin{aligned}
& H : \theta = 0 \\
\text{versus} \quad & K : \theta > 0 \qquad (\text{or } K : \theta \neq 0).
\end{aligned} \tag{6.1.1}
$$

The classical test for this situation is the two-sample pooled t-test. It is a test which is clearly not robust to outliers, based on experience gained with the one-sample t-tests in the last chapter.

Another difficulty arises as follows. Suppose that the X-population serves as control for some treatment given to the Y-population, and that θ measures the treatment effect. It would be most surprising if an effective treatment did not also affect the shape or variance of the Y-population. But this is precisely what we are assuming in our somewhat unrealistic model. Therefore it is important to know how robust the pooled t-test (or any other proposed test) is to variations in shape or variance.

Because normal populations are determined by only two parameters, it is easiest to study the case where $X \sim N(\mu_X, \sigma_X^2)$ and $Y \sim N(\mu_Y, \sigma_Y^2)$, and $\mu_Y = \mu_X + \theta$, $\sigma_Y = \sigma_X(1+h)$, for some $h > -1$. The parameter θ represents the unknown shift in the populations, while h is a measure of the heteroscedasticity. When $h = 0$ we have the classical two-sample problem and when $|h|$ is small, we have a reasonable model for departures from equal variances. If it is plausible that $1+h$ or $1/(1+h)$ is moderately large, then we need to be estimating the variances separately, and possibly using the two-sample t-test with Welch's approximation (1937) for degrees of freedom; see (6.2.5). When $h \gg 1$, there is real doubt as to whether it is reasonable to be comparing the means of the two populations; it is like comparing apples and oranges.

6.1.1 Examples

Example 1: Comparing Total Ridge Counts. The *total* ridge count of an individual is the sum of the fingerprint ridge counts on both hands. The (left, right) hand totals for two groups of individuals are given in Appendix C.1. The total ridge counts for these groups are summarized below in terms of means and variances:

$$
\begin{aligned}
m &= 27 \qquad \bar{x} = 125.96 \qquad s_x^2 = (50.20)^2 \\
n &= 27 \qquad \bar{y} = 154.00 \qquad s_y^2 = (29.04)^2.
\end{aligned}
$$

The X-population consists of members of a first-year statistics course at La Trobe University in Melbourne, which consists mainly of students from European and Asian background. The Y-population consists of members of an Aboriginal tribe. It is speculated a priori that the ridge counts for the

Aboriginal tribe are higher, on average, than those of the Europeans and Asians because of evolution toward greater hunting skills. The null hypothesis is that there is no difference between the TRCs. It seems reasonable to assume that an inherited characteristic such as total ridge count will show smaller variance among tribal members than among unrelated persons, and thus a two-sample test which estimates the variances seems in order.

It turns out that for the two-sample test with Welch's formula for d.f., [see (6.2.8) below] $t = 2.51$ yields a P-value of $p = .008$ at $\nu = 41$ d.f. If one naively applies the pooled t-test which assumes that the populations are normal with *equal* variances, then it turns out that the pooled test yields almost exactly the same results: $t_{\text{pool}} = 2.51$ at 52 d.f., which has P-value $= .0076$. But the fact that these tests yield the same results is simply a consequence of the fact that the sample sizes are equal and large.

If we use all 27 observations from the Y-population and select only the first eight data points from the X-population, and repeat the exercise, we find $\bar{x} = 102$ and $s_x^2 = 51^2$. Now the pooled test claims significance at level .0004, but the Welch t-test at level .012. The pooled test is surely overstating a claim, which frequently happens when the larger sample is taken from the population with the smaller variance, as explained below in Section 6.2. □

Example 2: Atomic Weight of Carbon. Brownlee [(1965), p. 251] illustrates the t-tests and the Wilcoxon test on a set of data on the measurements of the atomic weight of carbon. The nine measurements are taken from two preparations, which we call X and Y.

X	Y
11.9853	12.0072
11.9949	12.0064
11.9985	12.0054
12.0061	12.0016
	12.0077

In testing whether the weights are equal against the one-sided alternative that the Y preparation leads to a larger value, we find (later in Section 6.5) that the two-sample test which estimates the variances has a P-value of .08 and accepts the null hypothesis at level .05, while the pooled t-test rejects at level .05. Thus the decision as to which test to use should be made before looking at the results, or we are more likely to be biased in our selection. We will argue below that it is in almost all cases preferable to use the t-test which estimates the variances separately. This of course does not answer the question of how to deal with outliers; this is a question we will answer in Sections 6.5 and 6.6. □

6.1.2 Mathematical Formulation

We are given $N = m + n$ independent, normally distributed variables: in the first sample $X_1, X_2, \ldots, X_m$ each has mean 0 and variance σ_X^2; and in the second sample $Y_1, Y_2, \ldots, Y_n$ each has mean $\theta/\sqrt{N}$ and variance σ_Y^2, where $\sigma_Y = \sigma_X(1+h)$, some $h > -1$, and σ_X and h are both unknown. In the classical model $h = 0$, that is, the variances are equal. The parameter h is a measure of the heteroscedasticity.

We reject $H : \theta = 0$ in favor of $K : \theta > 0$ when a test statistic $T_{m,n}$ exceeds a critical value $C_{m,n;\alpha}$ which is chosen so that the probability of falsely rejecting H is

$$P_H\{T_{m,n} \geq C_{m,n;\alpha}\} = \alpha. \tag{6.1.2}$$

If we let

$$m, n \to \infty \qquad \text{with} \quad \frac{m}{m+n} \to \lambda, \qquad 0 < \lambda < 1 \tag{6.1.3}$$

then it is often the case that the power of $T_{m,n}$ against the alternative $\theta/\sqrt{N}$ converges to an asymptotic power function of the form

$$\Pi_T(\theta) = \Phi(a + bz_\alpha + ce_T\theta), \qquad \theta \geq 0 \tag{6.1.4}$$

where $z_\alpha = \Phi^{-1}(\alpha)$ and e_T is the efficacy of the test. The constants a, b, and c are called the test bias term, test bias factor, and test power factor, respectively; see Section 5.5.

Whether or not there are unequal variances, under the assumptions listed above, it turns out that for the tests we consider, the bias term $a = 0$. The asymptotic size of the test is therefore $\Phi(bz_\alpha)$, which is larger, equal to, or smaller than the nominal level α according to whether $b < 1$, $b = 1$, or $b > 1$. We want to avoid tests with $b < 1$, for they overstate the significance of a result. Clearly values of $c > 1$ are desired, for they increase the power against all alternatives.

6.2 ROBUSTNESS OF CLASSICAL *t*-TESTS

6.2.1 Asymptotic Power of the Pooled *t*-Test

Let $\overline{x}$, s_x^2 denote, respectively, the sample mean and the variance of the X-variables, and similarly define $\overline{y}$, s_y^2. The *pooled t-test* statistic is

$$T_{m,n} = \frac{(\overline{y} - \overline{x})}{S\sqrt{1/m + 1/n}} \tag{6.2.1}$$

where

$$S^2 = \frac{(m-1)s_x^2 + (n-1)s_y^2}{m+n-2}. \tag{6.2.2}$$

The one-sided test rejects H in favor of K when (6.1.1) holds, with the constant $C_{m,n,\alpha} = t_{m+n-2}(1-\alpha)$, the $1-\alpha$ quantile of the Student's t-distribution with $m+n-2$ degrees of freedom.

As m, n increase to infinity in the manner (6.1.3), the weighted variance S^2 converges in probability:

$$S^2 \xrightarrow{p} \sigma_X^2[1 + (1-\lambda)(2h + h^2)] \tag{6.2.3}$$

and the asymptotic power function is derived by the same series of steps used in the one-sample problem to find (5.2.10). The result is $\Pi_T(\theta) = \Phi(bz_\alpha + c(\theta/\sigma_X))$, where

$$\begin{aligned} b &= \left[\frac{1 + (1-\lambda)(2h + h^2)}{1 + \lambda(2h + h^2)}\right]^{1/2} \\ c &= \left[\frac{\lambda(1-\lambda)}{1 + \lambda(2h + h^2)}\right]^{1/2}. \end{aligned} \tag{6.2.4}$$

Remarks

1. Note that the test bias factor $b = 1$, and the pooled t-test has asymptotic level α, if *either*:
   (i) $h = 0$, that is, $\sigma_Y = \sigma_X$
   or
   (ii) $\lambda = \frac{1}{2}$, that is, $m/n \to 1$.
   Hence choosing equal sample sizes will protect the level against unequal variances.

2. Assuming that $h > 0$ ($\sigma_Y^2 > \sigma_X^2$) and $\lambda > \frac{1}{2}$, it follows that the bias factor $b < 1$. Thus in using the pooled t-test we tend to overstate the significance of a result if we happen to take a larger sample from the population with the smaller variance.

3. The test power factor c is maximized by choosing $\lambda = 1/(2+h)$, that is, by $n/m = \sigma_Y/\sigma_X$. In this case $b = [(1 + h + h^2)/(1+h)]^{1/2}$. The usual situation is that the standard deviations are unknown; but if one has any prior knowledge concerning their relative sizes, this fact would help in choosing sample sizes.

6.2.2 Comparison of the Two-Sample Pooled and the Welch t-Tests

In this section we will compare the above pooled t-test with the Welch t-test, which assumes that the variances may be unequal and uses the Welch (1937) formula for degrees of freedom (6.2.8). For normal data the pooled t-test is more efficient when the variances are equal, and otherwise the Welch test is preferred. Given the real possibilty that $\sigma_X^2 \neq \sigma_Y^2$ in any situation, and the small loss in power if the Welch t-test is used when $\sigma_X^2 = \sigma_Y^2$, it seems reasonable to dispense with the pooled test; see further remarks below.

The two-sample test statistic which estimates the variance for each sample separately is defined by

$$t = \frac{\overline{y} - \overline{x}}{s} \tag{6.2.5}$$

where

$$s^2 = \frac{s_x^2}{m} + \frac{s_y^2}{n}. \tag{6.2.6}$$

It will be shown in the problems that as $m/(m+n) \to \lambda$,

$$\begin{aligned} s^2 &\sim \sigma^2_{\overline{Y}-\overline{X}} \\ &\sim \frac{\sigma_X^2}{N}\left[\frac{1+2\lambda h+\lambda h^2}{\lambda(1-\lambda)}\right]. \end{aligned} \tag{6.2.7}$$

The Welch test rejects when the statistic t exceeds the $1-\alpha$ quantile from the Student's t-distribution with ν degrees of freedom. Welch suggested taking ν to be the integer nearest to

$$\nu = \frac{(A+B)^2}{A^2/(m-1) + B^2/(n-1)} \tag{6.2.8}$$

with $A = s_x^2/m$ and $B = s_y^2/n$.

The asymptotic power function of this test is readily derived and found to equal

$$\Pi_t(\theta) = \Phi\left(bz_\alpha + c\frac{\theta}{\sigma_X}\right) \tag{6.2.9}$$

with $b = 1$ and c as in (6.2.4) for the pooled t-test.

More Remarks

4. For large samples the two-sample t-test which estimates the variances has asymptotic level α, as advertised, regardless of h. Thus it has the same asymptotic level and power as the pooled t-test when the variances

are equal, or when the sample sizes are equal. However, if the variances are unequal and the (large) sample sizes are very different, then the two-sample test is preferable, because for $b < 1$ the pooled test overstates the significance of results, and for $b > 1$ the two-sample test is uniformly more powerful. These statements are based on the asymptotic results, so they need not hold for any finite samples. However, in the next section we will see that the asymptotics can be a very accurate guide even for small sample sizes.

5. When $m = n$, the two test statistics (6.2.1) and (6.2.5) have exactly the same value, but the Welch t-test will be based on fewer degrees of freedom. However, the loss in degrees of freedom is slight, only about 10%, if $\frac{1}{3} < s_x^2/s_y^2 < 3$. This last condition occurs with probability .95 for normal data if $\sigma_X = \sigma_Y$. Thus when the conditions for which the pooled t-test is valid hold, it is *usually* no great loss to use the two-sample test. Of course neither test is robust to outliers, so we must continue our search for a robust efficient test.

6. The pooled t-test is preferable when $m \gg n$, $\sigma_X = \sigma_Y$, and $s_x \gg s_y$. But this is a relatively rare event. [See Brownlee, (1965), p. 303, for further discussion.]

As a result of these considerations it seems reasonable to employ the Welch t-test rather than the pooled test when the normal model is justified.

6.3 FINITE SAMPLE RESULTS

In this section we make a comparison of asymptotic power functions with results from a simple Monte Carlo study based on 1000 repetitions of each experimental situation. The standard error of an estimate of power near p is therefore $\sqrt{p \times (1-p)/1000} \leq .016$.

6.3.1 Equal Variances

First consider the situation of equal variances where $\sigma_X = \sigma_Y = 4.2$, $m = 4$, $n = 8$, and $\alpha = .05$. Then $h = 0$, $\lambda = \frac{1}{3}$, and $c = \sqrt{2}/(3 \times 4.2) = .112$. The asymptotic power function for both tests is given by $\Phi(-1.645 + .112\theta)$. We will define μ_2 by $\theta = \sqrt{N}\mu_2 = \sqrt{12}\mu_2$, and calculate the power functions at various values of μ_2 (Table 6.1).

Table 6.1 Power Comparisons of Pooled and Welch t-Tests when Variances are Equal

μ_2	0	1	2	3	4	5	6
Asymptotic power for both tests	.05	.105	.193	.316	.464	.655	.754
Simulated pooled	.048	.117	.150	.290	.412	.559	.682
Simulated Welch	.041	.114	.154	.270	.382	.509	.648

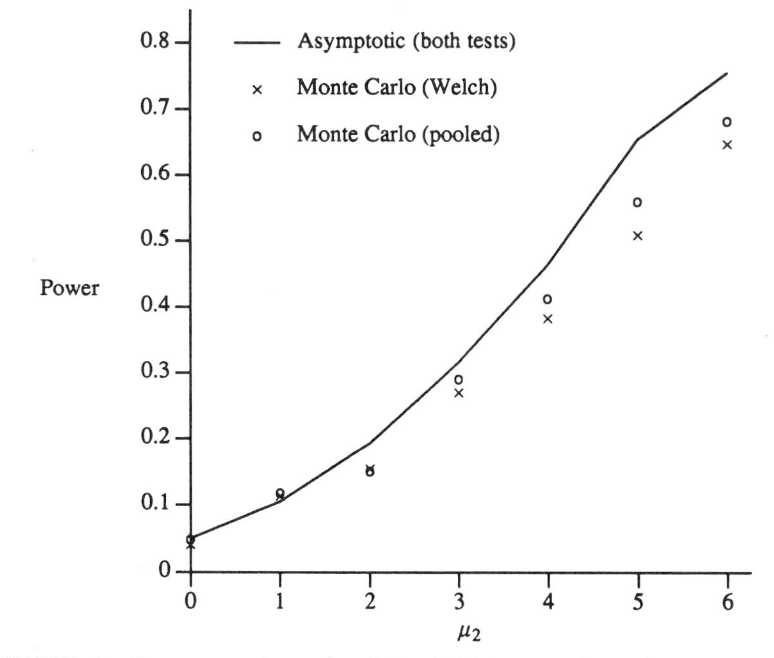

FIGURE 6.1 Power comparisons of pooled and Welch t-tests when variances are equal.

Remarks

In Figure 6.1 the asymptotic power function (solid line) is plotted for comparison with Monte Carlo approximations to the finite sample power functions of, respectively, the pooled test (circles) and the Welch test (crosses). The asymptotic power function is a useful guide to the finite sample power of these two tests, even though the sample sizes are only $m = 4$ and $n = 8$. The reason that the solid line appears to the left of the other

Table 6.2 Power Comparisons of Pooled and Welch t-Tests when Variances are Unequal

μ_2	0	1	2	3	4	5	6
	(a) Power Function for Pooled Test						
Asymptotic	.012	.043	.121	.265	.466	.677	.842
Monte Carlo	.024	.044	.114	.240	.396	.577	.753
	(b) Power Function for Welch Test						
Asymptotic	.05	.137	.289	.495	.702	.858	.947
Monte Carlo	.057	.116	.243	.436	.614	.803	.903

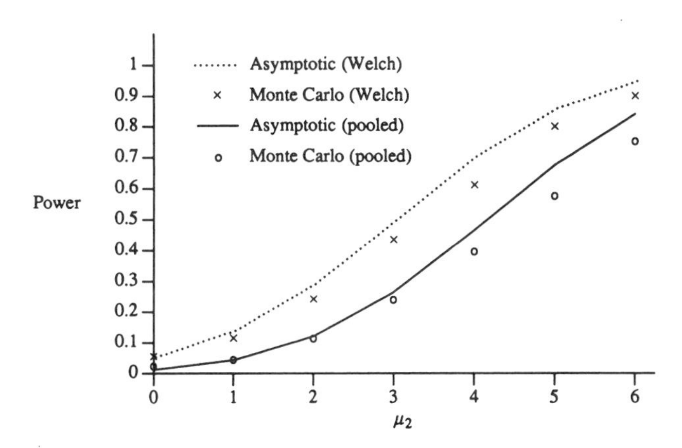

FIGURE 6.2 Power comparisons of pooled and Welch t-tests when variances are unequal.

two power functions is that it is plotted with the asymptotic critical point. If it is plotted instead with z_α replaced by the same critical point used for the finite sample tests, even better approximations result; see the problems. Examples such as this one encourage us to place faith in the asymptotic theory as a useful practical guide.

6.3.2 Unequal Variances

Next consider the case of unequal variances. $\sigma_X = 1$, $\sigma_Y = 5$, so $h = 4$. With all other parameters as above, the asymptotic power functions are $\Phi(-2.26 + .157\theta)$ for the pooled test and $\Phi(-1.645 + .157\theta)$ for the two-sample test (Table 6.2 and Figure 6.2).

6.4* RESULTS FOR TESTS BASED ON M-ESTIMATORS

Brown (1982) considers M-test statistics for the two-sample problem of the form $T = \sum_{i=1}^{m} \psi(X_i - \hat{\theta})$ where $\hat{\theta}$ is the solution to

$$0 = \sum_{i=1}^{m} \psi(X_i - \theta) + \sum_{j=1}^{n} \psi(Y_j - \theta)$$

and ψ is a nondecreasing odd function. Such tests are not always able to be implemented, because the ψ function may depend on the unknown F. However, they are useful for asymptotic purposes. It can be shown that various choices of ψ render tests with asymptotic power functions identical to otherwise defined tests. For example, if $\psi(x) = x$, we obtain a test equivalent to the t-test; if $\psi(x) = F(x) - \frac{1}{2}$, we obtain a test equivalent to the Wilcoxon test. Brown has found the asymptotic power function of M-tests and, in particular, formulas for the test bias factor b and test power factor c in terms of ψ, ψ', h, and λ. These formulas are difficult to evaluate for some ψ of interest, but he has expanded b in a power series in h, the first two terms of which are

$$b = 1 + h(1 - 2\lambda)\frac{\mathrm{E}[X\psi(X)\psi'(X)]}{\mathrm{E}[\psi^2(X)]} + o(h).$$

From this expression it is clear that quite generally choosing equal sample sizes ($\lambda = \frac{1}{2}$) will render a test with a robust significance level in the presence of slightly unequal variances (h small).

In addition, Brown's analysis of various tests leads him to conclude that:

(i) In the case of $m = n$, means, lightly trimmed means, and sign tests are all robust against small departures from "$\sigma_Y = \sigma_X$".

(ii) In the case of $m \neq n$, the robustness mentioned above deteriorates rapidly for all tests except the sign test.

Rousseeuw and Ronchetti (1981) developed an influence function for the level and for the power of tests which also leads to insights into the behavior of one- and two-sample tests. See also Chapter 3 of Hampel et al. (1986) for a systematic development.

6.5 THE TWO-SAMPLE TRIMMED t-TEST

So far we have learned that in sampling from two normal populations with possibly different means and variances, the pooled t-test is not level robust unless the sample sizes are equal, or the larger sample is taken from the population with the larger variance. This is not surprising in view of the fact that the pooled t-test was designed for normal populations with *equal* variances.

The Welch two-sample test does much better with regard to unequal variances, of course, because it is designed to accommodate unequal variances. Unfortunately, both tests will suffer in the presence of outliers, because they are based on sample means and standard deviations which are very sensitive to their presence.

In this section we look at the behavior of the two-sample trimmed t-test. This test was introduced by Yuen and Dixon (1973) and further studied by Yuen (1974). She showed via Monte Carlo sampling that:

1. For normal distributions a trimmed-t version of Welch's test (described below) had very accurate empirical levels when the nominal values were .01, .05, and .1, the sample sizes were 10 and/or 20, and the amount of trimming β was less than $\frac{1}{4}$. But, for sample size 5, the approximation is inadequate.
2. Furthermore, for a large variety of nonnormal distributions and the same sample sizes and proportions of trimming as above, Yuen's trimmed version of Welch's test maintains the approximate nominal levels except for very long tailed distributions, and then it is conservative. Generally speaking, her test has type I error probabilities closer to the nominal levels than the Welch test.
3. For exact normal distributions the power loss in using the trimmed t-test is small compared to the Welch test, but for long tailed distributions, the power gained is substantial.

These numerical results are obtained for specific cases of sample sizes, distributions, amounts of trimming, etc., so that we must tread warily until analytical confirmation is found for the general situtation. However, the asymptotic power function (6.5.4) for the two-sample trimmed test complements the extensive numerical experiments of Yuen (1974) in that it shows that the test is (asymptotically) level robust and furthermore has the same efficiency relative to the Welch test as the trimmed t-test does to the classical t-test in the one-sample problem.

To carry out the trimmed t-test, we need to calculate the two-sided trimmed mean and its standard error for each sample, and compare the Studentized difference of the trimmed means with the appropriate critical point.

The one-sample trimmed mean is introduced and studied in example 4 of Section 4.3.1 and example 4a of Section 4.3.2. Let $\overline{x}_{m,2\beta}$ be the mean of the m_β observations remaining after $[m\beta]$ observations are removed from each side of the ordered x's. Let $s_{x,m,\beta}$ denote the influence function estimate of its standard error; it is, from (4.3.5), the square root of

$$s_{x,m,\beta}^2 = \frac{s_{W_{x,m,2\beta}}^2}{m(1-2\beta)^2} \tag{6.5.1}$$

where $s_{W_{x,m,2\beta}}^2$ is the sample variance of the 2β symmetrically Winsorized X-sample.

Similarly define the symmetrically trimmed mean for the second population by $\overline{y}_{n,2\beta}$ and let

$$s_{y,n,\beta}^2 = \frac{s_{W_{y,n,2\beta}}^2}{n(1-2\beta)^2}.$$

The variance of the difference $\overline{y}_{n,2\beta} - \overline{x}_{m,2\beta}$ is naturally estimated by the sum $s_{m,n,\beta}^2 = s_{x,m,\beta}^2 + s_{y,n,\beta}^2$. The two-sample trimmed t-test statistic is

$$T_{m,n,\beta} = \frac{\overline{y}_{n,2\beta} - \overline{x}_{n,2\beta}}{s_{m,n,\beta}}. \tag{6.5.2}$$

The terminology "t-test" is justified only if the critical point $C_{m,n;\alpha}$, which defines the level-α test through (6.1.1), is determined by the Student's t-distribution for some number ν degrees of freedom. We will use a modification of Welch's approximation (6.2.8) for ν as follows. Let $m_\beta = m - 2[\beta m]$ and $n_\beta = n - 2[\beta n]$ be the putative sample sizes of the trimmed samples; and let $A_\beta = s_{x,m,\beta}^2$ and $B_\beta = s_{y,n,\beta}^2$. Then define ν_β by (6.2.8) with m replaced by m_β, A replaced by A_β, etc. Thus the appropriate degrees of freedom is

$$\nu_\beta = \frac{(A_\beta + B_\beta)^2}{A_\beta^2/(m_\beta - 1) + B_\beta^2/(n_\beta - 1)} \tag{6.5.3}$$

with $A_\beta = s_{x,m,\beta}^2$ and $B_\beta = s_{y,n,\beta}^2$.

As mentioned earlier near (5.2.12) there is to our knowledge no *theoretical* justification for using the t-distribution for the one-sample Studentized trimmed mean, but there is much empirical justification. The same situation holds here.

The asymptotic power function for the two-sample trimmed t-test is derived in a straightforward manner under the assumptions made in Section 6.1.3. By a proof outlined in the problems it is shown that

$$\Pi(\theta) = \Phi\left(z_\alpha + \frac{c\theta}{V_{x,\beta}^{1/2}}\right) \tag{6.5.4}$$

where c is the test power factor given by (6.2.4) and $V_{x,\beta}$ is the asymptotic variance of the one-sample symmetrically trimmed mean [the V_β defined by (5.2.14)]. This power function includes that of the Welch t-test (6.2.9) as a limiting case by letting $\beta \to 0$.

The asymptotic relative efficiency of the trimmed to the Welch t-test is the same as it was in the one-sample case: V_0/V_β. Some values of this efficiency are listed in Table 5.2. It may be shown that these efficiencies hold in the two-sample problem even when the variances are unequal. Moreover, there is robustness to up to proportion of β outliers on each side of each sample. Thus there is much to gain and little to lose by using Yuen's trimmed t-test. Let us see how it compares with other tests on some data sets.

Example 1: Comparing Total Ridge Counts (Continued from Section 6.1). In this problem we want to compare μ_X with μ_Y and reject at level .01, say, the null hypothesis of equality in favor of the alternative that $\mu_X < \mu_Y$. We will use the computer software package Minitab to facilitate the computation of various test statistics for comparing two populations. The dotplot shown below reveals the presence of numerous possible outliers which may create problems for tests that rely on the assumption of normal populations. Also note that the variances of the two samples are clearly different, which suggests that the pooled t-test is not appropriate. However, we will find the P-values of the pooled t-test, the Welch t-test, the Wilcoxon two-sample test (Mann-Whitney test), and Yuen's two-sample trimmed t-test. We will omit some computer output regarding confidence intervals for the mean difference of total ridge counts for the two groups. A summary of these intervals is given in the next section.

```
MTB > dotplot 'X' 'Y';
SUBC> same scale;
SUBC> start -20.

              :                    .
  .           :..  ... : ...::.    :: .          .
+---------+---------+---------+---------+---------+-------X
                           .   :.
                    .  ....::.:::. :.    ..   .
+---------+---------+---------+---------+---------+-------Y
0        50       100       150       200       250

MTB > twosample 'Y' 'X';
SUBC> pooled;
SUBC> alternative=1.
TWOSAMPLE T FOR Y VS X
-  N  MEAN STDEV SE MEAN
Y 27 154.0  29.0    5.59
X 27 126.0  50.2    9.66
```

```
POOLED STDEV=41.0
TTEST MU Y=MU X (VS GT): T= 2.51 P=0.0076 DF= 52
```

As we shall see below, the P-value for the Welch t-test, which estimates the variances separately, is similar to that of the pooled test. This is consistent with Remark 4 of Section 6.2.2, which points out that the two tests have similar behavior for large equal sample sizes.

```
MTB > twosample 'Y' 'X';
SUBC> alt 1.
TTEST MU Y=MU X (VS GT): T= 2.51 P=0.0080 DF= 41
```

Next we will use the Wilcoxon two-sample test (the Mann-Whitney test). It assumes that the first variable listed (the Y-variable) has a continuous, but not necessarily symmetric, distribution with median η_1, and that the second variable listed (our X-variable) has a distribution which is a shift of it with median η_2.

```
MTB > Mann-Whitney 1 'Y' 'X'
Mann-Whitney Confidence Interval and Test
Y  N=27 MEDIAN=153.00
X  N=27 MEDIAN=129.00
W=874.0
TEST OF ETA1=ETA2 VS. ETA1 G.T. ETA2 IS SIGNIFICANT AT 0.0117
```

Finally we use the Minitab macro tmean2, which is given in Appendix D. It finds the P-value of the trimmed two-sample t-test, as well as a 95% confidence interval for the mean difference in populations. We trimmed three observations off each side of the sample; this yields a $2 \times 3/27 = 2 \times \beta$ symmetrically trimmed mean, with $\beta \approx .1$.

The trimmed mean of the Y-population, plus or minus an estimate of its standard error, is 151.57 ± 4.78. For the X-population it is 125.5 ± 10.04. The difference in trimmed means, plus or minus an estimate of the standard error, is 26.05 ± 11.12. The estimated degrees of freedom for the Studentized difference is $\nu_\beta = 28.6$. [These results are based on formulae (6.5.1)–(6.5.3).]

The P-value of the one-sided trimmed t-test is .013, which is significant at the .05 level but not quite significant at level .01. Note that this P-value is close to that of the Mann-Whitney test, but both values are larger than those of the pooled and Welch tests. The latter two tests would just barely reject the hypothesis of no difference between the groups at level .01.

To summarize this example, there appear to be different population variances, and possible outliers, in the X-population, but the fact that the sample sizes are equal and moderately large irons out these wrinkles and renders the traditional t-tests roughly equal and comparable to the nonparametric and robust t-test. The former tests may be slightly overstating the significance of the data (see the Remarks in Section 6.2), and this can be avoided by using the latter tests. □

Example 2: Atomic Weight of Carbon (Continued from Section 6.1). We have rescaled the data by subtracting 12 and multiplying by 1000. Then the results of the four tests (pooled t, Welch, Mann-Whitney, and Yuen) for testing whether $\mu_X = \mu_Y$ against $\mu_X < \mu_Y$ were found by Minitab commands, as above.

```
MTB > dotplot 'X' 'Y';
SUBC> same scale;
            .                  .        .                     .
-----+---------+---------+---------+---------+---------+-X

                                       .           .  . . .
-----+---------+---------+---------+---------+---------+-Y
   -150      -100       -50         0        50       100

MTB > twosample 'Y' 'X';
SUBC> pooled;
SUBC> alternative=1.
TWOSAMPLE T FOR Y VS X
- N  MEAN STDEV SE MEAN
Y 5  56.6  24.3    10.9
X 4 -38.0  86.4    43.2
POOLED STDEV=  59.5
TTEST MU Y=MU X (VS GT): T=2.37 P=0.025 DF=7
MTB > twosample 'Y' 'X';
SUBC> alternative=1.
TTEST MU Y=MU X (VS GT): T= 2.12 P=0.062 DF= 3
MTB > Mann-Whitney alternative 1 'Y' 'X'
Mann-Whitney Confidence Interval and Test
Y  N=5 MEDIAN=  64.0
X  N=4 MEDIAN= -33.0
W=  33.0
TEST OF ETA1=ETA2 VS. ETA1 G.T. ETA2 IS SIGNIFICANT AT 0.0331
```

In applying the "tmean2" macro we trimmed one observation off the end of each sample, and obtained for the Y-variable a trimmed mean, plus or minus standard error, of 63.3 ± 6.7 and for the X-variable, -33 ± 20.8. The difference between the trimmed means was 96.3 ± 21.84 and the estimated degrees of freedom was $\nu = 1.21$. The Studentized difference has P-value .054.

It is tempting to argue that because both the parametrically based pooled t-test and the nonparametric Mann-Whitney test are in agreement—the data suggest rejection of the null hypothesis at level .05—we should therefore make this decision. However, the pooled test is based on assumptions which cannot be checked. The little evidence that is available for standard deviations suggests that the largest sample has been taken from the population with the smallest variance, a situation in which the pooled test tends to overstate the significance of a result (see Remark 2 of Section 6.2.1). Of course the Mann-Whitney test also assumes that the variances are equal, when they exist, and equality is doubtful here.

The Welch test and Yuen's trimmed version of it are in basic agreement that the data are not significant at level .05, although the latter test is closer to the .05 level. Of course the trimmed test is not affected by the possible outliers which appear in the two samples. In conclusion, the trimmed t-test is likely to be more reliable in the presence of unequal variances and/or outliers than the other three tests studied above. □

6.6 CONFIDENCE INTERVALS FOR THE DIFFERENCE IN POPULATION LOCATIONS

Confidence intervals for the difference θ between two population locations can be constructed using the results presented earlier in this chapter. In particular, if an estimate of θ, when Studentized, has a known t-distribution or an approximate normal distribution, then it is a routine problem to find an interval with the desired coverage probability. After discussing such methods briefly in Section 6.1.1, we will look at two different distribution-free approaches, one related to the Wilcoxon statistic and an alternative approach based on one-sample intervals.

6.6.1 Intervals Based on Studentized Statistics

Suppose that $(\hat{\theta} - \theta)/\widehat{\mathrm{SE}}[\hat{\theta}]$ converges in distribution to the standard normal distribution. Then an approximate $1 - \alpha$ confidence interval for θ is easily shown to be

$$\hat{\theta} - z_{1-\alpha/2}\widehat{\mathrm{SE}}[\hat{\theta}] < \theta < \hat{\theta} + z_{1-\alpha/2}\widehat{\mathrm{SE}}[\hat{\theta}]. \tag{6.6.1}$$

When the Studentized $\hat{\theta}$ has an exact t-distribution with ν degrees of freedom, as in the case of the pooled t-statistic given by (6.2.1), then an exact level $1-\alpha$ confidence interval is given by (6.6.1) with the z-quantiles replaced by t_ν-quantiles. If the Studentized statistic has an approximate t-distribution, as in the Welch t-test [see (6.2.5)] or Yuen's trimmed t-test statistic (6.5.2), then the confidence coefficient is only approximate. [*Remark*: in these three cases the Studentized estimate referred to is of the form $(\hat{\theta}-0)/\widehat{\mathrm{SE}}[\hat{\theta}]$ rather than $(\hat{\theta}-\theta)/\widehat{\mathrm{SE}}[\hat{\theta}]$ because in the context of (6.1.1) the null hypothesis value of θ was zero. However, it is clear that the Studentized estimates for nonzero θ have the same distributions.]

The confidence coefficient of such intervals depends on the distribution of the Studentized estimate; if this distribution is highly sensitive to outliers or unequal variances, then the coefficient is unreliable. In fact the breakdown point of the level is zero for the classical tests (pooled and Welch t-tests), and β for the 2β symmetrically trimmed means. The length of these confidence intervals is of the form $2|t_{\nu,\alpha/2}|\widehat{\mathrm{SE}}[\hat{\theta}]$, which will be robust to outliers only when the estimated standard error is robust. Again, this will not be true for the traditional intervals, but it will be the case for Yuen's trimmed mean (which uses the standard deviation of the 2β symmetrically Winsorized sample).

In Table 6.3 we collect the 95% confidence intervals for a difference in location for the data of Example 1 (total ridge counts) and for the data of Example 2 (atomic weight of carbon). In all cases, but of course under different assumptions, it is claimed that the lower and upper bounds on θ are realizations of statistics satisfying $P\{L<\theta<U\}=.95$. The trimming proportions for the third interval are the same used in the discussion of one-sided tests in Section 6.5 above. The validity of the claimed confidence coefficient for the first two intervals is highly dependent on the assumption of normality, while for the trimmed t interval it is less so. Intervals that have distribution-free confidence coefficients are those associated with the Mann-Whitney test and the one-sample sign statistics; see below.

In the first example the sample sizes were equal and moderately large ($m=n=22$), and there were only a few possible outliers. The intervals are all in basic agreement. In the second example sample sizes are small and unequal ($m=4, n=5$), and the intervals vary greatly in length.

6.6.2 Distribution-Free Confidence Intervals

Assume that the $X_1,\ldots,X_m$ are i.i.d. $F(x)$, where F is continuous, but not necessarily symmetric or having moments; and let $Y_1,\ldots,Y_n$ be i.i.d.

Table 6.3 95% Confidence Intervals for the Difference in Location in Examples 1 and 2

Method	Example 1		Example 2	
	L	U	L	U
Pooled	5.6	50.4	.3	188.9
Welch	5.4	50.6	−47.1	236
Mann-Whitney	5.0	51	−7	219
Trimmed	3.3	48.8	−88.8	281
One-sample sign	1.5	46.7	−6.5	203.1

$F(x-\theta)$. The problem is to estimate θ. This situation clearly contains the normal population two-sample problem stated in Section 6.1. Let $W_{X,Y}$ equal the number of (X_i, Y_j) pairs with $X_i < Y_j$. When $\theta = 0$, the distribution of this Wilcoxon (Mann-Whitney) statistic is free of F and can be found by a counting argument.

In addition, if $D_{(1)}, \ldots, D_{(mn)}$ denote the ordered differences of the form $D = Y - X$, then it may be shown that

$$P_\theta\{D_{(i)} \le \theta < D_{(i+1)}\} = P_0\{W_{X,Y} = i\}$$

so that by careful selection of L and U from these ordered differences a $1-\alpha$ confidence interval for θ may be constructed. For more details, see Lehmann (1975), Chapter 2. Insofar as the interval is distribution-free, the level is robust to outliers, although a large enough proportion of outliers could break down the length. What proportion is required is unknown to us.

Now let $x_{.5}$ and $y_{.5}$ be the medians of the X- and Y-distributions; then clearly $\theta = y_{.5} - x_{.5}$. We will give a distribution-free interval estimate of θ based on sign statistic confidence interval estimates of the individual medians. The latter are discussed is Section 4.6 and do not require the distribution F to be symmetric. Let $[L_x, U_x]$ and $[L_y, U_y]$ be the sign statistic interval estimates, each with confidence coefficient .84. Define $L = L_y - U_x$ and $U = U_y - L_x$. Then $[L, U]$ is the approximate .95 confidence interval for θ when m, n are large.

It is worth noting that a level .05 test for $H : \theta = 0$ versus $H : \theta \neq 0$ is given by rejecting H if $U_x > L_y$, that is, when the respective interval estimates for the two medians overlap [see Hettmansperger (1984b)]. This test is easily carried out on Minitab, and we illustrate it for the TRC data.

```
MTB > sint 84 'Y'
SIGN CONFIDENCE INTERVAL FOR MEDIAN
-              ACHIEVED
-   N MEDIAN CONFIDENCE CONFIDENCE INTERVAL POSITION
Y  27  153.0     0.7522     ( 144.0, 156.0)       11
-                0.8400     ( 142.8, 157.7)      NLI
-                0.8779     ( 142.0, 159.0)       10
MTB > sint 84 'X'
SIGN CONFIDENCE INTERVAL FOR MEDIAN
-              ACHIEVED
-  N MEDIAN CONFIDENCE CONFIDENCE INTERVAL POSITION
X 27  129.0     0.7522     ( 113.0, 139.0)       11
-               0.8400     ( 109.0, 141.3)      NLI
-               0.8779     ( 106.0, 143.0)       10
MTB > stack 'Y' 'X' c10;
SUBC > subscripts c20.
MTB > boxplot c10;
SUBC> by c20;
SUBC> notch 84.
```

```
C20
1                    --------[( +)]------     * *  *
2          ------------[    ( + )    ]---------------
       +---------+---------+---------+---------+---------+------C10
       0        50       100       150       200       250
```

The advantage of this last method is that it is based on statistics with high breakdown point; see Section 4.6.

6.7 PROBLEMS

Section 6.1

1. Under the two-sample assumptions given in Section 6.1.3 show that

   (a) $\mathrm{E}[\overline{Y}_n - \overline{X}_m] = \dfrac{\theta}{\sqrt{N}}$

   (b) $\mathrm{Var}[\overline{Y}_n - \overline{X}_m] = \sigma_X^2 \left[\dfrac{(1+h)^2}{n} + \dfrac{1}{m}\right]$

Section 6.2

2. Verify the limiting value of S^2 given in (6.2.3).

3. Derive the asymptotic power of the pooled t-test using (6.2.1) and (6.2.2) in (6.1.1), and show that it has the form given by (6.2.4).

4. Verify the claims made in Remarks 1 to 3.

5. Verify the limiting expression for s^2 given in (6.2.7).

6. Derive the asymptotic power function of the two-sample test given in (6.2.9).

7. Verify the details of Remark 5, which assumes that the condition of normality with equal variances holds exactly for the two populations:

(a) When $m = n$, the two-sample t-test statistic and the pooled t-test statistic have the same value.

(b) When $m = n$, show that the pooled t-test has more degrees of freedom than ν of the Welch t-test. *Hint*: Show that $A^2 + B^2 \geq \frac{1}{2}(A+B)^2$ with equality if and only if $A = B$; hence that ν given by (6.2.8) satisfies $\nu \leq 2(n-1)$.

(c) Show that loss of degrees of freeedom for the two-sample test is slight if $\frac{1}{3} < s_x^2/s_y^2 < 3$. In fact $\nu \geq 1.6(n-1)$.

(d) Also check that $P\{\frac{1}{3} \leq s_x^2/s_y^2 \leq 3\} \approx .9$ if $\sigma_x^2 = \sigma_Y^2$ and $m = n = 1$.

8. Let $y_1, \ldots, y_n$ have $\overline{y}_n$ and $s_{y,n}^2 = [1/(n-1)]\sum_{i=1}^{n}(y_i - \overline{y}_n)^2$ for the sample mean and sample variance. Then:

(a) $\overline{y}_n = \overline{y}_{n-1} + [y_n - \overline{y}_{n-1}]/n$ and $s_{y,n}^2 = s_{y,n-1}^2 + [y_n - \overline{y}_{n-1}]^2/n$.

(b) Hence for m, n fixed and S given by (6.2.2),

$$T = \frac{\overline{y}_n - \overline{x}_m}{S\sqrt{1/m + 1/n}}$$

$$\sim \frac{y_n/n}{\sqrt{1/m + 1/n}\sqrt{[(n-1)/(N-2)](1/n)y_n^2}} \quad \text{as} \quad y_n \to \infty$$

$$\to \sqrt{\frac{\lambda}{1-\lambda}} \quad \text{as} \quad m, n \to \infty \quad \text{with} \quad \frac{m}{m+n} \to \lambda.$$

Section 6.3

9. The asymptotic power function appears to the left of the finite sample power functions in Figure 6.1 because the asymptotic critical point of $z_{.95} = 1.645$ is used. Find and plot the same power function $\pi(\theta) = \Phi(z_\alpha + .112\theta)$ when z_α is replaced by:

(a) The .95 critical level for the pooled t-test $t_{m+n-2}(.95)$.

(b) The .95 critical level for the two-sample test using Welch's approximation, assuming that s_x^2 and s_y^2 are very close (equal) to the true values of σ_x^2 and σ_y^2, respectively.

10. Carry out a simple Monte Carlo experiment to check one of the results in Table 6.1. Estimate the standard error of your estimate of power.

11. Find the asymptotic power functions for the pooled t-test and the two-sample test for the case of unequal variances with $\sigma_X = 1$, $\sigma_Y = 5$ and with the formulas in Section 6.2 and compare the results with Table 6.2.

Section 6.5

12. Verify the four P-values for the different tests of the mean difference in drug effects given in Example 1, Section 6.5. Use a hand calculator and compare your results with those obtained by Minitab.

13. Verify the four P-values for the different tests of the mean difference in results using different methods of preparation given in Example 2, Section 6.5. Use a hand calculator and compare your results with those obtained by Minitab.

14. Consider the data from Appendix C. Find 95% trimmed-t confidence intervals for the mean difference between:

(a) Total ridge counts for the statistics class and the Aboriginal tribe.

(b) Total ridge counts on the left and right hands for the statistics class.

(c) Total ridge counts on the left and right hands for the Aborigine tribe.

15. This example illustrates quality control methods practiced in a clothing manufacturing plant. Levi-Strauss manufactures clothing from cloth supplied by several mills. The data used in this exercise are for two of these mills and were obtained from the quality control department of the Levi plant in Albuqerque, New Mexico. [See Koopmans (1987), p. 86.] In order to maintain the anonymity of these two mills we have coded them A and B. A measure of wastage due to defects in cloth and so on is called *run-up*. It is quoted as percentage of wastage per week and is measured relative to computerized layouts of patterns on the cloth. Since the people working in the plant can often beat the computer in reducing wastage by laying out the patterns by hand, it is possible for run-up to be negative. From the viewpoint of quality control, it is desirable not only that the run-up be small but that the quality from week to week be fairly consistent. There are 22 measurements on run-up for each of the two mills and they are:

Mill A		Mill B	
0.12	0.03	1.64	0.63
1.01	0.35	−0.60	0.90
−0.20	−0.08	−1.16	0.71
0.15	1.94	−0.13	0.43
−0.30	0.28	0.40	1.97
−0.07	1.30	1.70	0.30
0.32	4.27	0.38	0.76
0.27	0.14	0.43	7.02
−0.32	0.30	1.04	0.85
−0.17	0.24	0.42	0.60
0.24	0.13	0.85	0.29

(a) Are there any differences in the quality of the cloth supplied by Mills A and B? More precisely, using a hand calculator or a statistical package, find the P-values of the two-sided alternative to the null hypothesis for each of the five tests discussed in Section 6.5. [*Answers*: The P-value of the pooled t-test equals .28, Welch t-test .28, Mann-Whitney test .001, Yuen test (with $\beta = 2/22$) .034, Yuen test (with $\beta = 4/22$) .001.

(b) Which if any of the above results can be relied on? Explain your answer.

16. In this problem we derive the asymptotic power function (6.5.3).

(a) Under the assumptions given in (6.1.1) show that

$$N\,\mathrm{Var}[\overline{y}_{n,2\beta} - \overline{x}_{n,2\beta}] \rightarrow \frac{V_{x,\beta}}{\lambda} + \frac{V_{y,\beta}}{1-\lambda}$$

where $V_{x,\beta}$ is defined in (5.2.13).

(b) Show that with the same assumptions as in (a), $Ns^2_{x,m,\beta} \xrightarrow{p} V_{x,\beta}/\lambda$ and find a similar result for the y-sample.

(c) Show that $T_{m,n,\beta}$ is asymptotically normal, and combine this fact with the results from (a) and (b) to derive the power function (6.5.3).

Section 6.6

17. Verify the confidence interval entries in Table 6.3.

18. Find 95% confidence intervals for the mean difference in run-up for the mill data of Problem 15, using each of the five methods discussed in the text. Which method is appropriate in this context, and why?

6.8 COMPLEMENTS

Cressie and Whitford (1986) suggest modifications to the pooled t-test and Welch's version of it to take into account estimated skewness of the underlying distributions. They also relax the "identically distributed" assumption within each of the samples to allow for differing variances within the respective samples. They then propose an estimate of the appropriate degrees of freedom which generalizes Welch's suggestion.

Best and Rayner (1987) give more evidence supporting the use of Welch's version of the two-sample t-test for testing the difference of means in two normal populations whose variances may also differ. They compare Welch's test with the asymptotically optimal tests provided by the Wald, likelihood ratio, and score statistics for a variety of alternatives and find that the powers are comparable. They also find that the nominal .05 size is reliable under a wide range of parameter and sample size combinations. Even when the population variances are assumed equal, they recommend Welch's version of the t-test over the pooled t-test unless the estimated degrees of freedom is less than 5.

CHAPTER SEVEN

Regression

There are two main areas of research into linear regression—diagnostics and robust regression—which complement each other in the process of model building and verification as discussed in Chapter 1. After briefly reviewing some classical results we will first present the diagnostic material and then the algebraic and asymptotic justification for the robust methods of inference which follow afterward.

We now list some details of our choice of contents for this somewhat lengthy but important chapter. Regression analysis is often used in the construction phase of model building, and hence a variety of data analytic methods are available for determining a model which represents the scientific relationships. Transformations of variables, selection of variables, tests for correlated errors, and diagnostic tests for aberrant data are all part of the model building process. We will not discuss these topics except for diagnostic methods.

Once a linear regression model is adopted, it is desired to make inferences about its parameters, and a well-established theory is available. We assume the reader is familiar with the matrix formulation of the multiple linear regression model and the classical normal theory methods of inference as covered by Montgomery and Peck (1982) or Draper and Smith (1981), and which are summarized in Section 7.1. Unfortunately, outliers and other aberrations which appear to conflict with the model can arise and undermine the stated levels for confidence intervals and tests. Thus it is desirable to have robust methods which are still highly efficient in the presence of these aberrations. Such methods are sometimes also used as a reliable reference upon which to base diagnostics.

In Section 7.2 we present a detailed study of three diagnostics: Studentized residuals, leverage, and DFITS. In particular we derive the distribution of the Studentized residuals, explain the elementary properties of the hat matrix, and show how DFITS arises as a standardized influence of deletion on the fitted values. We illustrate how these diagnostics complement each other in the simple linear regression model, and in some examples of multiple linear regression.

We continue the study of leverage in Section 7.3 and derive some more properties of the hat matrix, which give some insights into the leverage points in multiple linear regression. We also include some new methods for constructing designs which have no points with large leverage. These methods may use a trial set of points, including some with large leverage, to roughly define the region over which the model is to be employed.

In Section 7.4 we consider the asymptotic properties of least squares estimates, and give two proofs of their asymptotic normality. The first is due to Huber (1973a) and assumes fixed carriers (independent variables). The second method of proof is for random carriers and uses a von Mises expansion to derive asymptotic normality. The influence function for the least squares estimators is derived along the way. From this the multiplicative effect of residual and position in the design space on the regression estimates may be readily ascertained.

Much of the asymptotic theory has been developed for M-estimators in regression. Since the M-estimating equations may be expressed as weighted normal equations where the weights depend on the data, we may equivalently study weighted least squares. We have chosen to present the asymptotic material in the weighted least squares formulation, because most statisticians are already familiar with the method of iteratively reweighted least squares (reviewed in Section 7.1.2). The influence function for each iterate toward a solution of the weighted normal equations is obtained. This reveals how the previous estimate affects the current one, and in particular shows how the initial (least squares, say) estimator can influence a one-step estimator.

These theoretical considerations are rounded off in Sections 7.6 by computational methods, standard errors of recommended estimators, and applications. In particular we compare Welsch's robust regression procedures based on DFITS with a few other robust estimators by finding the median absolute deviation and interquartile range of the residuals for the respective fitted values.

Thus there is enough material for an entire book on robust procedures in regression, and in this chapter we can only present the basic ideas. Our emphasis will be on methods which are simple to implement and which are supported at least by appealing heuristics, if not mathematical rigor. While the sections to follow are written in a logically ordered progression, the material in Sections 7.3–7.5 is more difficult and may be skipped without serious loss of continuity provided one accepts the asymptotic results which are later used as the basis for estimates of standard errors of robust regression estimates.

The reader may also pursue the topics of transformations, diagnostics, and robust regression in the recent books by Cook and Weisberg (1982), Atkinson (1985), Rousseeuw and Leroy (1987) and Carroll and Ruppert (1988).

7.1 BACKGROUND MATERIAL

7.1.1 The Classical Regression Problem

The classical regression problem is neatly symbolized by

$$\boldsymbol{Y} = \boldsymbol{X}\boldsymbol{\beta} + \boldsymbol{\epsilon} \tag{7.1.1}$$

where

$$\begin{aligned}
\boldsymbol{Y} &= [Y_1, \ldots, Y_n]^T \text{ is an } n \times 1 \text{ vector of observations} \\
\boldsymbol{X} &= [x_{ij}] \text{ is an } n \times (p+1) \text{ design matrix of full rank} \\
\boldsymbol{\beta} &= [\beta_0, \ldots, \beta_p]^T \text{ is a } (p+1) \times 1 \text{ vector of unknown parameters} \\
\boldsymbol{\epsilon} &= [\epsilon_1, \ldots, \epsilon_n]^T \text{ is an } n \times 1 \text{ vector of errors which satisfy}
\end{aligned}$$

$$\begin{aligned}
&\mathrm{E}[\boldsymbol{\epsilon}] = \boldsymbol{0} \\
\text{and} \quad &\mathrm{Cov}[\boldsymbol{\epsilon}] = \sigma^2 \boldsymbol{I}_n.
\end{aligned} \tag{7.1.2}$$

The constant $\sigma^2 > 0$ is also an unknown parameter. The *method of least squares* chooses an estimator $\boldsymbol{B}$ of $\boldsymbol{\beta}$ to minimize the length $\|\boldsymbol{r}\| = [\boldsymbol{r}^T\boldsymbol{r}]^{1/2}$ of the vector of *residuals* defined by

$$\boldsymbol{r} = \boldsymbol{Y} - \boldsymbol{X}\boldsymbol{B}. \tag{7.1.3}$$

Setting the derivative of $\|\boldsymbol{r}\|^2$ to zero leads to the *normal equations*

$$\boldsymbol{X}^T\boldsymbol{X}\boldsymbol{B} = \boldsymbol{X}^T\boldsymbol{Y} \tag{7.1.4}$$

and *least squares estimator*

$$\boldsymbol{B} = (\boldsymbol{X}^T\boldsymbol{X})^{-1}\boldsymbol{X}^T\boldsymbol{Y}. \tag{7.1.5}$$

Under the assumptions (7.1.2) it can be easily verified that $\boldsymbol{B}$ is unbiased and has covariance matrix proportional to the matrix $(\boldsymbol{X}^T\boldsymbol{X})^{-1}$:

$$\begin{aligned} \mathrm{E}[\boldsymbol{B}] &= \boldsymbol{\beta} \\ \mathrm{Cov}[\boldsymbol{B}] &= \sigma^2(\boldsymbol{X}^T\boldsymbol{X})^{-1}. \end{aligned} \tag{7.1.6}$$

The classical estimator of the parameter σ^2 is the *residual mean square*:

$$s^2 = \frac{\boldsymbol{r}^T\boldsymbol{r}}{n-p-1} = \frac{\boldsymbol{Y}^T\boldsymbol{Y} - \boldsymbol{B}^T\boldsymbol{X}^T\boldsymbol{Y}}{n-p-1}. \tag{7.1.7}$$

When it is further assumed that the errors are normally distributed

$$\boldsymbol{\epsilon} \sim N_n(0, \sigma^2\boldsymbol{I}_n) \tag{7.1.8}$$

then

1. $\boldsymbol{B}$ is also the MLE of $\boldsymbol{\beta}$, and hence denoted by $\hat{\boldsymbol{\beta}}$
2. $\hat{\boldsymbol{\beta}} \sim N_{p+1}(\boldsymbol{\beta}, \sigma^2(\boldsymbol{X}^T\boldsymbol{X})^{-1})$
3. $(n-p-1)s^2/\sigma^2 \sim \chi^2_{n-p-1}$
4. The standard error of $\hat{\boldsymbol{\beta}}_i$ is estimated by $s[(\boldsymbol{X}^T\boldsymbol{X})^{-1}]_{ii}^{1/2}$.

In the sequel we shall assume that (7.1.8) holds unless otherwise specified, and we will continue to identify $\boldsymbol{B}$ with $\hat{\boldsymbol{\beta}}$, even when the model is not normal.

There is usually a constant term in the model, which means that $\boldsymbol{X}$ will contain an $n \times 1$ column of 1's, denoted by $\mathbf{1}$, and usually placed first. It is then convenient to refer to this column by $X_0 = \mathbf{1}$. Other columns $X_1, X_2, \ldots, X_p$ will be called *explanatory variables* or *regressors* or *predictors*. Thus unless specifically stated otherwise, the letter p will denote the number of explanatory variables. If there is no constant term in the model, then the dimension of $\boldsymbol{\beta}$ is p, and all the results stated above for $p+1$ parameters still hold with $p+1$ replaced by p.

Some insight into the least squares methodology is given by a vector space interpretation. (See Appendix A.2 for a review of terminology.) The set of all linear combinations of the columns of $\boldsymbol{X}$ forms a vector subspace of the observation space R^n. The *fitted values* are defined to be $\hat{\boldsymbol{Y}} = \boldsymbol{X}\hat{\boldsymbol{\beta}}$, which is the projection of $\boldsymbol{Y}$ onto this subspace; in other words, $\hat{\boldsymbol{Y}}$ is the closest vector in the subspace to the observation vector $\boldsymbol{Y}$, where "closest" means in the Euclidean metric $\|\boldsymbol{Y} - \hat{\boldsymbol{Y}}\|$. The fitted values satisfy $\hat{\boldsymbol{Y}} = \boldsymbol{H}\boldsymbol{Y}$, where $\boldsymbol{H} = \boldsymbol{X}(\boldsymbol{X}^T\boldsymbol{X})^{-1}\boldsymbol{X}^T$. The matrix $\boldsymbol{H}$ carries with it useful algebraic information on the estimator $\hat{\boldsymbol{\beta}}$. A detailed analysis of this projection matrix $\boldsymbol{H}$, also dubbed the *hat matrix* since it maps $\boldsymbol{Y}$ into $\hat{\boldsymbol{Y}}$, is carried out in Sections 7.2.1, 7.2.2, and 7.3.

Table 7.1 Growth of Prices

Year x_i	40	41	42	43	44	45	46
Growth of prices y_i	1.62	1.63	1.90	2.64	2.05	2.13	1.94

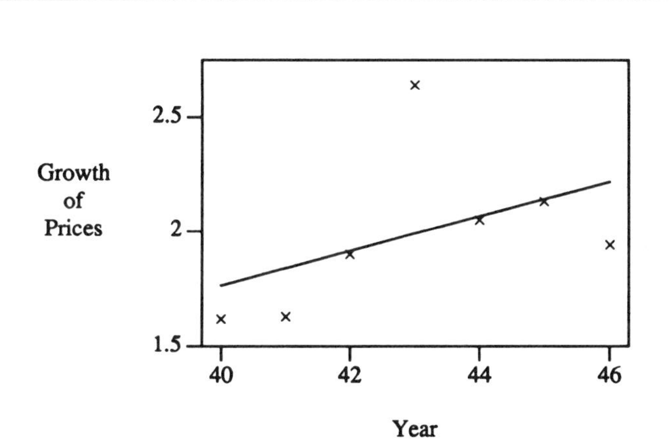

FIGURE 7.1 Growth of prices.

Example 1: Growth of Prices. The growth of city prices in Taiwan during the years 1940–1946 is shown in Table 7.1 and plotted in Figure 7.1. [See Simkin (1978).] We are interested in the average *rate* of growth over these years.

If a *simple* linear regression model is assumed, that is, if each year's growth in prices is assumed to arise from a model with only $p = 1$ explanatory variable (time) plus a constant term:

$$Y_i = \beta_0 + \beta_1 x_i + \epsilon_i, \qquad i = 1,\ldots,7$$

then $\beta = [\beta_0, \beta_1]^T$ and

$$X^T = \begin{bmatrix} 1 & 1 & \ldots & 1 \\ x_1 & x_2 & \ldots & x_7 \end{bmatrix}.$$

The reader may use (7.1.5) and (7.1.7) to find the least squares estimates and the residual mean square: $s^2 = (.3358)^2$. The least squares line $\hat{y} = -1.253 + .0754x$ is plotted in Figure 7.1. The residuals $r_i = y_i - \hat{y}_i$ are all negative except for one which is much larger in magnitude than the others. This observation during the year 1943 is suspect if the errors are assumed to be normal, as we shall see in the next section. How much faith can we have in the estimate of the growth rate $\hat{\beta}_1 = .0754$? The estimated standard error of this estimate is [according to the classical theory following (7.1.8)] found to be .06346, but if the model is in doubt, so is this estimate of error.

How should we proceed to estimate the rate of growth when the data and the model are in apparent conflict? The answer is not obvious, in this and many other problems, so we need to make a systematic study of estimators which will still give reasonable answers in the presence of suspect data.

We need to *identify* the data that may conflict with our model, partly because such data may signal a previously undiscovered phenomenon, and also because such data *may* be due to blunders or misrecording at some stage of the experiment. If after identification, the suspect data cannot be removed on qualitative grounds, then we need to be assured that our statistical methods are not overly affected by such data. This usually means employing a method which is known not to be the most efficient for the favored model. We would like the chosen procedure to be reasonably efficient for all models near the favored one; however, not much progress has been made on this complicated problem. There are at least three notions of relative efficiency for vector-valued estimators (see Section 7.4.5), and much more work needs to be done on the efficiency of robust estimators.

In Section 7.6 we illustrate how robust procedures handle data sets such as the growth of prices example, which includes an even more surprising outlier in the year 1948. This data set has also been analyzed by different robust methods in Rousseeuw, Daniels, and Leroy (1984) and by Rousseeuw and Leroy (1987), pp. 51–56. □

7.1.2 Weighted and Generalized Least Squares

Instead of choosing $\boldsymbol{B}$ to minimize the sum of squared residuals,

$$S(\boldsymbol{B}) = \boldsymbol{r}^T\boldsymbol{r} = \sum r_i^2 \tag{7.1.9}$$

we could more generally choose $\boldsymbol{B}_W$ to minimize the sum of weighted squared residuals,

$$S_W(\boldsymbol{B}) = \sum w_i r_i^2. \tag{7.1.10}$$

For example, in the fitting of a straight line to the growth of price data in Example 1, we may choose to downweight the residual arising from the suspect fourth observation.

Traditionally, weighted least squares has been employed in situations where the errors ϵ have been assumed to be uncorrelated with a variance that varies with $\boldsymbol{x}$ in a *known* way. More generally, generalized least squares has been used when the errors have been dependent, with a covariance structure that was known except for the multiplicative constant σ^2. More recently, variations of weighted least squares have been used in situations where there is the possibility of outliers in the observations or the design.

Such situations are handled by estimating the weights themselves; by choosing the weights to be prespecified functions of the data. Having obtained weighted least squares estimates of the unknowns $\boldsymbol{\beta}$, σ^2, the residuals arising from the fitted model are used to choose new weights, and the process is then repeated. This iterative procedure is especially effective in robust regression, and therefore essential to our discussion. A second reason for reducing a problem to the structure of weighted least squares is that it can in turn be solved by the methods of ordinary least squares, and computations are readily performed by standard computer package or on a hand calculator. We will quickly review this reduction to ordinary least squares to help fix the notation.

When the errors satisfy

$$\begin{aligned} \mathrm{E}[\boldsymbol{\epsilon}] &= 0 \\ \mathrm{Cov}[\boldsymbol{\epsilon}] &= \sigma^2 \boldsymbol{W}^{-1} \end{aligned} \tag{7.1.11}$$

for a known $n \times n$ symmetric positive-definite matrix $\boldsymbol{W}$, the problem $\boldsymbol{Y} = \boldsymbol{X}\boldsymbol{\beta} + \boldsymbol{\epsilon}$ can be transformed to one with errors satisfying (7.1.2). For under the conditions on $\boldsymbol{W}$ there exists a symmetric positive-definite square root $\boldsymbol{W}^{1/2}$ satisfying $\boldsymbol{W}^{1/2}\boldsymbol{W}^{1/2} = \boldsymbol{W}$. Let $\boldsymbol{Y}_W = \boldsymbol{W}^{1/2}\boldsymbol{Y}$, $\boldsymbol{X}_W = \boldsymbol{W}^{1/2}\boldsymbol{X}$, and $\boldsymbol{\epsilon}_W = \boldsymbol{W}^{1/2}\boldsymbol{\epsilon}$, so that (7.1.1) can be reexpressed

$$\boldsymbol{Y}_W = \boldsymbol{X}_W\boldsymbol{\beta} + \boldsymbol{\epsilon}_W. \tag{7.1.12}$$

It is easily seen that $\boldsymbol{\epsilon}_W$ satisfies the usual conditions (7.1.2) of ordinary least squares, and hence the ordinary least squares methodology for estimating $\boldsymbol{\beta}$ can be carried out in terms of $\boldsymbol{X}_W$ and $\boldsymbol{Y}_W$. The analogues of (7.1.4) to (7.1.6) are the normal weighted equations

$$(\boldsymbol{X}_W^T\boldsymbol{X}_W)\boldsymbol{B} = \boldsymbol{X}_W^T\boldsymbol{Y}_W$$

which clearly simplifies to

$$(\boldsymbol{X}^T\boldsymbol{W}\boldsymbol{X})\boldsymbol{B} = \boldsymbol{X}^T\boldsymbol{W}\boldsymbol{Y} \tag{7.1.13}$$

and has the *generalized least squares* solution

$$\boldsymbol{B}_W = (\boldsymbol{X}^T\boldsymbol{W}\boldsymbol{X})^{-1}\boldsymbol{X}^T\boldsymbol{W}\boldsymbol{Y}. \tag{7.1.14}$$

The mean and covariance of this estimator are:

$$\begin{aligned} \mathrm{E}[\boldsymbol{B}_W] &= \boldsymbol{\beta} \\ \mathrm{Cov}[\boldsymbol{B}_W] &= \sigma^2(\boldsymbol{X}^T\boldsymbol{W}\boldsymbol{X})^{-1}. \end{aligned} \tag{7.1.15}$$

The parameter σ^2 has an analogous estimator

$$s_W^2 = \frac{\boldsymbol{r}_W^T\boldsymbol{r}_W}{n-p-1}, \tag{7.1.16}$$

where $r_W = Y_W - X_W B_W$.

When the matrix of weights W is a *diagonal* matrix, the estimator given by (7.1.14) is called the *weighted least squares estimator*. This is the most commonly used form of generalized least squares and the simplest because the errors are assumed to be uncorrelated.

The terminology *iteratively reweighted least squares* describes the following process, which assumes that some method has been decided upon for choosing the weights $W_{ii} = w_i = w(x_i, r_i)$ after a trial fit.

1. Choose an initial estimate $B^{(0)}$ of β.
2. Define the residuals $r^{(j)} = Y - XB^{(j)}$ associated with a jth estimate and then find the weights to be used in the next (weighted) least squares estimate.
3. Use the weights obtained in 2 to solve for $B^{(j+1)}$ and return to 2, unless the estimates differ from $B^{(j)}$ by no more than the desired accuracy.

7.1.3 Goals of Robust Regression

In regression analysis we may want to test hypotheses regarding the unknown vector of coefficients β, and we require estimates of them plus the standard errors. There are a number of assumptions that may be violated: the distributional assumption that each error $\epsilon_i \sim N(0, \sigma^2)$; the independence of the errors; and the linear dependence on the explanatory variables.

As examples in Section 7.6 show, only a small violation in one or more of these assumptions can lead to a large change in the least squares estimator $\hat{\beta}$ or the estimates of standard errors based on the assumed covariance structure. Robust estimators of β ideally should satisfy the following goals:

1. Consistency, asymptotic normality, and high efficiency of the estimators if there are no model violations
2. Methods for forming confidence intervals for the unknown parameters and for testing hypotheses about them
3. Relative insensitivity of the properties in 1 and results in 2 to slight violations of the model
4. Simplicity of theory
5. Ease of computation, given a standard computer package.

7.2 DIAGNOSTICS FOR THE CLASSICAL MODEL

The linear regression model described in the last section is both useful and mathematically elegant, so that many statisticians want to use it whether or not the data appear to arise from a distribution which strictly satisfies the assumptions of the model (7.1.1), (7.1.2), and (7.1.8). Diagnostic methods help to identify cases which are inconsistent with the bulk of the data, or which have an unusually large affect on the estimates and fitted values. These cases are worth identifying for several reasons. First they may be anomolous because of some unexpected physical cause which is worthy of further scientific study. Second they may be due to blunders in recording or transcribing of results. Third they may have an undue effect on the least squares estimators, even though they have arisen by chance from distributions which satisfy the model assumptions. It may not always be possible to determine the reason for the unusual or suspect cases, and it is not always easy to identify them, especially in multiple linear regression. Thus good diagnostic procedures are extremely important.

In this section we assume that the data have already been gathered and presented to the statistician, who must make the best analysis possible with data which may not only contain outliers but which may also have arisen from a poor design. It will become apparent that many of the problems which arise in regression can be avoided by a proper choice of design matrix, and hence we will concentrate on methods for choosing it in Section 7.3.

Let $\boldsymbol{x}_i^T = [x_{i0},\ldots,x_{ip}]$ denote the ith row of the design matrix $\boldsymbol{X}$. Then we have $y_i = \boldsymbol{x}_i^T\boldsymbol{\beta} + \epsilon_i$, $i = 1,\ldots,n$, and the couple $[\boldsymbol{x}_i^T, y_i]$ will be referred as the ith datum point or ith case. In this section we systematically examine different diagnostics: first, the idea of leverage which is based entirely on $\boldsymbol{X}$ through the hat matrix; second, the Studentized residuals which depend on both the design and the dependent variable; and third, DFITS, a diagnostic which reveals the combined effect on the fitted values of both leverage and residuals. Particular attention is given to leverage, for in regression there is often the opportunity for choosing a good design. Sometimes the statistician is consulted only after the experimental results have been obtained; even then he or she may still examine only the design matrix to search for high leverage points, and suggest that they be either omitted or supplemented by further observations in the same area of the design space. Such recommendations are more objective in that they are not influenced by the values of the dependent variable, and they may well avoid some potential problems with analysis. If high leverage points must remain, at least the experimenter will be made aware of them.

Much of the material in this section and Section 7.3 is taken from Hoaglin and Welsch (1978), Huber (1981), Cook and Weisberg (1982), and Belsley, Kuh, and Welsch (1980). The latter book has an especially penetrating analysis of collinearity.

7.2.1 Design Diagnostics

The hat matrix (projection matrix) is defined in terms of the design matrix by $\boldsymbol{H} = \boldsymbol{X}(\boldsymbol{X}^T\boldsymbol{X})^{-1}\boldsymbol{X}^T$, and the fitted values $\hat{\boldsymbol{Y}}$ depend on the observations $\boldsymbol{Y}$ only through $\boldsymbol{H} = [h_{ij}]$, since $\hat{\boldsymbol{Y}} = \boldsymbol{X}\hat{\boldsymbol{\beta}} = \boldsymbol{H}\boldsymbol{Y}$ (see Section 7.1.1). Now by definition

$$\hat{Y}_i = \sum_{j=1}^{n} h_{ij} Y_j$$

so the fitted value $\hat{Y}_i$ at $\boldsymbol{x}_i^T$ changes at the rate h_{ij} with respect to a change in the jth observed value Y_j. The jth column of the hat matrix gives the entire vector of these rates of response $\boldsymbol{H}_j = [h_{1j} \ldots h_{nj}]^T$ of the fitted values to a change in the jth observed value. It will be called the jth *leverage vector*. The length squared $\|\boldsymbol{H}_j\|^2$ of this vector gives a measure of the overall leverage applied by the jth observation on the fitted values, and will simply be called the jth *leverage*. One of the properties (see the list below) of the hat matrix is that the length squared of the jth column is equal to the jth diagonal element h_{jj}, so that the leverage of the jth observation is simply h_{jj}. We emphasize that the leverage vectors depend only on the design matrix $\boldsymbol{X}$ and not at all on the observations $\boldsymbol{Y}$. Furthermore an error in an observation at a design point with a large leverage is likely to badly affect the fit at many of the other design points.

Example 1: Growth of Prices (Continued). The hat matrix and the leverages are given in Table 7.2. We see that $h_{17} = -.18$, so (not surprisingly) a positive error in observation at x_7 can have a large negative effect on the fit at x_1. (The reader may find it helpful to look again at Figure 7.1.) Note also that the leverages are increasing as x_i moves away from $\overline{x}$. Since the average of the leverages is about .28, only the outermost points have much greater leverage than the average, and hence the potential to greatly influence the fit. This potential does not appear to be realized in this example since the corresponding y-values appear to be consistent with the bulk of the data; see Figure 7.1. However, the observation at x_7 has a much larger effect than meets the eye, as will be seen in the Section 7.2.4. □

Table 7.2 Hat Matrix and Leverages for the Growth of Prices Example

$$
H = \begin{bmatrix}
.46 & .36 & .25 & .14 & .04 & -.07 & -.18 \\
.36 & .29 & .21 & .14 & .07 & .00 & -.07 \\
.25 & .21 & .18 & .14 & .11 & .07 & .04 \\
.14 & .14 & .14 & .14 & .14 & .14 & .14 \\
.04 & .07 & .11 & .14 & .18 & .21 & .25 \\
-.07 & .00 & .07 & .14 & .21 & .29 & .36 \\
-.18 & -.07 & .04 & .14 & .25 & .36 & .46
\end{bmatrix}
$$

x_i	40	41	42	43	44	45	46
h_{ii}	.46	.29	.18	.14	.18	.29	.46

7.2.2 The Basic Properties of H

1. The hat matrix H is an (orthogonal) *projection* matrix with domain R^n and range spanned by the columns of X. All such projection matrices are *symmetric* and *idempotent*, i.e., $H = H^T$ and $H^2 = H$, as the reader may readily verify.[†]
2. It follows from the properties in 1 that the length squared of the ith leverage vector H_i, that is the leverage of x_i^T, is equal to h_{ii}.
3. The diagonal elements of H lie in the interval [0,1]. This also follows from the properties given in 1: $h_{ii} = h_{ii}^2 + \sum_{j \neq i} h_{ij}^2 \geq h_{ii}^2$.
4. The trace of H is $\text{tr}(H) = \sum h_{ii} = p + 1$, which implies that the *average leverage* is $(1/n \sum_{i=1}^n h_{ii} = (p+1)/n$. To verify this result, show that if the products AB and BA are defined, $\text{tr}(AB) = \text{tr}(BA)$; then apply this result to the matrices $A = X(X^TX)^{-1}$ and $B = X^T$.

Now let us see what the hat matrix tells us about the relationship between residuals, observations, and fitted values. The residual vector $r = Y - \hat{Y}$ satisfies

$$r = (I - H)Y = (I - H)\epsilon \tag{7.2.1}$$

In particular

$$r_i = (1 - h_{ii})\epsilon_i - \sum_{j \neq i} h_{ij}\epsilon_j.$$

This identity shows that if $h_{ii} \approx 1$, an unexpected error in Y_i (large error ϵ_i) might not be reflected in r_i, even though it might affect some other

[†]It can be shown that an orthogonal projection matrix is uniquely determined by these two properties in that there is a unique symmetric, idempotent linear transformation from R^n onto a given vector subspace.

residual. Thus the magnitude of a residual r_i is not necessarily related to an error in the observation Y_i; it may be due to an error in another observation with a large value of h_{ij}.

Another point follows from the fact that

$$\text{Cov}[\hat{\boldsymbol{Y}}] = \sigma^2 \boldsymbol{H} \qquad \text{and} \qquad \text{Cov}[\boldsymbol{r}] = \sigma^2(\boldsymbol{I} - \boldsymbol{H}). \tag{7.2.2}$$

Comparing the corresponding diagonal terms ($\text{Var}[\hat{Y}_i]$ with $\text{Var}[r_i]$), we see that a point with a large leverage will produce a fitted value with a large variance and a corresponding residual with a small variance. The relationship is reversed if the leverage is small.

To summarize this subsection, the hat matrix carries much useful information with it concerning the design matrix, especially because it identifies points with high leverage which have the potential to greatly affect the fitted regression model. We will look more closely at the hat matrix in Section 7.3.

7.2.3 Standardized and Studentized Residuals

The variance of each residual is proportional to the unknown parameter σ^2, so some form of standardization is required to judge the extent to which a residual is "outlying." The *standardized residual* is suggested by the equality $\text{Cov}[\boldsymbol{r}] = (\boldsymbol{I} - \boldsymbol{H})\sigma^2$, which in turn follows from the fact that $\boldsymbol{r} = (\boldsymbol{I} - \boldsymbol{H})\boldsymbol{\epsilon}$. Let

$$r_i^* = \frac{Y_i - \hat{Y}_i}{s\sqrt{1 - h_{ii}}} \approx \frac{r_i}{\text{SE}[r_i]}. \tag{7.2.3}$$

It may be shown that $(r_i^*)^2/(n - p - 1)$ has a β distribution, but tables are not so readily available, and it is not easy to determine the significance of an outlier among the residuals using r_i^*. Therefore we consider a different method of standardizing residuals.

Let $(\boldsymbol{X}_{(i)}, \boldsymbol{Y}_{(i)})$ denote the data with the ith case $(\boldsymbol{x}_i^T, Y_i)$ removed, and compute $\hat{\boldsymbol{\beta}}_{(i)} = (\boldsymbol{X}_{(i)}^T \boldsymbol{X}_{(i)})^{-1} \boldsymbol{X}_{(i)}^T \boldsymbol{Y}_{(i)}$. Then we may compare Y_i with the predicted value $\hat{Y}_{(i)} = \boldsymbol{x}_i^T \hat{\boldsymbol{\beta}}_{(i)}$ at $\boldsymbol{x}_i^T$ based on the remaining observations. Define the ith *predicted residual* by $r_{(i)} = Y_i - \hat{Y}_{(i)}$. Figure 7.2 shows the relationship between r_i and $r_{(i)}$ in the special case of three data points in simple linear regression.

The *Studentized* residual will be defined below as the ith predicted residual divided by an appropriate estimate $\widehat{\text{SE}}$ of its standard error:†

$$r_{(i)}^* = \frac{Y_i - \hat{Y}_{(i)}}{\widehat{\text{SE}}[Y_i - \hat{Y}_{(i)}]}. \tag{7.2.4}$$

†By *appropriate* we mean one which leads to a diagnostic statistic with the Student's t-distribution.

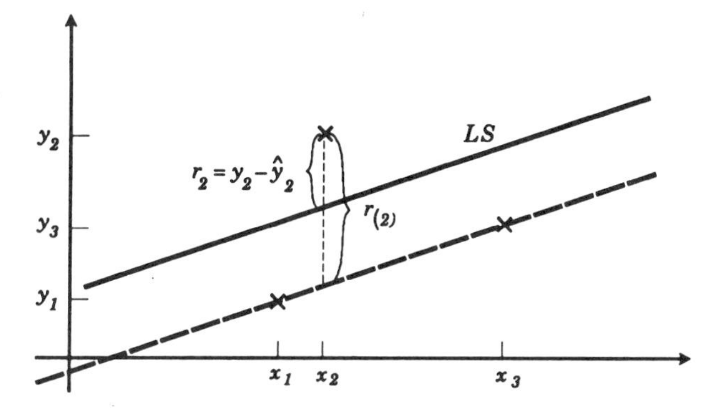

FIGURE 7.2 Example of the ith predicted residual for simple linear regression with three observations.

Leaving details to the problems, we may show with the help of (7.2.8) that under the classical assumptions (7.1.8),

$$r_{(i)} = Y_i - \hat{Y}_{(i)} = \frac{Y_i - \hat{Y}_i}{1 - h_{ii}} = \frac{r_i}{1 - h_{ii}} \qquad (7.2.5)$$

and

$$\mathrm{Var}[Y_i - \hat{Y}_{(i)}] = \frac{\sigma^2}{1 - h_{ii}}.$$

This last result shows that $\mathrm{SE}[Y_i - \hat{Y}_{(i)}] = \sigma/\sqrt{1 - h_{ii}}$. We estimate σ not by s, but by the corresponding $s_{(i)}$ for the deleted sample. This $s_{(i)}$ may be conveniently computed with the relationship

$$(n - p - 2)s_{(i)}^2 = (n - p - 1)s^2 - \frac{r_i^2}{1 - h_{ii}}. \qquad (7.2.6)$$

We therefore define the *Studentized residual* or *t-residual* by

$$r_{(i)}^* = \frac{r_i}{(1 - h_{ii})^{1/2} s_{(i)}}. \qquad (7.2.7)$$

This outlier diagnostic is sensitive to an outlier Y_i in that it is linearly increasing in Y_i, and hence will reveal it. Furthermore, under the assumption of normal errors, $r_{(i)}^* \sim t_{n-p-2}$. To see this, note that the three variables $(Y_i, \hat{Y}_{(i)}, s_{(i)})$ are mutually independent. Hence $r_i = (Y_i - \hat{Y}_{(i)})(1 - h_{ii})$ is independent of $s_{(i)}$ and we may express $r_{(i)}^*$ as the familiar ratio

$$r_{(i)}^* = \frac{r_i/\sigma(1 - h_{ii})^{1/2}}{s_{(i)}/\sigma} \sim \frac{N(0,1)}{\sqrt{\chi_{n-p-2}^2/(n - p - 2)}} \sim t_{n-p-2}.$$

Table 7.3 Residuals, Predicted Residuals, and Their Standardized Counterparts for the Growth of Prices Example

Year x_i	40	41	42	43	44	45	46
Growth of prices y_i	1.62	1.63	1.90	2.64	2.05	2.13	1.94
r_i	−.14	−.21	−.01	.65	−.01	−.01	−.27
$r_{(i)}$	−.26	−.30	−.01	.76	−.01	−.01	−.50
r_i^*	−.57	−.73	−.04	2.10	−.04	−.03	−1.11
$r_{(i)}^*$	−.53	−.69	−.03	5.47	−.04	−.02	−1.15

The standardized residuals r_i^* and the Studentized residual $r_{(i)}^*$ are sometimes referred to as internally and externally Studentized residuals, respectively. We reserve the term Studentized only for $r_{(i)}^*$, which follows the Student's t-distribution. Note that r_i^* and $r_{(i)}^*$ are related by (7.2.6) and (7.2.7).

The Studentized residuals are *not* independent of each other so that rejection of an observation (x_i^T, Y_i) because it has $|r_{(i)}^*| > t_{n-p-2}(1-\alpha/2)$ for some α does not carry a level α when many such dependent comprisons are made. Of course one can compare $\max_i |r_{(i)}^*|$ with $t_{n-p-2}(1-\alpha/2n)$ to obtain at least a level α rejection test for the observation with maximum absolute Studentized residual. We prefer to use $|r_{(i)}^*|$ first as a pointer to possibly misquoted or otherwise bungled data, and then to use robust methods or nonparametric methods if no qualitative reason exists to discard them.

Example 1: Growth of Prices (Continued). In Table 7.3 we list the residuals $r_i = Y_i - \hat{Y}_i$, the predicted residuals $r_{(i)} = Y_i - \hat{Y}_{(i)} = r_i/(1-h_{ii})$, and their standardized versions r_i^* and $r_{(i)}^*$. Note that the last of these quantities identifies very clearly the observation for 1943 as unlikely to come from normal data, if the simple linear regression model is appropriate.
We note in passing that a t-test for significance of any standardized residual r_i^* is not legitimate, even if the errors are normally distributed; on the other hand the Studentized residuals $r_{(i)}^*$ may be so tested subject to the qualifications given in the last paragraph. For example, when $\alpha = .1$, the critical point for a test of the maximum absolute Studentized residual is

$$t_{n-p-2}\left(1-\frac{\alpha}{2n}\right) = t_4\left(1-\frac{1}{140}\right) < t_4(.995) = 4.60.$$

Since 5.47 exceeds this value, it is rejected at level .1 as having come from normal data. □

7.2.4 Combining Design and Residual Effects on Fit—DFITS

We will make use of the following matrix identity due to Gauss (1821). Let A be $p \times p$ symmetric of rank p, and suppose that $\boldsymbol{B}, \boldsymbol{C}$ are $q \times p$ of rank q. Then assuming inverses exist,

$$(A + B^T C)^{-1} = A^{-1} - A^{-1} B^T (I + CA^{-1} B^T)^{-1} CA^{-1}. \tag{7.2.8}$$

This formula will enable us to determine the effect of deletion of a design point on the fit. First write

$$X = \begin{bmatrix} X_{(i)} \\ x_i^T \end{bmatrix}$$

and note that $X_{(i)}^T X_{(i)} = X^T X - x_i x_i^T$, so by (7.2.8)

$$(X_{(i)}^T X_{(i)})^{-1} = (X^T X)^{-1} + \frac{(X^T X)^{-1} x_i x_i^T (X^T X)^{-1}}{1 - h_{ii}}$$

where $h_{ii} = x_i^T (X^T X)^{-1} x_i$. Then after some algebra we find

$$\begin{aligned} \hat{\beta} - \hat{\beta}_{(i)} &= \frac{(X^T X)^{-1} x_i r_i}{1 - h_{ii}} \\ \hat{Y}_i - \hat{Y}_{(i)} &= \frac{h_{ii}}{1 - h_{ii}} r_i. \end{aligned} \tag{7.2.9}$$

This relationship shows clearly the effect which the ith observation (x_i^T, Y_i) has on the fitted value. Kuh and Welsch (1980) standardize this "influence of deletion" by the standard error of $\hat{Y}_i$, namely, $\text{SE}[\hat{Y}_i] = h_{ii}^{1/2}\sigma \approx h_{ii}^{1/2} s_{(i)}$, to obtain a standardized difference of fitted values DFITS_i:

$$\text{DFITS}_i = \frac{\hat{Y}_i - \hat{Y}_{(i)}}{h_{ii}^{1/2} s_{(i)}} = \left[\frac{h_{ii}}{1 - h_{ii}}\right]^{1/2} r_{(i)}^*. \tag{7.2.10}$$

Since the average value of $h_{ii} = (p+1)/n$ and values of $|r_{(i)}^*| > 2$ are "significant" at level .05, it is often suggested that any observation be checked for validity if

$$|\text{DFITS}_i| > 2\left[\frac{(p+1)/n}{1 - (p+1)/n}\right]^{1/2} = 2\sqrt{\frac{p+1}{n-p-1}}. \tag{7.2.11}$$

A simple-to-remember rule of thumb given by Belsley, Kuh, and Welsch (1980) is $2\sqrt{(p+1)/n}$. In our experience these bounds are too optimistic in that we often find questionable points with a value of DFITS less than them. For diagnostic work we suggest replacing 2 by 1.5 in the bound (7.2.11).

Table 7.4 Diagnostics for the Growth of Prices Example: Leverages, Studentized or t-Residuals, and DFITS

x_i	40	41	42	43	44	45	46
h_{ii}	.46	.29	.18	.14	.18	.29	.46
$r^*_{(i)}$	−.53	−.69	−.03	5.47	−.04	−.02	−1.15
DFITS$_i$	−.49	−.43	−.02	2.23	−.02	−.02	−1.07

Example 1: Growth of Prices (Continued). The diagnostic DFITS is calculated using (7.2.10) and listed in Table 7.4. The rule of thumb $2[(p+1)/n]^{1/2}$ is 1.07, so this criterion draws our attention to observations 4 and 7 as worthy of further investigation. Although the 7th observation was not considered a problem by the Studentized residual diagnostic, when this residual is weighted by the leverage through DFITS, it appears to warrant further attention. □

7.2.5 Simple Linear Regression

This example illustrates the relative effects of one datum $(\boldsymbol{x}_i^T, Y_i) = [1\ x\ y]$ on the above three diagnostic measures. Let

$$Y^T = (y_1, y_2, \ldots, y_n, y)$$

$$\boldsymbol{X}^T = \begin{bmatrix} 1 & 1 & \ldots & 1 & 1 \\ x_1 & x_2 & \ldots & x_n & x \end{bmatrix}$$

and assume $\boldsymbol{Y} = \boldsymbol{X}\boldsymbol{\beta} + \boldsymbol{\epsilon}$, where $\boldsymbol{\epsilon} \sim N(0, \sigma^2 \boldsymbol{I}_{n+1})$. For ease of presentation, we denote the leverage $h_{n+1,n+1}$ of x by h.

Then the reader is asked to show in the problems that:

1. Regardless of the value of y, the leverage at x satisfies $1/(n+1) \leq h$ and $h \to 1$ as $x \to \infty$. Hence for large n, h effectively takes on all values from 0 to 1 as x varies from $\bar{x}_n = \sum_1^n x_i/n$ to $+\infty$.
2. The t-residual of (x, y) is

$$r^*_{(n+1)} = \frac{[y - (\hat{\beta}_0 + \hat{\beta}_1 x)]}{s_n}(1-h)^{1/2} \qquad (7.2.12)$$

   where $\hat{\beta}_0$, $\hat{\beta}_1$, and s_n are the classical estimates based on $x_1, \ldots, x_n$. Hence for fixed x, $r^*_{(n+1)}$ takes on all real values as y varies from $-\infty$ to $+\infty$.
3. The observation (x, y) will be tagged for special consideration by DFITS$_{n+1}$ according to (7.2.11) when it lies far from $(\bar{x}_n, \hat{Y})$ based

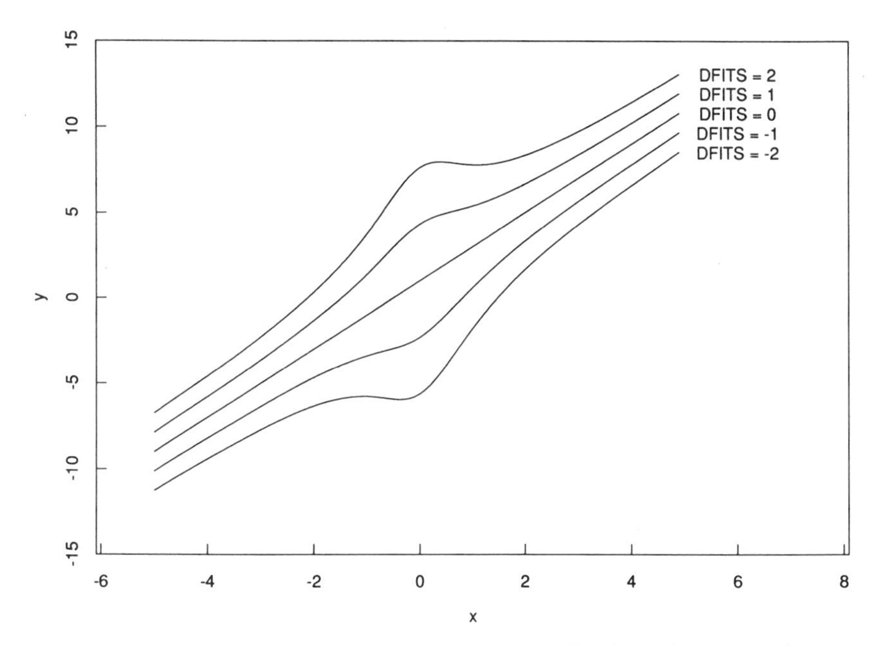

FIGURE 7.3 Curves denoting points of equal DFITS for simple linear regression.

on the first n observations, where "far" means outside boundaries, as shown in Figure 7.3. These boundaries of equal DFITS = d may be obtained from the definition (7.2.10) and (7.2.12) by setting

$$d = \left[\frac{h}{1-h}\right]^{1/2} r^*_{(n+1)} \tag{7.2.13}$$

and solving for y as a function of x. Note that h depends on x and n via the formula given in Problem 7; for $\overline{x} = 0$, it is

$$h = \frac{1}{n+1} + \frac{1}{c_n s^2/x^2 + 1 + 1/n^2}$$

where $c_n = (n^2 - 1)(n + 1)/n^2$.

Example 2: Brownlee's Stack Loss Data. Table 7.5 lists the stack loss data of Brownlee [(1965), Section 13.12]. The data have been subsequently analyzed by numerous statisticians, including Draper and Smith [(1966), pp. 204–215], Andrews (1974), and Daniel and Wood [(1980), Chapter 5].

Table 7.5 Brownlee's Stack Loss Data, Fitted Values, and the Diagnostics t-Residuals, Leverages and DFITS

Case i	Y_i	X_{i1}	X_{i2}	X_{i3}	$\hat{Y}_i$	$r^*_{(i)}$	h_{ii}	DFITS$_i$
1	42	80	27	89	38.77	1.21	0.30	0.79
2	37	80	27	88	38.92	−0.71	0.32	−0.48
3	37	75	25	90	32.44	1.62	0.17	0.74
4	28	62	24	87	22.30	2.05	0.13	0.79
5	18	62	22	87	19.71	−0.53	0.05	−0.12
6	18	62	23	87	21.01	−0.96	0.08	−0.28
7	19	62	24	93	21.39	−0.83	0.22	−0.44
8	20	62	24	93	21.39	−0.47	0.22	−0.25
9	15	58	23	87	18.14	−1.05	0.14	−0.42
10	14	58	18	80	12.73	0.43	0.20	0.21
11	14	58	18	89	11.36	0.88	0.16	0.38
12	13	58	17	88	10.22	0.97	0.22	0.51
13	11	58	18	82	12.43	−0.47	0.16	−0.20
14	12	58	19	93	12.05	−0.02	0.21	−0.01
15	8	50	18	89	5.64	0.80	0.19	0.39
16	7	50	18	86	6.09	0.29	0.13	0.11
17	8	50	19	72	9.52	−0.60	0.41	−0.50
18	8	50	19	79	8.46	−0.15	0.16	−0.07
19	9	50	20	80	9.60	−0.20	0.17	−0.09
20	15	56	20	82	13.59	0.44	0.08	0.13
21	15	70	20	91	22.24	−3.33	0.28	−2.10

We have changed Brownlee's notation to ours in presenting the data and paraphrase his description of the experiment. The data were obtained from 21 days of operation of a plant for the oxidation of ammonia to nitric acid. The nitric acids which are produced are absorbed in a countercurrent absorption tower. The dependent variable Y is 10 times the percentage of the ingoing ammonia that escapes from the absorption tower; the more that escapes, the less efficient the plant. We take $X_0 = 1$. The three regressor variables—which presumably can be controlled—are X_1, the rate of operation of the plant; X_2, the temperature of cooling water which is circulated through coils in the absorption tower; and X_3, the (rescaled) concentration of acid circulating.

The least squares solution $\hat{Y} = X\hat{\beta}$ is

$$\hat{Y} = -39.9X_0 + .716X_1 + 1.3X_2 - .152X_3.$$

The estimated value of the error standard deviation defined by (7.1.7) is $s = 3.243$. The matrix $(X^TX)^{-1}$ is

$$(X^TX)^{-1} = \begin{bmatrix} 13.453 & .027 & -.062 & -.159 \\ & .002 & -.004 & -.001 \\ & & .013 & .000 \\ & & & .002 \end{bmatrix}$$

The classical t-test for significance of regression of any particular variable may now be carried out. For example, the third explanatory variable X_3 does not seem to be an important variable for the model since the estimate of its coefficient $\hat{\beta}_3 = -.152$ has estimated standard error given by the square root of

$$s^2 \times (X^TX)_{33}^{-1} = (3.243)^2 \times .002 = (.16)^2$$

namely, $\text{SE}[\hat{\beta}_3] = .16$. (The t-ratio $= -.152/.16 \approx -1$, which is far from significant at the usual .05 level.) A summary of all estimates and t-ratios follows:

Variable	Coefficient	SE	t
X_0	−39.92	11.90	−3.4
X_1	.716	.135	5.3
X_2	1.295	.368	3.5
X_3	−.152	.156	−.97

Note that the fitted values and the three diagnostics studied in this chapter are listed in Table 7.5. The reader should try to identify troublesome points with (or without) the aid of these diagnostics, and also to speculate on the validity of the classical assumptions upon which the above results are obtained. After an extensive and careful analysis, Daniel and Wood (1980) found four points (1, 3, 4, and 21) to be anomolous and suggested their removal from the analysis. These four points are precisely those which have the largest value of DFITS. They would not be identified by the rule (7.2.11), unless the 2 is replaced by 1.5. In this connection we emphasize that *no* rule can divide the data into "good" and "bad" cases; rather the rule of thumb is a practical guide which is meant to reduce the number of cases that the statistician must examine individually. Note that only points 4 and 21 are identified by the rule that compares the t-residuals ($r^*_{(i)}$) with the usual cut-off value of 2.

For those readers who want to gain some familiarity with robust regression estimates before plunging into the mathematically more demanding Sections 7.3–7.5, we suggest reading Section 7.5.1 and Sections 7.6 and 7.7, wherein the motivation and methods for computing robust estimates are found. Using the (robust weighted least squares) methods which downweights cases with values of DFITS that are greater than $2\sqrt{(p+1)/n}$, the following estimates were found for the stack loss data. We note that only case 21 was downweighted ($w_{21} = .33$), which shows that this particular estimation method would not have considered cases 1, 3, and 4 as worthy of special treatment.

Variable	Coefficient	$\widehat{\text{SE}}$	z
X_0	−42.13	5.75	−7.3
X_1	.817	.148	5.4
X_2	1.016	.403	2.5
X_3	−.126	.0785	−1.6

For the sake of comparison, we also list the results obtained when downweighting observations with values of DFITS that are greater than $\sqrt{(p+1)/n}$. In this case the downweighted points have weights $w_1 = .49$, $w_3 = .59$, $w_4 = .46$ and $w_{21} = .16$.

Variable	Coefficient	$\widehat{\text{SE}}$	z
X_0	−40.93	5.25	−7.8
X_1	.84	.137	6.1
X_2	.80	.276	2.9
X_3	−.105	.067	−1.6

We will return to this example again in the next section. □

7.3* EQUILEVERAGE DESIGNS

In this section we look in more depth at the design matrix. In particular we examine the geometry of the design with an eye toward eventually choosing designs that avoid large leverage points. The reader may find it helpful to review the material on matrix algebra in Appendix A before reading this section. We begin by adding four more properties of the hat matrix to the list started in Section 7.2. Then we will consider a large number of examples of equileverage designs, and explain some methods for generating them. Often there is an elliptical region in the range of the explanatory variables which we particularly want to include in our experiment—this region may be precisely outlined or only vaguely specified in terms of a few trial points. We will illustrate how to prescribe a design for various values of n and p which lies in this region, and which does not include points of high leverage.

7.3.1 More Properties of the Hat Matrix (continued from 7.2.2)

5. The hat matrix $\boldsymbol{H}$ is invariant under nonsingular reparametrizations. That is, for a nonsingular square matrix $\boldsymbol{A}$, let $\boldsymbol{\alpha} = \boldsymbol{A}^{-1}\boldsymbol{\beta}$ and $\boldsymbol{W} = \boldsymbol{X}\boldsymbol{A}$, so $\boldsymbol{Y} = \boldsymbol{X}\boldsymbol{\beta} + \boldsymbol{\epsilon} = \boldsymbol{W}\boldsymbol{\alpha} + \boldsymbol{\epsilon}$. Then $\boldsymbol{H}_W = \boldsymbol{W}(\boldsymbol{W}^T\boldsymbol{W})^{-1}\boldsymbol{W}^T = \boldsymbol{H}$. Belsley, Kuh, and Welsch (1980) note that if $\boldsymbol{X}$ is ill-conditioned in the sense that two columns are roughly parallel, then a reparametrization which leads to more nearly orthogonal columns in $\boldsymbol{W}$ is appropriate *provided* that one is interested in the new parameter $\boldsymbol{\alpha}$. However, if one is mainly interested in $\boldsymbol{\beta}$ and transforms back to $\hat{\boldsymbol{\beta}} = \boldsymbol{A}\hat{\boldsymbol{\alpha}}$, the ill-conditioning problem is reintroduced via $\boldsymbol{A}$. Property 5 shows that $\boldsymbol{H}$ is independent of attempts to resolve these collinearity problems.
6. We may *decrease* the leverage of the ith point by replication. In fact if $\boldsymbol{x}_i^T$ is replicated m times, $h_{ii} \leq 1/m$. For when $\boldsymbol{x}_i^T = \boldsymbol{x}_j^T$, $h_{ij} = h_{ii}$, and hence $h_{ii} = \sum_{j=1}^n h_{ij}^2 \geq m h_{ii}^2$. The practical significance of this result is that the statistician may lessen the undue leverage of one datum without removing it, provided that more observations may be taken in the same region of the design space.
7. By partitioning the independent variables (including the constant term, if any), the hat matrix may be decomposed into the sum of mutually orthogonal hat matrices. To see this, we exploit a decomposition given in Cook and Weisberg [(1982), p. 12]. Partition $\boldsymbol{X} = (\boldsymbol{X}_A, \boldsymbol{X}_B)$, where $\boldsymbol{X}_A$ is $n \times q$ of rank q, with $q < p+1$. Let H, H_A denote the hat matrices corresponding to $\boldsymbol{X}, \boldsymbol{X}_A$, respectively. Then define $\boldsymbol{X}_B^* = \boldsymbol{X}_B - H_A\boldsymbol{X}_B$, which is orthogonal to $\boldsymbol{X}_A$ ($\boldsymbol{X}_A^T\boldsymbol{X}_B^* = 0 = \boldsymbol{X}_B^{*T}\boldsymbol{X}_A$).

   Let $\boldsymbol{H}_B^*$ be the hat matrix corresponding to $\boldsymbol{X}_B^*$. Then $\boldsymbol{H}_A + \boldsymbol{H}_B^*$ is easily seen to be symmetric and idempotent. It has the same range as $\boldsymbol{H}$, since any vector in the range of $\boldsymbol{H}$ can be written as a linear combination of the columns of $\boldsymbol{X}_A$ and $\boldsymbol{X}_B$. By the uniqueness of projection matrices, we have

$$H = H_A + H_B^*. \tag{7.3.1}$$

In particular, take $X_A = X_0 = \mathbf{1}$ and X_B to be the submatrix of explanatory variables. Throughout this section we denote X_B^* by Z. Then Z is the matrix of centered explanatory variables; that is, $Z_{ij} = X_{ij} - \overline{X}_{\cdot j}$, where $\overline{X}_{\cdot j} = (1/n)\sum_{i=1}^{n} X_{ij}$. Letting H_X,H_Z denote the respective hat matrices of X,Z, the decomposition (7.3.1) may be written

$$H_X = \frac{1}{n}\mathbf{1}\mathbf{1}^T + H_Z. \tag{7.3.2}$$

The matrices Z and H_Z will be important in the remainder of this section.

8. We note for further reference that not only does the hat matrix H_Z of centered variables Z satisfy properties 1–7 listed above, but it also has row sums equaling 0. This in turn implies that the row sums of H_X are all equal to 1. We also note in passing that if the rows of Z are given the empirical distribution (which puts mass $1/n$ on each z_i^T), then $(1/n)Z^TZ$ is the sample or empirical covariance matrix associated with the rows of Z.

A geometric interpretation of (7.3.2) will yield further insight. Recall that any linear transformation of R^p onto R^p can be expressed as matrix multiplication by a nonsingular $p \times p$ matrix C. An *eigenvalue* of C is any real λ_j such that there is a nonzero vector U_j (called an *eigenvector* associated with λ_j), such that $CU_j = \lambda_j U_j$. If U_j is an eigenvector associated with λ_j, so is any nonzero multiple of U_j; hence there is an eigenvector of length 1. Hereafter we reserve the notation U_j for this unit eigenvector. Under certain conditions, the set of unit eigenvectors associated with the eigenvalues of C forms an orthonormal basis for R^p. Then the properties of C are often more simply expressed in terms of these eigenvalues and eigenvectors, because the linear transformation is represented by a diagonal matrix with respect to the basis of these eigenvectors.

By Theorem A.5 in the Appendix, any symmetric matrix may be diagonalized. That is, if the matrix C is $p \times p$ symmetric, there exists an orthogonal matrix $U(UU^T = I)$ such that $U^TCU = \Lambda$, where $\Lambda = [\lambda_j]$ is a diagonal matrix. Since $UU^T = I_p$, we have $CU = U\Lambda$. Writing U_j for the jth column of U, we have $CU_j = \lambda_j U_j$, so U_j is an eigenvector of unit length corresponding to the jth eigenvalue λ_j of C. Rewriting the diagonalization $U^TCU = \Lambda$ as $C = U\Lambda U^T$ and substituting $C = Z^TZ$, we find

$$(H_Z)_{ii} = z_i^T(Z^TZ)^{-1}z_i = (z_i^TU)\Lambda^{-1}(U^Tz_i) = \sum_{k=1}^{p} \frac{(U_k^Tz_i)^2}{\lambda_k} \tag{7.3.3}$$

where z_i^T is the ith row of $\mathbf{Z}$.

Now the cosine of the angle between $\boldsymbol{U}_k$ and z_i^T is the ratio of their inner product to the product of their lengths:

$$\cos[\theta_{ki}] = \frac{\boldsymbol{U}_k^T z_i}{\|\boldsymbol{U}_k\|\,\|z_i\|} \tag{7.3.4}$$

so by (7.3.2) we have

$$h_{ii} = \frac{1}{n} + z_i^T z_i \sum_{k=1}^{p} \frac{\cos^2(\theta_{ki})}{\lambda_k}. \tag{7.3.5}$$

From (7.3.5) we can see that the ith point $\boldsymbol{x}_i^T$ will have large leverage if z_i^T has a relatively large length, and if $\cos^2(\theta_{ki})$ is large while simultaneously λ_k is small for some k. Now $\cos^2(\theta_{ki})$ is large when z_i^T is roughly parallel to $\boldsymbol{U}_k$, so the ith point has large leverage if it has a large component along an eigenvector with small associated eigenvalue.

Now for a fixed centered design matrix $\mathbf{Z}$ and an arbitrary vector z in R^p, define the quadratic form

$$q_Z(z) = z^T(\mathbf{Z}^T\mathbf{Z})^{-1}z. \tag{7.3.6}$$

Then by definition

$$\mathcal{E}_Z(k) = \{z : q_Z(z) = k\} \tag{7.3.7}$$

is an ellipsoid with axes determined by the eigenstructure of $C = \mathbf{Z}^T\mathbf{Z}$, and the constant k: the axes of the ellipsoid are along the axes determined by the eigenvectors, and the lengths of the ellipsoidal axes are proportional to the square roots of the eigenvalues.

It follows from (7.3.2) that for a design point $\boldsymbol{x}_i^T$ with corresponding z_i^T (row of $\mathbf{Z}$), we have

$$q_Z(z_i) = (H_Z)_{ii} = (H_X)_{ii} - \frac{1}{n} = h_{ii} - \frac{1}{n}$$

so points of equal leverage will have centered explanatory variables which lie on the same ellipsoid. In particular any points whose centered explanatory variables lie on the ellipsoid determined by $k = p/n$ will have the average leverage $(p+1)/n$.

7.3.2 Data Examples (Continued)

Example 1: Growth of Prices (Continued). Recall that in this example of simple linear regression $p = 1$ and

$$\boldsymbol{X}^T = \begin{bmatrix} 1 & 1 & \ldots & 1 \\ 40 & 41 & \ldots & 46 \end{bmatrix}$$

so that $Z^T = [-3 \ -2 \ -1 \ 0 \ 1 \ 2 \ 3]$ and $Z^TZ = [28]$. The ellipsoids defined via (7.3.7) are simply the boundaries of the intervals centered at 0. Design points whose corresponding centered versions which lie on the ellipsoid $\mathcal{E}_Z(\frac{1}{7}) = \{z : z^2/28 = \frac{1}{7}\} = \{-2, +2\}$ would have the average leverage of $\frac{2}{7}$. As it turns out, points $x_2 = 41$ and $x_6 = 45$ have this average leverage, and their corresponding Z-values do lie on the ellipsoid $\mathcal{E}_Z(\frac{1}{7})$. Points within this ellipsoid of course have smaller leverage, while those outside it have larger leverage. While the geometric interpretation is unnecessary in this example because of the low dimensionality of the design space, in higher dimensions it will become a useful aide. □

Example 2: Stack Loss Data (Continued). We drop the third explanatory variable X_3 (acid concentration) from the model and define Z to be the matrix of centered variables which remain (rate of production and water temperature); then

$$Z^TZ = \begin{bmatrix} 1681.14 & 453.14 \\ 453.14 & 199.81 \end{bmatrix}$$

which has eigenvalues $\lambda_1 = 1808.77$ and $\lambda_2 = 72.19$, with corresponding eigenvectors

$$U = [U_1 U_2] = \begin{bmatrix} .963 & -.271 \\ .271 & .963 \end{bmatrix}.$$

Using these results we may illustrate the geometric interpretation of leverage given by (7.3.5). In Figure 7.4 we plot the centered explanatory variables,

$$z_i^T = [X_{i1} - \overline{X}_{\cdot 1}, X_{i2} - \overline{X}_{\cdot 2}], \qquad i = 1, \ldots, 21.$$

We also plot the ellipse of average leverage in Figure 7.4 and list some values of h_{ii} corresponding to points of high leverage; note that these values are different from those given in Table 7.5 since variable X_3 has now been dropped from the model.

Note that while it is easy to spot x_i^T having large leverages h_{ii} when they are parallel to the axis of the eigenvector with the largest eigenvalue—such as cases 1 and 2 which have $h_{11} = .281$—it is not obvious that case 21 has almost the same leverage $h_{21\ 21} = .276$. Cases with the largest leverage values are the $h_{11} = .281 = h_{2\ 2} = .281$; $h_{21\ 21} = .276$; $h_{12\ 12} = .203$; $h_{3\ 3} = .174$; and $h_{19\ 19} = .150$. In examples with more explanatory variables it will be even more difficult to pick out points of high leverage without the aid of a computer. □

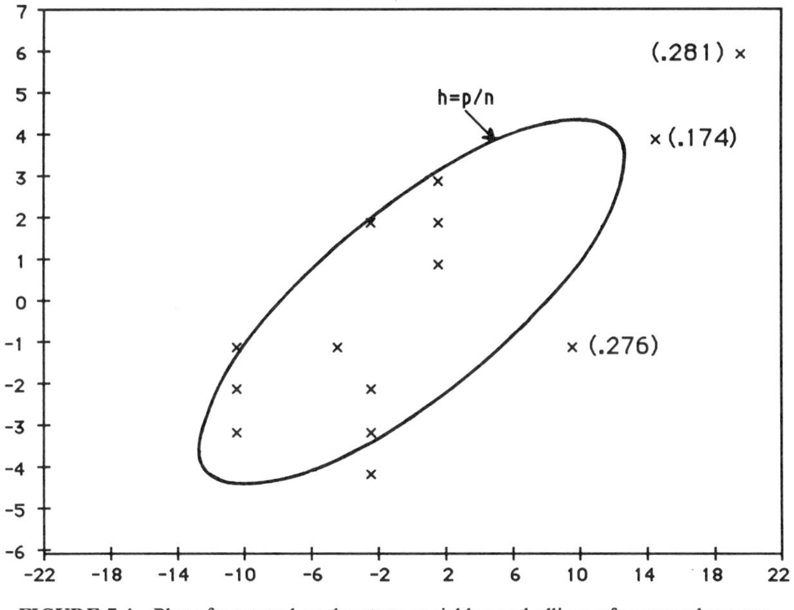

FIGURE 7.4 Plot of centered explanatory variables and ellipse of average leverage.

7.3.3 Properties of Equileverage Designs

We define an *equileverage design* as one having $h_{ii} = (p+1)/n$, $i = 1,\ldots,n$. The advantage of such designs is that they minimize the maximum leverage that any one point may have. The disadvantage is that they may not allow for adequate testing for lack of fit. We will discuss this problem and others of implementation after listing some of the properties of such designs and giving a number of methods for generating them.

It will be convenient and sufficient to restrict our discussion to the $n \times p$ matrix of centered explanatory variables $\mathbf{Z}$, for as pointed out earlier, X will be equileverage if and only if $\mathbf{Z}$ is equileverage. The ellipsoids $\mathcal{E}_Z(k)$ generated by $\mathbf{Z}$ via (7.3.7) may be transformed via nonsingular transformations to the unit sphere, without affecting the leverages. For example, if $\mathbf{Z}$ has the eigenstructure described following property 8 above, let $\mathbf{W} = \mathbf{Z}\mathbf{U}$; then $\mathbf{W}^T\mathbf{W} = \mathbf{\Lambda}$, the diagonal matrix of eigenvalues of $\mathbf{Z}^T\mathbf{Z}$. Hence the ellipse defined by the design $\mathbf{W}$ has its axes along the coordinate axes. Next we make a change of scale in each coordinate so as to obtain a sphere of radius 1. Let $\mathbf{V} = \mathbf{W}(k\mathbf{\Lambda})^{-1/2}$; then $\mathbf{V}^T\mathbf{V} = (1/k)\mathbf{I}_p$, and the corresponding ellipsoid

$$\mathcal{E}_V(k) = \{v^T : v^T(V^TV)^{-1}v = k\}$$

is the desired sphere of radius 1.

An equileverage design $\boldsymbol{Z}$ with n points in $\mathbb{R}^p$ will have all of its points (row vectors) lying on the ellipse $\mathcal{E}_Z(p/n)$. Since the leverages are unaffected by nonsingular transformations, (see property 5 of hat matrices), these points must be transformed to the unit sphere by the above nonsingular transformations.

Conversely given any design on the origin-centered unit sphere, we may apply nonsingular transformations to obtain a design on an arbitrary ellipsoid. Thus without loss of generality we may study equileverage designs on the origin-centered unit sphere.

If a design $\boldsymbol{Z}$ has orthogonal columns, then $\boldsymbol{Z}^T\boldsymbol{Z}$ is a diagonal matrix. If all of these diagonal elements are the same, we will call the design a *scalar design*, since $\boldsymbol{Z}^T\boldsymbol{Z}$ is a scalar multiple of the identity matrix. Scalar designs $\boldsymbol{Z}$ have an especially simple form which makes it possible to combine them easily. The following results may be found in Dollinger and Staudte (1990a).

Theorem 7.1

All designs are assumed to be centered; that is, column sums are zero.

(i) Let $\boldsymbol{Z}$ be an $n \times p$ scalar design. Then $\boldsymbol{Z}$ is an equileverage design if and only if the rows of $\boldsymbol{Z}$ have equal length. If, in addition, $\boldsymbol{Z}$ lies on the *unit* sphere, then $\boldsymbol{Z}^T\boldsymbol{Z} = (n/p)\boldsymbol{I}_p$.

(ii) Conversely, if $\boldsymbol{Z}$ is an equileverage design on an origin-centered sphere, and if this sphere is the only origin-centered ellipsoid in $\mathbb{R}^p$ which passes through all the points (rows) of $\boldsymbol{Z}$, then $\boldsymbol{Z}$ is a scalar design.

(iii) If $\boldsymbol{Z}$ is an $n \times p$ scalar design and $\boldsymbol{U}$ is a $p \times p$ orthogonal matrix, then $\boldsymbol{ZU}$ is a scalar design. Thus scalar designs are preserved under arbitrary rotations and inversions. The result is also true for equileverage designs, since they are invariant under nonsingular transformations (property 5 of hat matrices).

(iv) Let $\boldsymbol{V}$ and $\boldsymbol{W}$ be $m \times p$ and $n \times p$ scalar designs on the same origin centered sphere. Then the combined design $\boldsymbol{Z}^T = [\boldsymbol{V}^T\boldsymbol{W}^T]$ is an $(m+n) \times p$ scalar equileverage design on the same sphere.

(v) Let $\boldsymbol{V}$ and $\boldsymbol{W}$ be $m \times p$ and $n \times q$ equileverage designs. Assume that the leverages p/m and q/n are equal. Then the matrix

$$\boldsymbol{Z} = \begin{bmatrix} \boldsymbol{V} & 0 \\ 0 & \boldsymbol{W} \end{bmatrix}$$

defines an equileverage design of dimension $(m+n) \times (p+q)$. If in addition V and W are scalar equileverage designs on spheres of the same radius, then Z is also a scalar equileverage design.

(vi) Let V and W be $m \times p$ and $n \times q$ equileverage designs on unit spheres. Let $0 < r < 1$ and $s = (1-r^2)^{1/2}$. For each $i = 1, \ldots, m$ and $j = 1, \ldots, n$ define a row of Z by $z^T = [rv_i^T \;\; sw_j^T]$. The rows of Z may be arranged in the following block form:

$$Z = \begin{bmatrix} \begin{pmatrix} rv_1^T & sw_1^T \\ rv_2^T & sw_1^T \\ \vdots & \vdots \\ rv_m^T & sw_1^T \end{pmatrix} \\ \vdots \\ \begin{pmatrix} rv_1^T & sw_n^T \\ rv_2^T & sw_n^T \\ \vdots & \vdots \\ rv_m^T & sw_n^T \end{pmatrix} \end{bmatrix}.$$

Then Z is an equileverage design of dimension $(mn) \times (p+q)$. In addition, if V and W are scalar designs, then Z is a scalar design if and only if $r = [p/(p+q)]^{1/2}$.

(vii) Let W be an $m \times (p-1)$ scalar design with $W^T W = kI_{p-1}$. For each row w_i^T of W define $c_i = pk/m - \|w_i^T\|^2$, and suppose that each $c_i \geq 0$. For each i define two points in R^p by $z^T = [w_i^T, \pm c_i^{1/2}]$. The resulting matrix Z is a $2m \times p$ scalar equileverage design on the sphere of radius $(pk/m)^{1/2}$.

Proof

(i) Let $Z^T Z = kI$. Then the ith leverage is $z_i^T (kI)^{-1} z_i = k^{-1}\|z_i\|^2$. It follows that Z is equileverage if and only if the rows have equal length. Furthermore, if each row has length 1, then the leverages are all $1/k$; and hence, from Section 7.2, $k = n/p$.

(ii) Because Z is equileverage, the points of Z lie on the ellipsoid $\mathcal{E}_Z(p/n)$. By the uniqueness assumption, this ellipsoid coincides with an origin-centered sphere. Hence the quadratic form $q(z) = z^T(Z^T Z)^{-1} z$ on R^p has spherical level contours. This can only occur if $(Z^T Z)^{-1}$ and hence $Z^T Z$ is a scalar matrix.

(iii) The proof is immediate.

(iv) Because $Z^T Z$ is the sum $V^T V + W^T W$ of two scalar matrices, it is also a scalar matrix. The result now follows directly from part (i).

(v) The result follows directly from a calculation of the leverages of Z.

(vi) The design Z is on the origin-centered unit sphere in $\mathbb{R}^{p+q}$, since $r^2 + s^2 = 1$. The (i,k)th element of $Z^T Z$ is $(Z^T Z)_{ik} = Z_i^T Z_k$, where Z_i denotes the ith column of Z. Since the ith column of Z is composed of either n repetitions of a column of V (times a constant r) or n repetitions of a column of W (times s),

$$(Z^T Z)_{ik} = \begin{cases} nr^2 (V^T V)_{ik}, & \text{if } i,k \le m \\ ms^2 (W^T W)_{ik}, & \text{if } i,k > m \\ 0, & \text{if } i \le m < k. \end{cases}$$

The last line of the above display follows from the fact that the column sums of V and W are zero. Hence for all i,

$$\begin{aligned} (H_Z)_{ii} &= \frac{1}{n}(H_V)_{ii} + \frac{1}{m}(H_W)_{ii} \\ &= \frac{p}{nm} + \frac{q}{mn} = \frac{p+q}{mn} \end{aligned}$$

that is, Z is equileverage. To see the result concerning scalar designs, note that if V and W are scalar designs, then from the above expression for $(Z^T Z)_{ik}$ and part (i), $Z^T Z$ is a diagonal matrix with entries

$$(Z^T Z)_{ii} = \begin{cases} nr^2 \left(\dfrac{m}{p}\right) & i \le m \\ ms^2 \left(\dfrac{n}{q}\right) & i > m. \end{cases}$$

The conclusion follows by equating these values.

(vii) By construction the rows of Z all have length $(pk/m)^{1/2}$. So by part (i) it suffices to check that Z is a scalar design. Denote the columns of W and Z by W_j and Z_j, respectively. Then we have

$$\|Z_j\|^2 = 2\|W_j\|^2 = 2k, \qquad j = 1, \ldots, p-1.$$

Also

$$\begin{aligned} \|Z_p\|^2 &= 2\sum_{i=1}^{m} c_i = 2\left[m\left(\frac{pk}{m}\right) - \sum_{i=1}^{m} \|w_i^T\|^2\right] \\ &= 2\left[pk - \sum_{j=1}^{p-1} \|W_j\|^2\right] = 2k. \end{aligned}$$

The orthogonality of the columns of $\boldsymbol{Z}$ follows directly from the orthogonality of the columns of $\boldsymbol{W}$ and from the symmetry of the plus and minus signs in the last column of $\boldsymbol{Z}$. □

Remarks on Theorem 7.1.

One reason for the emphasis on scalar equileverage designs is part (iv) of the theorem. For, as will be seen in the examples below, a combination of two nonscalar equileverage designs need not be equileverage. From the proof of part (iv), we observe that such a combination of scalar designs is scalar.

We call the construction in part (vii) the *inverse projection design* since the design $\boldsymbol{W}$ is projected up and down onto the surface of a sphere from its equatorial hyperplane. It should be noted that $\boldsymbol{W}$ need not be an equileverage design and that the radius of the resulting sphere is uniquely determined by $\boldsymbol{W}$. Examples below illustrate this method.

Although parts (iv)–(vi) of the theorem are stated for combinations of two designs, they easily extend to three or more. Finally we note that in order to apply parts (iii)–(vii) to construct equileverage designs, some elementary building blocks are needed. We supply some of these blocks in the following examples.

7.3.4 Examples of Equileverage Designs

In simple linear regression ($p = 1$) there is a paucity of useful equileverage designs. The only equileverage designs for simple linear regression are those for which the sample size $n = 2m$ is even, and m points are placed on each side of and equidistant from the origin. Such designs do not allow for any test of the hypothesis of linearity, since what happens between these points is unknown.

Examples in $\mathbb{R}^2$ for Two Explanatory Variables

Example 1: Regular Polygons. In $\mathbb{R}^2$ there is a scalar design for every $n \geq p + 1 = 3$, namely, the points of the regular polygon with n vertices which is inscribed in the unit circle. The points may be defined counterclockwise on the unit circle starting say with $(1,0)$:

$$\left[\cos\left(\frac{2\pi j}{n}\right), \sin\left(\frac{2\pi j}{n}\right)\right], \qquad j = 0, \ldots, n-1. \tag{7.3.8}$$

The regular polygons for $n = 3$ and $n = 4$ are shown in Figure 7.5(a) (b). The design with only $n = 3$ points would be inadequate to fit the $p + 1 = 3$ linear parameters as well as the unknown error variance σ^2, but it is

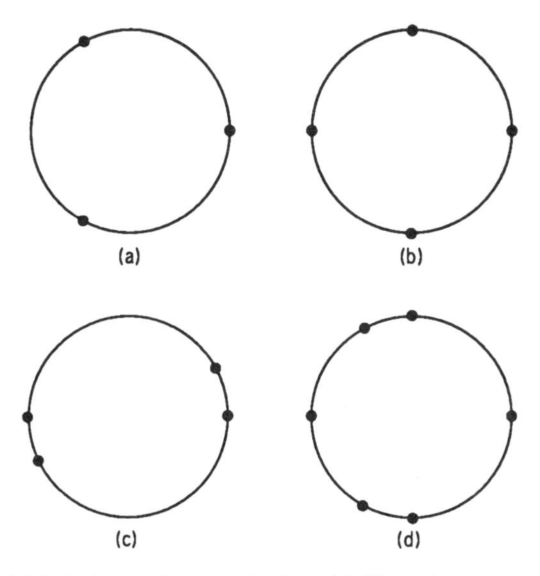

FIGURE 7.5 (a),(b) Scalar equileverage designs. (c) Nonscalar average leverage design. (d) Non-equileverage design.

included as a possible building block which could be combined with other designs, or with itself to obtain replications of the three points. The regular polygon design with $n = 4$ is often referred to as the $2 \times 2 = 2^2$ factorial design, for it allows two factors (variables) to each be tested at two levels.

Figure 7.5(c) shows an average leverage design that is not a scalar design.

The regular polygons may be combined using Theorem 7.1(iv) to obtain other scalar designs, with larger n. For example, the designs of Figure 7.5(a) and (b) could be so combined to obtain a seven-point scalar design. Figure 7.5(d) shows the seven-point design that combines designs (a) and (b) in the manner described in Theorem 7.1(iv). This example shows that a scalar design combined with an equileverage design need not be equileverage.

Example 2: Inverse Projection in $\mathbb{R}^2$. Consider a situation where the first centered explanatory variable $\mathbf{Z}_1$ must be selected at $n = 2m$ symmetrically distributed points denoted by $\pm a_1, \ldots, \pm a_m$, where without loss of generality we assume that $0 < a_1 \leq a_2 \leq \cdots \leq a_m$. The variance of these $n = 2m$ points is $\sigma_a^2 = (1/m)\sum a_i^2$. Suppose that we are free to choose the values $b_1, \ldots, b_n$ of the second explanatory variable and want to obtain an equileverage design on an origin-centered circle. It is clear (see Figure 7.6) that there is really no choice but to take the points $(a_i, \pm b_i)$ above and below

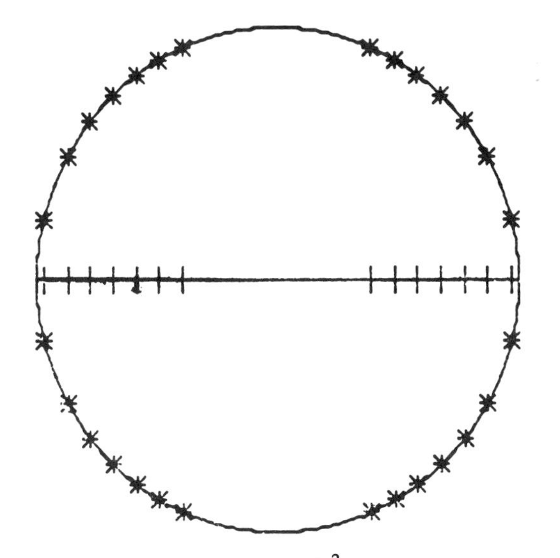

FIGURE 7.6 An inverse projection design in $\mathbb{R}^2$ generated by a scalar design in $\mathbb{R}^1$.

the a_i's, with $b_i = \sqrt{r^2 - a_i^2}$, where r is the radius of a circle. But the choice of r is restricted by the following considerations. First, r must be at least as big as $\max_i |a_i|$. Second, the points cannot be too spread out. For note that the columns of $\mathbf{Z}$ are orthogonal, and hence $\mathbf{Z}^T\mathbf{Z}$ is a diagonal matrix. The diagonal elements are the same if and only if the lengths of the columns are equal; that is, we require $\sigma_a^2 = \sigma_b^2$. This can be arranged only by choosing $r^2 = 2\sigma_a^2$. But then b_i is well-defined only if $\max_i |a_i| \leq \sqrt{2}\sigma_a$. Thus this method will not work for an arbitrary choice of a_i's, but requires that they not include isolated points far from the origin, or too many points very close to the origin.

The above design does not necessarily lie on the *unit* circle, but it may be rescaled by dividing each a_i and each b_i by $\sqrt{2}\sigma_a$. The rescaled design lies on the unit circle and is a scalar design and hence by Theorem 7.1(i) an equileverage design. Since a scalar multiple of an equileverage design is equileverage, so is the original design. It is called an *inverse projection design* because it is the set of all points on the circle whose projection onto the first coordinate axis lies in the generating set $\{\pm a_1, \ldots, \pm a_n\}$.

To illustrate this method consider the design generated by $a_i = i$, $i = 1, \ldots, m$. Then $\sigma_a^2 = (m+1)(2m+1)/6$, and the condition $\max_i |a_i| \leq \sqrt{2}\sigma_a$ is violated. It may seem that the points of largest magnitude should be removed, but doing so again yields a grid centered on the origin. However, if we omit the points $\pm 1, \ldots, \pm j$, where $j < m$ is chosen so that the remaining points satisfy the necessary condition, then the design can be completed. Elementary computations show that $j \approx m/4$ is too small, but $j \approx 2m/5$ is sufficiently large. The design resulting from $j = 3$ and $m = 10$ is generated by the values $a_i = 4, 5, 6, 7, 8, 9, 10$. After symmetrization and rescaling, the design has 28 points on the unit circle which are plotted in Figure 7.6. The design matrix is

$$Z^T = \begin{bmatrix} .389 & .486 & .583 & .680 & .777 & .874 & .971 & \ldots \\ .921 & .874 & .813 & .733 & .629 & .486 & .238 & \ldots \end{bmatrix}. \quad \square$$

7.3.5 Methods for Generating Equileverage Designs in R^p

It is possible to construct a large variety of equileverage (or scalar) designs in R^p with the aid of Theorem 7.1 and some basic designs. Some of these designs were introduced in Section 7.3.4 in the context of two explanatory variables. A few basic scalar designs on the origin-centered unit sphere in R^p are:

1. The 2^p *factorial* designs, which have all the rows of $\boldsymbol{Z}$ of the form $\boldsymbol{z}^T = 1/\sqrt{p}[\pm 1, \ldots, \pm 1]$. Of course, the number of points is very large: $n = 2^p$.
2. The *star* designs, which have only $n = 2p$ points of the form

$$\boldsymbol{z}^T = [\pm 1, 0, \ldots, 0], [0, \pm 1, 0, \ldots, 0], \ldots, [0, 0, \ldots, 0, \pm 1].$$

   The 2^p factorial designs and the star designs are often combined with points at the origin in response surface analysis [see, for example, John (1971), Section 10.5].
3. The *simplicial*† designs, which have only $n = p + 1$ points, and which are defined recursively as follows. In R^1 take $\boldsymbol{Z}_1 = [1 \;\; -1]^T$, the two-point scalar design. Then take $c = \sqrt{1 - \frac{1}{4}}$ and define in R^2 the design

$$\boldsymbol{Z}_2 = \begin{bmatrix} -\frac{1}{2} & \\ -\frac{1}{2} & c\boldsymbol{Z}_1 \\ 1 & 0 \end{bmatrix} = \begin{bmatrix} -\frac{1}{2} & \frac{\sqrt{3}}{2} \\ -\frac{1}{2} & -\frac{\sqrt{3}}{2} \\ 1 & 0 \end{bmatrix}.$$

†The *simplex* in R^{p+1} is the convex hull of the $p + 1$ unit vectors, and lies in a p-dimensional subspace. Its vertices are a translation and scale change of the designs given in 3.

Then Z_2 is a three-point scalar design in $\mathbb{R}^2$. For $p \geq 3$, define

$$Z_p = \begin{bmatrix} -1/p & \\ -1/p & \\ \vdots & c_p Z_{p-1} \\ -1/p & \\ 1 & 0 \quad 0 \quad \dots \quad 0 \end{bmatrix}$$

where $c_p = \sqrt{1-p^{-2}}$. It is straightforward to show that Z_p is a scalar design on the unit sphere.

The Method of Inverse Projection in $\mathbb{R}^p$

The inverse projection method was introduced in the last section for $p = 2$. It extends readily, again subject to certain conditions, to more variables. Consider the case $p = 3$. Suppose that a set of points (vectors) is chosen in the (a,b)-plane so that their sum is the origin and so that the sum of squares of their first coordinates equals the sum of squares of their second coordinates: $\sum_i a_i = 0 = \sum_i b_i$ and $\sum_i a_i^2 = \sum_i b_i^2$. Also assume that $\sum_i a_i b_i = 0$. Then letting σ_a^2 denote the variance of the first coordinates, assume that $a_i^2 + b_i^2 \leq \sqrt{3}\sigma_a$, for all i. (If this condition fails for any i, then the original set of points must be modified.) Next define for each point (a_i, b_i) two points on a sphere (of radius $\sqrt{3}\sigma_a$), namely, $(a_i, b_i, \pm c_i)$, where $c_i = \sqrt{3\sigma_a^2 - a_i^2 - b_i^2}$; see Figure 7.7(b) for an example. The points are then rescaled by dividing each coordinate by $\sqrt{3}\sigma_a$. It is clear that the matrix Z with rows $z_i^T = (a_i, b_i, \pm c_i)$ is a scalar design and hence by Theorem 7.1(i) an equileverage design.

The same argument works in p dimensions, as proved in Theorem 7.1(vii). We summarize the steps required in this *inverse projection method*:

1. Choose an $m \times (p-1)$ matrix W which has column sums equal to zero and orthogonal columns of equal length. Denote the length by $\sqrt{m}\sigma$.
2. Check that the rows w^T of W each satisfy the condition

$$\|w^T\|^2 \leq p\sigma^2.$$

   If this condition is not satisfied, another choice for W must be found.
3. For each row w^T of W define two points in $\mathbb{R}^p : z^T = [w^T \pm c]$, where

$$c = \sqrt{p\sigma^2 - \|w\|^2}.$$

If all the points w^T in R^{p-1} are on a sphere of radius r, and the conditions in 1 are satisfied, then W is a scalar design and $r^2 = (p-1)\sigma^2$, so that condition 2 is satisfied. Thus if an equileverage design of m points can be found in R^{p-1}, an inverse projection design of $n = 2m$ points can be defined in R^p. Generally speaking, the experimenter will want the points in R^p to not be so restricted.

One method of obtaining a greater spread of points is to choose several scalar designs in R^{p-1} on concentric spheres, and to combine them to satisfy conditions 1 and 2 before obtaining the inverse projection 3.

For further examples and discussion related to the literature on optimal designs, see Dollinger and Staudte (1990a). Also see Box and Draper (1975), and Huber (1973a), (1983).

7.3.6 Applications to Experimental Conditions

To illustrate how an equileverage design may be constructed in a problem where explanatory variables are subject to certain constraints, we consider again Brownlee's stack loss problem of Sections 7.2.5 and 7.3.2. Supposing first that we want to choose only the first two explanatory variables in the same range of values given in Table 7.5, but also yielding an equileverage design. The results of a straightforward procedure, explained below, are shown in Table 7.6.

In the original design, points were centered at (60, 21). After recentering at these points (see Figure 7.4) the major and minor axes of the equileverage ellipse were roughly of lengths $13 \approx \sqrt{1808p/n}$ and $2.55 \approx \sqrt{72p/n}$, where $p/n = 2/21$. The angle of rotation (counterclockwise) from the axes was roughly $\tan^{-1}(.27/.96)$. This information is also stored in the eigenvalues and eigenvectors of $\mathbf{Z}^T\mathbf{Z}$ given in Example 2, Section 7.3.2, and restated below.

After obtaining an equileverage design on the unit sphere, the above results can be used to transform the design back to the desired location, by reversing the steps listed above Theorem 7.1. They were translation ($Z_{ij} = X_{ij} - \overline{X}_{.j}$); rotation and/or inversion ($\mathbf{W} = \mathbf{ZU}$); and rescaling ($\mathbf{V} = \mathbf{W}((p/n)\mathbf{\Lambda})^{-1/2}$). Thus given an equileverage design $\mathbf{V}$ on the unit sphere, the transformation $\mathbf{Z} = \mathbf{V}[(p/n)\mathbf{\Lambda}]^{1/2}\mathbf{U}^T$ will carry it onto the ellipsoid $\mathcal{E}_Z(p/n)$, from where it may be translated back to the desired region.

Now we will illustrate this procedure. A 21-point equileverage design on the unit circle is given by the vertices of the regular polygon [see (7.3.8)], and stored in a design matrix $\mathbf{V}$:

$$\mathbf{V}^T = \begin{bmatrix} .96 & .83 & .62 & .37 & \dots \\ .29 & .56 & .78 & .93 & \dots \end{bmatrix}.$$

Table 7.6 Original and Equileverage Designs for Two Explanatory Variables of Stack Loss Data

80	27	78.6	25.6
80	27	76.5	25.9
75	25	73.4	25.7
62	24	69.6	25.1
62	22	65.3	24.2
62	23	61.0	23.0
62	24	57.1	21.6
62	24	53.9	20.1
58	23	51.6	18.7
58	18	50.6	17.6
58	18	50.8	16.7
58	17	52.3	16.2
58	18	54.9	16.2
58	19	58.4	16.5
50	18	62.5	17.3
50	18	66.9	18.4
50	19	71.0	19.7
50	19	74.6	21.2
50	20	77.4	22.6
56	20	79.1	23.9
70	20	79.5	24.9

The appropriate rescaling and rotation is then $\boldsymbol{Z} = \boldsymbol{V}[(p/n)\boldsymbol{\Lambda}]^{1/2}\boldsymbol{U}^T$, where

$$p = 2, \qquad n = 21, \qquad \boldsymbol{\Lambda} = \begin{bmatrix} 1808 & 0 \\ 0 & 72 \end{bmatrix}, \qquad \boldsymbol{U} = \begin{bmatrix} .96 & -.27 \\ .27 & .96 \end{bmatrix}.$$

Finally, the design is translated back to (60, 21) to obtain the equileverage design in Table 7.6.

It appears that this methodology is based on the existence of a data set $\boldsymbol{X}$ which suggests the $\boldsymbol{\Lambda}, \boldsymbol{U}, p, n$; but of course all that is needed is $\boldsymbol{\Lambda}$, $\boldsymbol{U}$, p, and n. These arise precisely by specification of the desired ellipse in the plane or, more informally, by the specification of location, major and minor axes, and rotation. It may be simpler, especially in higher dimensions, to suggest a trial design $\boldsymbol{X}$, which is not necessarily equileverage. Then use a computer package to find $\boldsymbol{\Lambda}$ and $\boldsymbol{U}$, and to construct an equileverage design on the equileverage ellipsoid suggested by the original $\boldsymbol{X}$. This is what we have in fact done above.

Next we suppose that three independent variables X_1, X_2, X_3 may be controlled, and that we would like an equileverage design. Of course there are many possibilities. We demonstrate two of them, and outline a third in the

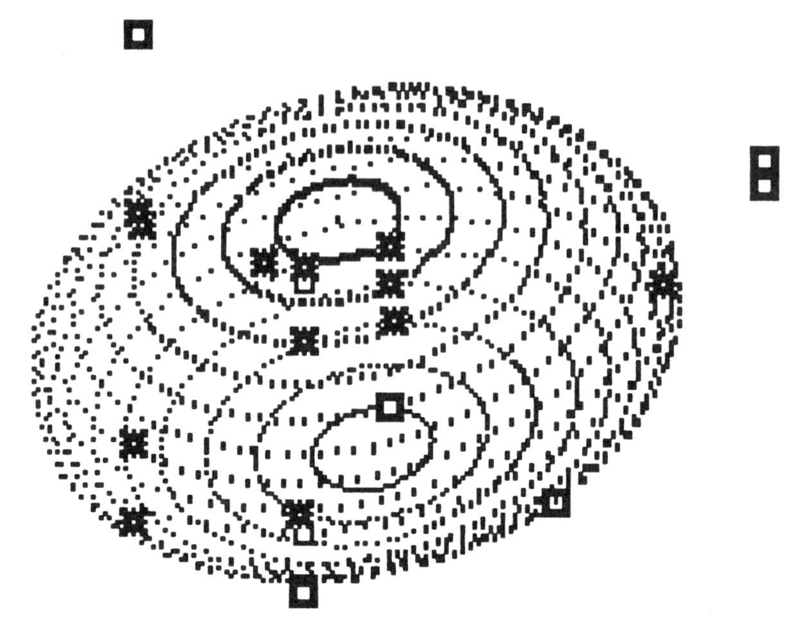

FIGURE 7.7 (a) Stack loss data with sphere of average leverage.

problems. Consider the three stack loss variables of Section 7.2.5; center them on 0 and let the resulting 21×3 matrix be Z. These 21 points and the ellipsoid of equileverage $\mathcal{E}_Z(p/n) = \mathcal{E}_Z(\frac{3}{21})$ are plotted in Figure 7.7(a). The eigenvalues and eigenvectors of $Z^T Z$ are given in

$$\Lambda = \begin{bmatrix} 1991.52 & 0 & 0 \\ 0 & 391.62 & 0 \\ 0 & 0 & 72.09 \end{bmatrix}, \qquad U = \begin{bmatrix} -.906 & .321 & .276 \\ -.254 & .108 & -.961 \\ -.338 & -.941 & -.016 \end{bmatrix}.$$

As discussed earlier in the two-variable case, an equileverage design V on the unit sphere may be transformed to $\mathcal{E}_Z(p/n)$ via the transformation:

$$W = V \left[\frac{p}{n}\Lambda\right]^{1/2} U^T. \tag{7.3.9}$$

The problem is to choose V. The simplest design to construct is the inverse projection (Section 7.3.5) of the vertices of a regular polygon as shown in Figure 7.7(b). This method yields an even number of points, so for $n = 20 = 2m$, define V_1 by the inverse projection method applied to the 10 points defined by (7.3.8). These points are plotted in Figure 7.7(c). After transformation to W_1 by (7.3.9), these points are shown in Figure 7.7(d).

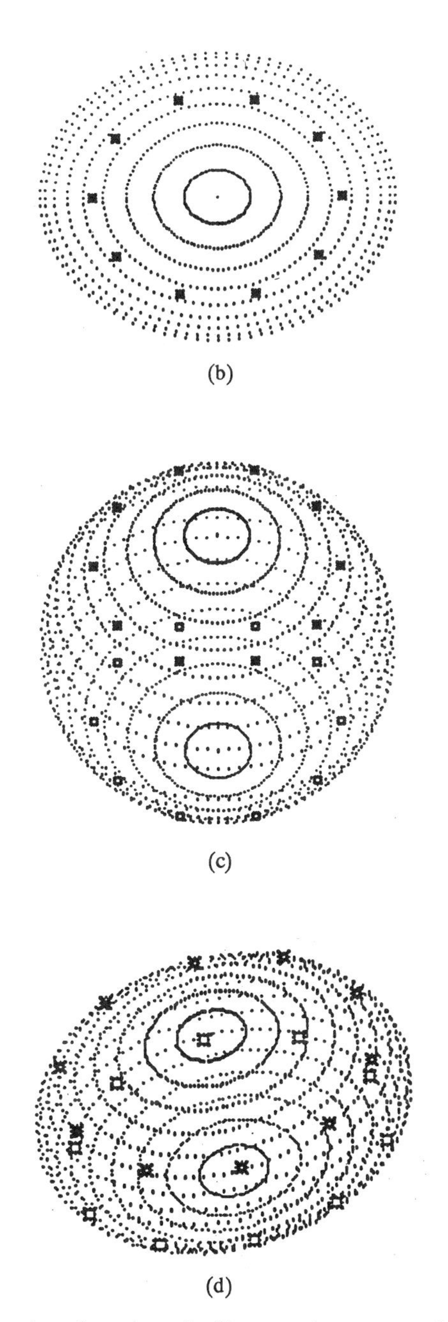

FIGURE 7.7 Vertices of regular polygon in (b) are used to generate design on the sphere in (c), which is then transformed to the ellipsoid of equal leverage in (d).

An alternative equileverage design of 20 points is generated by inverting a regular tetrahedron on the unit sphere about each of its vertices. More precisely let $\boldsymbol{T}$ be the tetrahedron defined by the simplicial design in three dimensions:

$$\boldsymbol{T} = \begin{bmatrix} -\frac{1}{3} & -\frac{\sqrt{2}}{3} & \sqrt{\frac{2}{3}} \\ -\frac{1}{3} & -\frac{\sqrt{2}}{3} & -\sqrt{\frac{2}{3}} \\ -\frac{1}{3} & \frac{2\sqrt{2}}{3} & 0 \\ 1 & 0 & 0 \end{bmatrix}.$$

When this tetrahedron is inverted about each vertex, and the resulting 16 points are joined to it, a symmetric 20-point design V_2 is the result. After the transformation (7.3.9), $\boldsymbol{W}_2$ lies on the ellipsoid shown in Figure 7.7(a).

These examples only illustrate some of the many possibilities for equileverage designs. The main advantage in using them is that the leverage of all points is the same, and the effect of outliers in the observations will not be reinforced. An examination of the Studentized residuals (t-residuals) for such designs will be equivalent to an examination of DFITS. Thus diagnosis is simpler and quicker.

One of the disadvantages of equileverage designs is that they cannot be used to check whether the model is indeed linear. If this is desired, then a compromise design can be created, which includes points both inside and outside the ellipsoid $\mathcal{E}_Z(p/n)$. Provided those points which have the largest leverage are numerous, the resulting design will still prevent an extreme leverage point. One possibility, for example, is to choose designs on concentric spheres in $\mathbb{R}^p$.

7.4* ASYMPTOTIC THEORY FOR LEAST SQUARES

Under the classical assumptions of uncorrelated, mean zero normal errors [(7.1.2) and (7.1.8)], the least squares estimator $\hat{\boldsymbol{\beta}} = (\boldsymbol{X}^T\boldsymbol{X})^{-1}\boldsymbol{X}^T\boldsymbol{Y}$ is a linear transformation of a multivariate normal vector and hence has a multivariate normal distribution itself. Dropping the assumption of normal errors, we seek conditions under which there is *asymptotic* normality:

$$n^{1/2}[\hat{\boldsymbol{\beta}} - \boldsymbol{\beta}] \xrightarrow{d} N_p(\mathbf{0}, \sigma^2\Sigma^{-1}) \quad \text{as} \quad n \to \infty \tag{7.4.1}$$

where Σ is a positive definite matrix. It is often the case that $\Sigma = \lim(1/n)(\boldsymbol{X}^T\boldsymbol{X})$. If (7.4.1) holds, then the classical methods such as formation of confidence ellipsoids for $\boldsymbol{\beta}$ will still be possible, at least with approximate levels for large n. In addition, we want to prove asymptotic normality for weighted least squares, where the weights are prespecified functions of the observations, so that we may obtain asymptotic normality for robust regression methods.

The problem is complicated by the fact that the rows of X may arise as the realizations of observations on a p-dimensional random vector. In this case the values of the explanatory variables are not chosen by design, but arise at random. However, the regression of Y on x is carried out conditionally on the realized values, for each fixed n, so that the same methods derived for the case of fixed explanatory variables are used in this situation.

Despite the similarity of the methodology in the two cases of fixed and random explanatory variables, there are good reasons for treating them separately. First, there are certain cases where the explanatory variables are clearly not the result of any random process, such as the "growth of prices" example in Section 7.1.1. Second, if it is possible that random errors have occurred in the calculation of the explanatory variables, then these errors may be included naturally in a model which assumes that the variables themselves are random, with the same distribution for each row. We will look at this situation in Section 7.4.2.

In Section 7.4.3 we derive the influence function for least squares estimators, and in Section 7.4.4 we prove the asymptotic normality of them using a von Mises expansion. In Section 7.4.5 we briefly discuss some criteria for comparing multivariate estimators.

7.4.1 Fixed Explanatory Variables

To begin with, let us consider simple linear regression and gradually introduce the complicating factors.

Example 1: Simple Linear Regression. Suppose that we may take observations on $y_i = \beta_0 + \beta_1 x_i$, $i = 1, 2, \ldots, n$, for arbitrarily large n. Let $\overline{x}_n = \sum_{i=1}^n x_i/n$, $\overline{x^2}_n = \sum_{i=1}^n x_i^2/n$, $s_{x,n}^2 = \overline{x_n^2} - \overline{x}_n^2$, and $s_{xy,n} = \sum_{i=1}^n x_i y_i/n - \overline{x}_n \overline{y}_n$. Then for each n we have

$$X^T = \begin{bmatrix} 1 & \cdots & 1 \\ x_1 & \cdots & x_n \end{bmatrix}, \qquad \text{and} \qquad \frac{1}{n}(X^T X) = \begin{bmatrix} 1 & \overline{x}_n \\ \overline{x}_n & \overline{x_n^2} \end{bmatrix}.$$

Assuming that each of $\overline{x}_n \to \mu_1$ and $\overline{x_n^2} \to \mu_2$ converge to finite limits and $\mu_2 - \mu_1^2 > 0$, the resulting limit $\Sigma = \lim(1/n)(X^T X)$ exists and is positive definite. We also want conditions under which the least squares estimators $\hat{\beta}_1 = s_{xy}/s_x^2$ and $\hat{\beta}_0 = \overline{y} - \hat{\beta}_1 \overline{x}$ are jointly asymptotically normal. (We have dropped the subscript n for simplicity.)

We will see in Theorem 7.2 below that the least squares estimator will be asymptotically normal provided that the largest diagonal element of the hat matrix (the largest leverage value) tends to zero with increasing n; that is, $\max_i h_{ii} \to 0$. In simple linear regression the leverages are

$$h_{ii} = \frac{1}{n} + \frac{(x_i - \overline{x})^2}{\sum_{i=1}^{n}(x_i - \overline{x})^2}$$

$$\sim \frac{1}{n}\left[1 + \left(\frac{x_i - \overline{x}}{s_x}\right)^2\right]. \qquad (7.4.2)$$

Choosing, for example, each $x_i = i$ yields $\max h_{ii} \sim 4/n$, so the least squares estimators of β_0 and β_1 are jointly asymptotically normal. An example where $\max_i h_{ii}$ does not converge to zero is the design given by $x_i = 2^i$; this yields $\max h_{ii} \sim \frac{3}{4}$. See Problem 22 for details. □

Huber (1973a) states and proves the following basic theorem on the asymptotic normality of least squares estimators [See also Eicker (1963).] This theorem treats the independent variables as part of a sequence which need not have arisen as the realization of a random procedure, and is hence applicable to deterministic sequences such as regular time intervals. The proof uses the Lindeberg central limit theorem for double arrays, and thus we may allow for more flexibility by letting the sample size n_N and the number of explanatory variables p_N to be indexed by $N = 1, 2, \ldots$. The common case of $n_N = N$ is called the *triangular array*. Note that even if there is a constant term in the model, we will use p_N (rather than $p_N + 1$) to represent the dimension of the vector of unknown parameters in this section and the next on asymptotic theory.

Theorem 7.2

For each $N = 1, 2, \ldots$ suppose that there is a regression equation

$$\boldsymbol{Y}_N = \boldsymbol{X}_N \boldsymbol{\beta}_N + \boldsymbol{\epsilon}_N \qquad (7.4.3)$$

where $\boldsymbol{X}_N$ is a design matrix of size $n_N \times p_N$, and the components of $\boldsymbol{\epsilon}_N$ are i.i.d. with mean 0 and variance $\sigma^2 > 0$. Assume further that the hat matrix $\boldsymbol{H}_N = \boldsymbol{X}_N(\boldsymbol{X}_N^T \boldsymbol{X}_N)^{-1}\boldsymbol{X}_N^T = [h_{ij}]$ satisfies:

$$M_N = \max_{1 \le i \le n_N} h_{ii} \to 0 \quad \text{as} \quad n_N \to \infty. \qquad (7.4.4)$$

Then $\hat{\boldsymbol{\beta}}_N$ is asymptotically normal.†

†Huber (1981) explains why the condition $\max_i h_{ii} \to 0$ is also *necessary* for joint asymptotic normality of the least squares estimator.

Proof. To show that the least squares estimator $\hat{\beta}_N$ is asymptotically normal, it suffices by the Cramer-Wold device (Theorem B.12 of the Appendix) to show that every linear combination of its components is asymptotically normal. Let $\boldsymbol{a}_N = [a_1, \ldots, a_{p_N}]^T$ be any vector of length 1, and consider $\hat{\alpha}_N = \boldsymbol{a}_N^T \hat{\boldsymbol{\beta}}_N$. Now the property of asymptotic normality of a seqence of vectors does not depend on which basis they are expressed in. For each N we may therefore without loss of generality choose any basis in R^{p_N}, so that $\boldsymbol{X}_N^T \boldsymbol{X}_N = \boldsymbol{D}_N^{-1}$, where $\boldsymbol{D}_N^{-1}$ is a diagonal matrix. Hence $\hat{\boldsymbol{\beta}}_N = \boldsymbol{D}_N \boldsymbol{X}_N^T \boldsymbol{Y}_N$ and $\hat{\alpha}_N = \boldsymbol{a}_N^T \boldsymbol{D}_N \boldsymbol{X}_N^T \boldsymbol{Y}_N$. Letting $\boldsymbol{t}_N = \boldsymbol{X}_N \boldsymbol{D}_N \boldsymbol{a}_N$, we may reexpress $\hat{\alpha}_N$ as a linear combination of the observations $\boldsymbol{Y}_N$:

$$\hat{\alpha}_N = \boldsymbol{t}_N^T \boldsymbol{Y}_N = \sum_{i=1}^{n_N} t_{Ni} Y_{Ni}.$$

To apply the Lindeberg version of the central limit theorem (see Theorem B.7 of the Appendix) to $\hat{\alpha}_N$, we need to consider the double array:

$$\begin{pmatrix} t_{11}Y_{11} & t_{12}Y_{12} & \cdots & t_{1n_1}Y_{1n_1} & & \\ \vdots & \vdots & & & \ddots & \\ t_{N1}Y_{N1} & t_{N2}Y_{N2} & \cdots & \cdots & \cdots & t_{Nn_N}Y_{Nn_N} \\ \vdots & \vdots & & & & \ddots \end{pmatrix}. \tag{7.4.5}$$

The entries in each row are independent, the Nth row sum is $\hat{\alpha}_N$, and the number of terms in the Nth row is $n_N \to \infty$. This last condition follows from (7.4.4) and the fact that $p_N/n_N = (1/n)\sum h_{ii} \leq \max_{1 \leq i \leq n_N} h_{ii} = M_N$.

Hereafter we drop the subscript N. In the proof we will need the following bound on the size of the t_i's, which is an application of the Cauchy-Schwarz inequality (Theorem A.10 of the Appendix.):

$$t_i^2 = \left(\sum_{j=1}^{p} a_j d_j x_{ij} \right)^2 \leq \sum_{j=1}^{p} a_j^2 d_j \sum_{j=1}^{p} x_{ij}^2 d_j = \|\boldsymbol{t}\|^2 h_{ii}. \tag{7.4.6}$$

Now define

$$\mu_i = \mathrm{E}[t_i Y_i] = t_i [\boldsymbol{X}\beta_i]$$

and

$$B^2 = \sum_{i=1}^{n} \mathrm{Var}[t_i Y_i] = \sum_{i=1}^{n} t_i^2 \sigma^2 = \sigma^2 \|\boldsymbol{t}\|^2.$$

Then the Lindeberg condition requires that for every $\delta > 0$, the truncated variances of the Nth row converge to 0:

$$\frac{1}{B^2}\sum_{i=1}^{n} \mathrm{E}[(t_iY_i - \mu_i)^2 I\{|t_iY_i - \mu_i| > \delta B\}] \to 0 \qquad \text{as} \quad N \to \infty.$$

Now $B^2 = \sum_{i=1}^{n} t_i^2\sigma^2 = \|t\|^2\sigma^2$, so with the aid of (7.4.6) we may place a bound on the truncated variances:

$$\frac{1}{\sigma^2\|t\|^2}\sum_{i=1}^{n} t_i^2 \mathrm{E}[\epsilon_i^2 I\{|t_i\epsilon_i| > \delta\sigma\|t\|\}] \le \frac{1}{\sigma^2\|t\|^2}\sum_{i=1}^{n} t_i^2 \mathrm{E}\left[\epsilon^2 I\left\{|\epsilon| > \frac{\delta\sigma\|t\|}{\sqrt{M}\|t\|}\right\}\right]$$

$$= \frac{1}{\sigma^2}\mathrm{E}\left[\epsilon^2 I\left\{|\epsilon| > \frac{\delta\sigma}{\sqrt{M}}\right\}\right] \to 0$$

as $N \to \infty$ since $M \to 0$. Thus the row sums of (7.4.5) are asymptotically normal, and $\hat{\beta}$ is asymptotically multivariate normal. □

7.4.2 Random Explanatory Variables

Assume that the rows of X are i.i.d. observations from a p-dimensional distribution F_x. Let F be the *joint* distribution for the $(p+1)$-dimensional vector $[x^T y]$, where x^T is a random row of X and y is the associated dependent variable defined by $y = x^T\beta + \epsilon$. Previously we have used capital Y to indicate that the dependent variable (unlike its independent counterpart) was a random variable. However, now both are random, and we choose to use lowercase y for both. It is useful to think of F as determined by the marginal distribution of x^T, and the conditional distribution $F_{y|x}$ of y given x^T. The latter is in turn determined by the distribution of ϵ, say F_ϵ, and the definition

$$y - x^T\beta = \epsilon \sim F_\epsilon.$$

The *regression problem* is characterized by the assumption that $y - x^T\beta$ and x^T are *independent*. The reason for allowing x^T to be random is that we may then model chance errors in the X matrix. Moreover, we will have the option of interpreting results unconditionally or, as usual, conditionally on $x_1^T, \ldots, x_n^T$.

Now let F_n denote the empirical distribution which puts mass $1/n$ on each of the n rows $[x_i^T, y_i]$, $i = 1, \ldots, n$, of the matrix $[X \mid Y]$. We want to express the estimator $\hat{\beta} = T(F_n)$, where $T(F)$ is a vector-valued functional on an appropriate class of distributions, and F_n is an empirical distribution in this class. To this end we use the approach of Hinkley (1977) and define the vector-valued functional $\gamma(F) = E_F[y\boldsymbol{x}]$. It maps the class of all distributions on R^{p+1} for which the vector of expectations exists, into R^p. When evaluated at F_n,

$$\gamma(F_n) = \frac{1}{n}\sum_{i=1}^{n} y_i \boldsymbol{x}_i = \frac{1}{n}\boldsymbol{X}^T\boldsymbol{Y}. \tag{7.4.7}$$

In a similar way one may define $\boldsymbol{\Sigma}(F) = E_F[\boldsymbol{x}\boldsymbol{x}^T]$ and show that $\boldsymbol{\Sigma}(F_n)$ satisfies

$$\boldsymbol{\Sigma}(F_n) = \frac{1}{n}\sum_{i=1}^{n} \boldsymbol{x}_i\boldsymbol{x}_i^T = \frac{1}{n}\boldsymbol{X}^T\boldsymbol{X}. \tag{7.4.8}$$

Hence the *least squares functional* $\boldsymbol{T}(F) = \boldsymbol{\Sigma}^{-1}(F)\gamma(F)$ yields

$$\boldsymbol{T}(F_n) = (\boldsymbol{X}^T\boldsymbol{X})^{-1}\boldsymbol{X}^T\boldsymbol{Y}. \tag{7.4.9}$$

Example 1: Simple Linear Regression (Continued from Section 7.4.1). Assume that the "design" matrix is determined by a random process. For example, let $x_i \sim \text{Poisson}(\lambda)$. Such a situation arises frequently in applications: the x_i's may be the number of traffic accident claims during successive months and y_i is the associated total monthly insurance payouts; or the x_i's may be the number of organisms in different samples from the same solution, while y_i is the mass associated with these x_i organisms. Note that

$$\epsilon_i = y_i - \boldsymbol{x}_i^T\boldsymbol{\beta} = y_i - [1 \quad x_i]\begin{bmatrix}\beta_0\\ \beta_1\end{bmatrix} = y_i - \beta_0 - \beta_1 x_i \sim F_\epsilon$$

where we assume F_ϵ is symmetric about 0 with unknown variance $\sigma^2 > 0$. Thus

$$\begin{aligned}\gamma(F) &= E_F[y\boldsymbol{x}]\\ &= E_F\begin{bmatrix} y\\ yx\end{bmatrix} = \begin{bmatrix}\beta_0 + \beta_1\lambda\\ (\beta_0+\beta_1)\lambda + \beta_1\lambda^2\end{bmatrix}.\end{aligned}$$

Also

$$\boldsymbol{\Sigma}(F) = E_F[\boldsymbol{x} \quad \boldsymbol{x}^T] = \begin{bmatrix}1 & \lambda\\ \lambda & \lambda+\lambda^2\end{bmatrix}.$$

Hence

$$\boldsymbol{T}(F) = \boldsymbol{\Sigma}^{-1}(F)\gamma(F) = \begin{bmatrix}\beta_0\\ \beta_1\end{bmatrix}.$$

Now if $x_1, \ldots, x_n$ are the realized values of n independent observations on the Poisson(λ) distribution, and $\epsilon_1, \ldots, \epsilon_n$ are the corresponding errors in observing $y_1, \ldots, y_n$, then the empirical distribution F_n puts mass $1/n$ on each point $(1 \;\; x_i \;\; y_i)$. It follows that

$$T(F_n) = \begin{bmatrix} \hat{\beta}_0 \\ \hat{\beta}_1 \end{bmatrix}$$

where $\hat{\beta}_1 = s_{xy}/s_x^2$ and $\hat{\beta}_0 = \overline{y} - \hat{\beta}_1 \overline{x}$. These least squares estimates need not be normally distributed because F_ϵ need not be. However, for large n we may treat them as though they were normally distributed. This follows from the Corollary to Theorem 7.3. It also follows from Theorem 7.2 for fixed x proved in the last section. For the hat matrix has diagonal elements (see Example 1 in Section 7.2.3)

$$h_{ii} = \frac{1}{n} + \frac{(x_i - \overline{x}_n)^2}{\sum_{i=1}^n (x_i - \overline{x}_n)^2} \sim \left[1 + \frac{(x_i - \lambda)^2}{\lambda}\right] \frac{1}{n}$$

so $\max_i h_{ii} \to 0$ almost surely. Hence for almost all random sequences condition (7.4.4) holds, and asymptotic normality of the least squares estimator follows. □

7.4.3 The Influence Function for Least Squares

In this section we derive the influence function for the least squares functional $T(F) = \Sigma^{-1}(F)\gamma(F)$. This function will reveal how outliers in the dependent and independent variables may combine to affect the least squares estimates of β. It will also be essential to our proof of asymptotic normality given in the next section. By definition, the influence function gives for each $[x^T y]$ the directional derivative of T at F in the direction of $\Delta - F$, where $\Delta = \Delta_{[x^T y]}$. It is the limit as $t \to 0$ of $[T((1-t)F + t\Delta) - T(F)]/t = [T(F + t(\Delta - F)) - T(F)]/t$. We will in fact find more generally the Gateaux derivative (see Appendix B) of T at F in the direction of H, express this derivative as an integral with respect to H, and obtain the influence curve from the kernel of this derivative. Let $a(t) = T(F + tH)$, where H is possibly the difference of two distributions. We will show that for any such H, $a'(0)$ may be written as an integral with respect to H of some kernel function.

For convenience we express the inverse $(\Sigma_F)^{-1}$ of Σ_F by Σ_F^{-1}. Using the formula for differentiation of the inverse of a matrix given in Appendix A.1,

$$\begin{aligned} \frac{d}{dt}\left[(\Sigma_F + t\Sigma_H)^{-1}\right]_{t=0} &= -\Sigma_F^{-1} \frac{d}{dt}[(\Sigma_F + t\Sigma_H)]_{t=0} \Sigma_F^{-1} \\ &= -\Sigma_F^{-1}\Sigma_H\Sigma_F^{-1}. \end{aligned} \tag{7.4.10}$$

Hence

$$T(F+tH) = \Sigma_{F+tH}^{-1}(\gamma_{F+tH})$$
$$= (\Sigma_F + t\Sigma_H)^{-1}(\gamma_F + t\gamma_H)$$

has derivative at $t = 0$

$$\frac{d}{dt}[T(F+tH)]_{t=0}$$
$$= \left\{\left[\frac{d}{dt}(\Sigma_F + t\Sigma_H)^{-1}\right](\gamma_F + t\gamma_H) + (\Sigma_F + t\Sigma_H)^{-1}\gamma_H\right\}_{t=0}$$
$$= \Sigma_F^{-1}\gamma_H - \Sigma_F^{-1}\Sigma_H\Sigma_F^{-1}\gamma_F$$
$$= \Sigma_F^{-1}(\gamma_H - \Sigma_H T(F)). \qquad (7.4.11)$$

Now the left-hand side of (7.4.11) is $a'(0)$, which is better described by the notation $T_F'(H)$. The equality in (7.4.11) may then be reexpressed,

$$T_F'(H) = \Sigma_F^{-1}\mathrm{E}_H[xy - xx^T T(F)] = \mathrm{E}_H[IF(x^T, y; F)]$$

where

$$IF(x^T, y; F) = \Sigma_F^{-1}[x(y - x^T T(F))]. \qquad (7.4.12)$$

The function IF is the kernel of our linear approximation to $T(F)$ and is hence the *influence function* of $T(F)$ in the direction $[x^T, y]$. Note that this (total) influence factors into an influence of position $\Sigma_F^{-1}x$ and the influence of residual $(y - x^T T(F))$. Also note that $\mathrm{E}_F[IF(x, y; F)] = \mathbf{0}$ if the error at x is independent of x, which is an assumption of the regression formulation.

7.4.4 Asymptotic Normality via the von Mises Expansion

The idea is to expand the vector-valued function $a(t) = T(F + tG)$ in a Taylor series about 0:

$$a(t) = a(0) + ta'(0) + \cdots.$$

Then taking $t = n^{-1/2}$ and $G = n^{1/2}(F_n - F)$ in the expansion for $a(t) = T(F + tG)$, we obtain the von Mises expansion

$$\hat{\beta} = T(F_n) = T(F) + \frac{1}{n}\sum_{i=1}^{n} IF(x_i^T, y_i; F) + R_n$$

or

$$n^{1/2}[\boldsymbol{T}(F_n) - \boldsymbol{T}(F)] = n^{-1/2}\sum_{i=1}^{n} \boldsymbol{IF}(\boldsymbol{x}_i^T, y_i; F) + n^{1/2}\boldsymbol{R}_n \qquad (7.4.13)$$

where $\boldsymbol{R}_n$ represents the error in the one-term expansion. Assuming in (7.4.13) that $\|n^{1/2}\boldsymbol{R}_n\| \xrightarrow{p} 0$, we have by the standard central limit theorem

$$n^{1/2}[\boldsymbol{T}(F_n) - \boldsymbol{T}(F)] \xrightarrow{d} N_p(\boldsymbol{0}, \text{Cov}(F)) \qquad (7.4.14)$$

where

$$\begin{aligned}\text{Cov}(F) &= \text{E}_F[\boldsymbol{IF}\ \boldsymbol{IF}^T] \\ &= \Sigma_F^{-1}\text{E}_F[\boldsymbol{x}\epsilon\ \epsilon^T \boldsymbol{x}^T]\Sigma_F^{-1} \\ &= \Sigma_F^{-1}\text{E}_F[\epsilon^2] = \sigma^2\Sigma_F^{-1} \end{aligned} \qquad (7.4.15)$$

where we have used the assumed independence of $\boldsymbol{x}$, ϵ under F. Thus the problem of proving aysmptotic normality reduces to showing that $\|n^{1/2}\boldsymbol{R}_n\|$ converges to 0 in probability.

Theorem 7.3

Let F_n be the empirical distribution on the points $[\boldsymbol{x}_i^T, y_i]$, $i = 1, \ldots, n$, which are assumed to be i.i.d. from a distribution F which is determined by a distribution $F_{\boldsymbol{x}}$ for each p-dimensional row vector $\boldsymbol{x}^T$, a distribution F_ϵ which is symmetric about 0, the linear relationship $y = \boldsymbol{x}^T\boldsymbol{\beta} + \epsilon$, and the assumption that ϵ and $\boldsymbol{x}$ are independent. Assume that $\Sigma(F) = E_F[\boldsymbol{x}\boldsymbol{x}^T]$ exists and is invertible, and that $\gamma(F) = E_F[y\boldsymbol{x}]$ exists. Further assume that

$$n^{1/4}[\gamma(F_n) - \gamma(F)] = O_p(1) \qquad (7.4.16)$$

and

$$n^{1/4}[\Sigma(F_n) - \Sigma(F)] = o_p(1). \qquad (7.4.17)$$

(These conditions may seem rather unintuitive, but they appear to be the weakest conditions for which this method of proof goes through. Moment conditions which imply them are given in the corollary.) This last condition (7.4.17) implies that (X^TX/n) converges to $\Sigma(F)$ with probability approaching one. This fact together with the assumed invertibility of $\Sigma(F)$ guarantees that with probability approaching one $\Sigma(F_n)$ is invertible and $\boldsymbol{T}(F_n)$ is well defined by (7.4.9). Under these conditions the remainder term in the von Mises expansion (7.4.13) satisfies

$$n^{1/2}\|\boldsymbol{R}_n\| \xrightarrow{p} 0 \quad \text{as} \quad n \to \infty \qquad (7.4.18)$$

so that

$$n^{1/2}[\boldsymbol{T}(F_n)-\boldsymbol{T}(F)] \xrightarrow{d} N(0,\boldsymbol{V}(F)) \tag{7.4.19}$$

with $\boldsymbol{V}(F)=\sigma^2\boldsymbol{\Sigma}_F^{-1}$. □

Proof. Substituting the influence function (7.4.12) into the von Mises expansion and using (7.4.7) and (7.4.8) we obtain the following relationship in $\boldsymbol{R}_n$:

$$\boldsymbol{T}(F_n)-\boldsymbol{T}(F)=\boldsymbol{\Sigma}_F^{-1}\left[\frac{1}{n}\boldsymbol{X}^T\boldsymbol{Y}-\frac{1}{n}\boldsymbol{X}^T\boldsymbol{X}\boldsymbol{T}(F)\right]+\boldsymbol{R}_n. \tag{7.4.20}$$

We next rearrange the terms in $\boldsymbol{R}_n$ after using the identities (7.4.7) and (7.4.8),

$$\begin{aligned}R_n &= \left(\frac{1}{n}\boldsymbol{X}^T\boldsymbol{X}\right)^{-1}\left(\frac{1}{n}\boldsymbol{X}^T\boldsymbol{Y}\right)-\boldsymbol{\Sigma}_F^{-1}\gamma_F-\boldsymbol{\Sigma}_F^{-1}\left(\frac{1}{n}\boldsymbol{X}^T\boldsymbol{Y}\right)\\ &\quad+\boldsymbol{\Sigma}_F^{-1}\left(\frac{1}{n}\boldsymbol{X}^T\boldsymbol{X}\right)\boldsymbol{\Sigma}_F^{-1}\gamma_F\\ &=\boldsymbol{\Sigma}_{F_n}^{-1}\gamma_{F_n}-\boldsymbol{\Sigma}_F^{-1}\gamma_F-\boldsymbol{\Sigma}_F^{-1}\gamma_{F_n}+\boldsymbol{\Sigma}_F^{-1}\boldsymbol{\Sigma}_{F_n}\boldsymbol{\Sigma}_F^{-1}\gamma_F\\ &=(\boldsymbol{\Sigma}_{F_n}^{-1}-\boldsymbol{\Sigma}_F^{-1})(\gamma_{F_n}-\gamma_F)+(\boldsymbol{\Sigma}_{F_n}^{-1}-\boldsymbol{\Sigma}_F^{-1})\boldsymbol{\Sigma}_{F_n}(\boldsymbol{\Sigma}_{F_n}^{-1}-\boldsymbol{\Sigma}_F^{-1})\gamma_F.\end{aligned} \tag{7.4.21}$$

The second term in this last expression twice contains the factor $\boldsymbol{\Sigma}_{F_n}^{-1}-\boldsymbol{\Sigma}_F^{-1}$. Now in view of the fact that

$$\boldsymbol{\Sigma}_{F_n}^{-1}-\boldsymbol{\Sigma}_F^{-1}=\boldsymbol{\Sigma}_{F_n}^{-1}(\boldsymbol{\Sigma}_F-\boldsymbol{\Sigma}_{F_n})\boldsymbol{\Sigma}_F^{-1}$$

and condition (7.4.17) it is clear that $n^{1/2}$ times the second term in (7.4.21) converges to zero in probability. The first term, times $n^{1/2}$, is a product of a factor that is converging to zero in probability, and a factor that, by (7.4.16), is bounded in probability. The product converges to zero in probability by Slutsky's lemma, which completes the proof. □

Corollary to Theorem 7.3. If the row vectors $\boldsymbol{x}^T$ have fourth moments existing (for each component), then the conditions (7.4.16) and (7.4.17) of Theorem 7.3 are satisfied, and hence the normality of the least squares estimator follows.

Proof of the Corollary. As in (7.4.7) we may write $\gamma(F_n)=(1/n)\sum_{i=1}^n y_i\boldsymbol{x}_i$. Thus $\gamma(F_n)$ is an average of i.i.d. vectors and we may apply a standard central limit theorem for averages of random vectors [such as Theorem B, p. 28, of Serfling (1980)]. It requires only that the covariance matrix of the vector $y\boldsymbol{x}$ exist to conclude that $n^{1/2}[\gamma(F_n)-\gamma(F)]$ is asymptotically normal, and hence bounded in probability, which implies (7.4.16). It suffices to show that the second moment matrix of $y\boldsymbol{x}$ exists. Let $\boldsymbol{x}^T=[x_1,x_2,\ldots,x_p]$ denote the components of a vector selected at random. (These components must not be confused with different realizations $\boldsymbol{x}_i,\boldsymbol{x}_j$ of $\boldsymbol{x}$.) Then $(y\boldsymbol{x})_i=(\boldsymbol{x}^T\boldsymbol{\beta}+\epsilon)x_i$. Thus the diagonal components of the second moment matrix of $y\boldsymbol{x}$ are

$$\begin{aligned}\mathrm{E}[(y\boldsymbol{x})_i]^2 &= \mathrm{E}[(\boldsymbol{x}^T\boldsymbol{\beta})^2x_i^2]+2\mathrm{E}[\epsilon x_i^2(\boldsymbol{x}^T\boldsymbol{\beta})]+\mathrm{E}[\epsilon^2x_i^2]\\ &= \mathrm{E}\left[\left(\sum\beta_jx_j\right)^2x_i^2\right]+\sigma^2\mathrm{E}[x_i^2]\end{aligned}$$

where we have used the independence of $\boldsymbol{x}$ and ϵ and the properties of the moments of ϵ. Terms of the form $\mathrm{E}[x_i^2x_jx_k]$ exist for all i,j,k by the Cauchy-Schwarz Inequality and the assumed fourth moments of each component:

$$|\mathrm{E}[x_i^2x_jx_k]|^2\le\mathrm{E}[x_i^4]\mathrm{E}[x_j^2x_k^2]\le\mathrm{E}[x_i^4]\{\mathrm{E}[x_j^4]\mathrm{E}[x_k^4]\}^{1/2}.$$

The cross product terms $\mathrm{E}[(y\boldsymbol{x})_i(y\boldsymbol{x})_j]$ required for the off-diagonal terms of the second-moment matrix of $y\boldsymbol{x}$ exist because the diagonal terms do, again by Cauchy-Schwarz. Hence $n^{1/2}[\gamma(F_n)-\gamma(F)]$ is bounded in probability, and condition (7.4.16) holds. □

A similar argument leads to (7.4.17) and is therefore left to the problems.

7.4.5 Criteria for Comparing Multivariate Estimators

For future reference we list three commonly used criteria for comparing the effectiveness of estimators of vector parameters. Suppose that we are given two estimators $\hat{\boldsymbol{\beta}}_1,\hat{\boldsymbol{\beta}}_2$ of the vector $\boldsymbol{\beta}$ (based on a finite or infinite sample) which satisfy unbiasedness

$$\mathrm{E}[\hat{\boldsymbol{\beta}}_1]=\boldsymbol{\beta}=\mathrm{E}[\hat{\boldsymbol{\beta}}_2].$$

The following three criteria have been employed to compare their corresponding covariance matrices $\boldsymbol{\Sigma}_1=\mathrm{Cov}[\hat{\boldsymbol{\beta}}_1]$ and $\boldsymbol{\Sigma}_2=\mathrm{Cov}[\hat{\boldsymbol{\beta}}_2]$; in each case $\hat{\boldsymbol{\beta}}_1$ is said to be more efficient than $\hat{\boldsymbol{\beta}}_2$ for the given criterion.

1. $\mathrm{tr}(\boldsymbol{\Sigma}_1)<\mathrm{tr}(\boldsymbol{\Sigma}_2)$
2. $\boldsymbol{\Sigma}_2-\boldsymbol{\Sigma}_1$ is positive semidefinite
3. $\det(\boldsymbol{\Sigma}_1)<\det(\boldsymbol{\Sigma}_2)$

The motivation for employing criterion 1 is that

$$\mathrm{tr}(\Sigma) = \sum_i \mathrm{E}[\hat{\beta}_i - \beta_i]^2 = \mathrm{E}[\|\hat{\boldsymbol{\beta}} - \boldsymbol{\beta}\|]^2$$

which is the expected mean squared error of $\hat{\beta}$ in R^p.

The justification for criterion 2 follows from the definition, for $\Sigma_2 - \Sigma_1$ is positive semidefinite if and only if for all $\boldsymbol{a} \neq \boldsymbol{0}$, $\boldsymbol{a}\Sigma_2\boldsymbol{a}^T \geq \boldsymbol{a}\Sigma_1\boldsymbol{a}^T$. To estimate *any* linear combination of the parameters, say $\boldsymbol{a}^T\boldsymbol{\beta}$, it is natural to use $\boldsymbol{a}^T\hat{\beta}_1$, which is unbiased for $\boldsymbol{a}^T\boldsymbol{\beta}$, with

$$\mathrm{Var}[\boldsymbol{a}^T\hat{\beta}_1] = \boldsymbol{a}^T\Sigma_1\boldsymbol{a}.$$

Thus criterion 2 is very strong in the sense that $\boldsymbol{a}^T\hat{\beta}_1$ has smaller variance than $\boldsymbol{a}^T\hat{\beta}_2$, for *any* choice of $\boldsymbol{a}$.

The third criterion is justified when the estimators have a multivariate normal distribution, and confidence ellipsoids of the form

$$\{\boldsymbol{\beta} : (\hat{\boldsymbol{\beta}} - \boldsymbol{\beta})^T\Sigma^{-1}(\hat{\boldsymbol{\beta}} - \boldsymbol{\beta}) \leq c\}$$

are formed for the unknown vector $\boldsymbol{\beta}$. The volume of the confidence ellipsoid can be shown (Appendix, Theorem A.4) to be proportional to the square root of det (Σ), and hence the estimator that yields the smaller volume confidence ellipsoid (at the same level) is to be preferred.

7.5* ASYMPTOTIC THEORY FOR WEIGHTED LEAST SQUARES

In this section we will briefly review the development of M-estimators for regression and their connection with weighted least squares. Then we will show in Section 7.5.2 that if a Fisher-consistent estimator of the regression coefficients is used to choose the weights, then the first iterate of weighted least squares—also called the *explicit* weighted least squares estimator—will also be Fisher-consistent. This will lead to weak consistency of the associated weighted least squares estimates. In Section 7.5.3 we will derive the influence function for both explicitly and implicitly defined weighted least squares estimators, the latter being the limit of continued iteration by weighted least squares. Much of this material and further results for weighted least squares are given in Dollinger and Staudte (1990b).

7.5.1 From M-Estimates to Weighted Least Squares

The least squares estimator is the value of $\hat{\boldsymbol{\beta}}$ minimizing the sum of the *squared* residuals:

$$\sum_i r_i^2 = \sum_i (y_i - \boldsymbol{x}_i^T\hat{\boldsymbol{\beta}})^2.$$

The existence of $\hat{\boldsymbol{\beta}}$, its ease of derivation, and its efficiency for normal errors are all positive consequences of squaring the residuals, but there is a negative consequence; the observations with large residuals have a much larger contribution to the above sum of squares, and hence more effect on the resulting solution. An alternative to least squares which treats each residual equally is the least absolute deviations estimator $\hat{\boldsymbol{\beta}}_{\text{LAD}}$; it minimizes

$$\sum_i |r_i| = \sum_i |y_i - \boldsymbol{x}_i^T \hat{\boldsymbol{\beta}}_{\text{LAD}}|.$$

Unfortunately, there is a substantial loss of efficiency for these estimators when the errors are normally distributed, and there are problems with large leverage points. See Bloomfield and Steiger (1983) for an excellent introduction to least absolute deviation estimators.

More generally, Huber (1973a) suggested choosing $\hat{\boldsymbol{\beta}}$ which minimizes

$$\sum_i \rho(r_i) \tag{7.5.1}$$

where ρ is a convex function. The minimum of (7.5.1) can often be found by setting the partial derivatives of ρ to zero to obtain

$$\sum_i \psi(r_i)\boldsymbol{x}_i = \sum_i \psi(y_i - \boldsymbol{x}_i^T \hat{\boldsymbol{\beta}}_\psi)\boldsymbol{x}_i = \boldsymbol{0} \tag{7.5.2}$$

with $\psi = \rho'$. Solutions $\hat{\boldsymbol{\beta}}_\psi$ of (7.5.2) are called *classical M-estimators* or Huber estimators because they were the first extensions of the location M-estimators to the regression case. As in the location case theory they are not scale equivariant but may be modified to be so by replacing r_i by r_i/σ in (7.5.2), where σ is a scale factor which also needs to be estimated. (See the material on location–scale M-estimators in Section 4.4.) By judicious choice of the ψ function, the M-estimators of $\boldsymbol{\beta}$ may attain high efficiency relative to least squares and also be robust against large residuals. However, they are not robust to outliers in the design space. This is no problem if the $\boldsymbol{x}$'s are in fact *chosen* by design to have equileverage, or at least a large proportion of points with the maximum leverage. But if the $\boldsymbol{x}$'s themselves are random or otherwise subject to errors, then the classical M-estimators may be unreliable, for the equations in (7.5.2) are directly affected by the $\boldsymbol{x}$'s. Thus several statisticians have proposed alternatives to (7.5.2) of the form

$$\sum_i \eta(\boldsymbol{x}_i, r_i)\boldsymbol{x}_i = \sum_i \eta(\boldsymbol{x}_i, y_i - \boldsymbol{x}_i^T \hat{\boldsymbol{\beta}}_\eta)\boldsymbol{x}_i = \boldsymbol{0}. \tag{7.5.3}$$

In other words the domain of the ψ function has been enlarged to include the design points, as well as the residuals. See Hampel, et al. [(1986), Section 6.3] and Hettmansperger (1987) for further references and history. Solutions to (7.5.3) are also called GM-estimators or generalized M-estimators, but we will avoid this terminology because M-estimators are generalized maximum likelihood estimators. The M-equations in (7.5.3) are similar to those of weighted least squares (7.1.13), at least when the latter are rewritten in the form

$$\sum_i (y_i - x_i^T\hat{\beta})w_i\boldsymbol{x}_i = \mathbf{0} \tag{7.5.4}$$

with $w_i = w(\boldsymbol{x}_i, y_i - x_i^T\hat{\beta})$.

Thus given an η function defining (7.5.3) which satisfies the condition $\eta(\boldsymbol{x}, 0) = 0$, there is an equivalent weight function defining (7.5.4) with the same solutions, if any. The equivalence is given by

$$\eta(\boldsymbol{x}, r) = w(\boldsymbol{x}, r)r. \tag{7.5.5}$$

Maronna and Yohai (1981) derive asymptotic normality of $\hat{\beta}_\eta$ under general conditions on $\eta(\boldsymbol{x}, r)$.

7.5.2 Consistency of Weighted Least Squares Estimators

In the next two sections we will define an explicit weighted least squares functional $T_w(G)$ and derive its influence function. The influence function will provide insight into the estimator $T_{\hat{w}}(F_n) = \hat{\boldsymbol{\beta}}_{\hat{w},n}$ of $\boldsymbol{\beta}$ and show how the initial estimator of the unknown parameters affects the explicit weighted least squares estimator. More generally, it will reveal how each iteration to an implicitly defined weighted least squares solution depends on the previous iterate.

We begin with ordinary (unweighted) least squares. In this section we assume for convenience that whether or not $\boldsymbol{x}$ contains a constant term, it lies in R^p, and hence so does the vector of unknowns $\boldsymbol{\beta}$.

Let $[\boldsymbol{x}^T, y]$ have a distribution G on R^{p+1} for which the following expectations exist:

$$\begin{aligned} \boldsymbol{\gamma}(G) &= \mathrm{E}_G[y\boldsymbol{x}] \\ \boldsymbol{\Sigma}(G) &= \mathrm{E}_G[\boldsymbol{x}\boldsymbol{x}^T]. \end{aligned} \tag{7.5.6}$$

When $\boldsymbol{\Sigma}(G)$ is invertible, we may define the functional

$$\boldsymbol{T}(G) = \boldsymbol{\Sigma}^{-1}(G)\boldsymbol{\gamma}(G). \tag{7.5.7}$$

A sufficient condition for $\Sigma(G)$ to be invertible [and hence for the existence of $\boldsymbol{T}(G)$] is that the distribution of the nonconstant part of $\boldsymbol{x}$ be nonsingular. This is shown in the problems.

We assume that the components $\boldsymbol{x}^T, y$ of $[\boldsymbol{x}^T y]$ satisfy the *regression structure*, which means that there exists $\boldsymbol{\beta} \in \mathrm{R}^p$ such that

$$\begin{array}{ll} \text{(i)} & \mathrm{E}_G[y \mid \boldsymbol{x}] = \boldsymbol{x}^T\boldsymbol{\beta} \\ \text{(ii)} & \epsilon = y - \boldsymbol{x}^T\boldsymbol{\beta} \text{ is independent of } \boldsymbol{x}^T \\ \text{(iii)} & G_\epsilon \text{ is symmetric about 0 with finite variance } \sigma^2 > 0. \end{array} \qquad (7.5.8)$$

When the moment conditions (7.5.6) and the regression structure (7.5.8) hold for a distribution G, we denote it by $F = F_{\beta,\sigma}$. If $\boldsymbol{T}(G)$ is well defined, it is Fisher-consistent at the model $F_{\beta,\sigma}$. To see this, note that $\gamma(F_{\beta,\sigma}) = \Sigma(F_{\beta,\sigma})\boldsymbol{\beta}$, and thus $\boldsymbol{T}(F_{\beta,\sigma}) = \boldsymbol{\beta}$. The $\boldsymbol{T}$ defined by (7.5.7) is called the *least squares functional*, because when it is applied to the empirical disribution which puts mass $1/n$ on each of $[\boldsymbol{x}_i^T y_i]$, $i = 1,\ldots,n$, the result is

$$\boldsymbol{T}(F_n) = (\boldsymbol{X}^T\boldsymbol{X})^{-1}\boldsymbol{X}^T\boldsymbol{Y} = \hat{\boldsymbol{\beta}}_n \qquad (7.5.9)$$

the least squares estimator of $\boldsymbol{\beta}$.

Next define the least squares *residual* of y, given $\boldsymbol{x}$ by

$$r(\boldsymbol{x}^T, y, G) = y - \boldsymbol{x}^T\boldsymbol{T}(G). \qquad (7.5.10)$$

Since $\boldsymbol{T}$ is Fisher-consistent, the residual when $G = F_{\beta,\sigma}$ satisfies the regression structure (7.5.8) is $r = y - \boldsymbol{x}^T\boldsymbol{\beta} = \epsilon$. We note that the least squares estimator may be replaced by any other Fisher-consistent estimator in defining the residuals, so that our results will hold more generally. We use least squares only for definiteness.

Until Section 7.6 we will for the sake of simplicity let $\sigma^2 = 1$. We are interested in weight functions of the form $w = w(\boldsymbol{x}, r)$, where r is given by (7.5.10). We emphasize that the weight attached to a point $[\boldsymbol{x}^T, y]$ depends not only on the point, but on the distribution G as well, because the position of the point is relative to G and the residual at the point is dependent on an estimate $\boldsymbol{T}(G)$.

When G is the empirical distribution F_n on n rows $[\boldsymbol{x}_i^T, y_i]$, the empirical estimate of (7.5.10) is the residual

$$r(\boldsymbol{x}_i^T, y_i, F_n) = y_i - \boldsymbol{x}_i^T\hat{\boldsymbol{\beta}} = r_i. \qquad (7.5.11)$$

We assume that G is nice enough to define the least squares functional $\boldsymbol{T}(G)$ given in (7.5.7). We want to define weighted analogues of (7.5.6) and (7.5.7) for a weight function which depends on each row $[\boldsymbol{x}^T, y]$ and its distribution G through the residual defined by (7.5.10). Thus our first need is to define

$$\begin{aligned}\gamma_w(G) &= \mathrm{E}_G[wy\boldsymbol{x}]\\ \Sigma_w(G) &= \mathrm{E}_G[w\boldsymbol{x}\boldsymbol{x}^T] \\ \boldsymbol{T}_w(G) &= \Sigma_w^{-1}(G)\gamma_w(G).\end{aligned} \tag{7.5.12}$$

We will give sufficient conditions for the existence of the weighted least squares functional in the following theorem, as well as the consistency of the associated estimators.

It will be important to distinguish between the variables

$$\gamma_w(F_n) = \frac{1}{n}\sum w(\boldsymbol{x}_i, y_i - \boldsymbol{x}^T\boldsymbol{\beta})y_i\boldsymbol{x}_i$$

and

$$\gamma_{\hat{w}}(F_n) = \frac{1}{n}\sum w(\boldsymbol{x}_i, y_i - \boldsymbol{x}^T\boldsymbol{T}(F_n))y_i\boldsymbol{x}_i.$$

The former is not a statistic because it depends on the unknown $\boldsymbol{\beta}$, while the latter is a statistic because the unknown is estimated by an initial estimate $\boldsymbol{T}(F_n)$. The subscript w represents the appropriate weight if the parameter is known, and the subscript $\hat{w}$ is the estimated weight when it is unknown. We similarly distinguish between $\Sigma_w(F_n)$ and $\Sigma_{\hat{w}}(F_n)$; and between $\boldsymbol{T}_w(F_n)$ and $\boldsymbol{T}_{\hat{w}(F_n)}$. It is left for the reader to show that:

$$\begin{aligned}\gamma_{\hat{w}}(F_n) &= \frac{1}{n}(\boldsymbol{X}^T\hat{\boldsymbol{W}}\boldsymbol{Y})\\ \Sigma_{\hat{w}}(F_n) &= \frac{1}{n}(\boldsymbol{X}^T\hat{\boldsymbol{W}}\boldsymbol{X}) \\ \boldsymbol{T}_{\hat{w}}(F_n) &= (\boldsymbol{X}^T\hat{\boldsymbol{W}}\boldsymbol{X})^{-1}\boldsymbol{X}^T\hat{\boldsymbol{W}}\boldsymbol{Y}\end{aligned} \tag{7.5.13}$$

where $\hat{\boldsymbol{W}}$ is the diagonal matrix with

$$\hat{\boldsymbol{W}}_{ii} = w(\boldsymbol{x}_i, y_i - \boldsymbol{x}_i^T\boldsymbol{T}(F_n)).$$

Let $\hat{\boldsymbol{\beta}}_{\hat{w},n} = \boldsymbol{T}_{\hat{w}}(F_n)$ define the *explicit weighted least squares estimator.* The initial estimate of $\boldsymbol{\beta}$ is the (unweighted) least squares estimate $\hat{\boldsymbol{\beta}}_n = \boldsymbol{T}(F_n)$. Now to the proof of consistency of this estimator.

Theorem 7.4

(i) Assume that the conditions for the existence of $\gamma(G)$, $\Sigma(G)$ and $\boldsymbol{T}(G)$ in (7.5.6) and (7.5.7) are satisfied. Let r be defined by (7.5.10), and assume that $w = w(\boldsymbol{x}, r)$ is nonnegative, bounded and measurable in $(\boldsymbol{x}, r)$. Then $\gamma_w(G) = \mathrm{E}_G[wy\boldsymbol{x}]$ and $\Sigma_w(G) = \mathrm{E}_G[w\boldsymbol{x}\boldsymbol{x}^T]$ are well defined.

(ii) We have already assumed in (i) that $w \geq 0$. If for all non-zero $\boldsymbol{a}$ the set

$$\{[\boldsymbol{x}^T, y] : \boldsymbol{a}^T\boldsymbol{x} \neq 0 \text{ and } w(\boldsymbol{x}, y - \boldsymbol{x}^T\boldsymbol{\beta}) > 0\}$$

has positive probability under G, then the existence of $\boldsymbol{T}(G)$ implies the existence of $\boldsymbol{T}_w(G)$.

(iii) If in addition to the earlier assumptions in (i) and (ii), $G = F_\beta$ has the regression structure (7.5.8), and $w(\boldsymbol{x}, r) = w(\boldsymbol{x}, -r)$ for all $\boldsymbol{x}, r$, then $\boldsymbol{T}_w(F_\beta) = \boldsymbol{\beta}$; i.e. $\boldsymbol{T}_w$ is Fisher consistent for $\boldsymbol{\beta}$.

(iv) In addition to conditions (i)–(iii) assume that the components of $\boldsymbol{x}$ have fourth moments existing and that the weight function satisfies $|w(x,q) - w(x,r)| \leq c|q - r|$ for some positive constant c and all q, r. Then we have weak consistency of the weighted least squares estimator:

$$\hat{\boldsymbol{\beta}}_{\hat{w},n} \xrightarrow{p} \boldsymbol{\beta}$$

Proof

(i) Since $\gamma(G) = \mathrm{E}_G[y\boldsymbol{x}]$ exists and w is a bounded measurable function, $wy\boldsymbol{x}$ also has moments existing for each component; thus $\gamma_w(G) = \mathrm{E}_G[wy\boldsymbol{x}]$ exists. A similar argument yields the existence of the matrix $\Sigma_w(G)$.

(ii) Since $w \geq 0$, Σ_w is non-negative definite. It remains to show that Σ_w is invertible. If it is not positive definite there exists a vector $\boldsymbol{a} \neq \boldsymbol{0}$ such that

$$0 = \boldsymbol{a}^T\Sigma_w(G)\boldsymbol{a} = \mathrm{E}_G[\boldsymbol{a}^T\boldsymbol{x}w\boldsymbol{x}^T\boldsymbol{a}] = \mathrm{E}_G[w^{1/2}\boldsymbol{a}^T\boldsymbol{x}]^2.$$

Now if a non-negative function $[w^{1/2}\boldsymbol{a}^T\boldsymbol{x}]^2$ has expected value 0 under G, it must be identically zero with probability one under G. Thus there exists $\boldsymbol{a} \neq \boldsymbol{0}$ such that with probability one under G,

$$\{[\boldsymbol{x}^T, y] : \boldsymbol{a}^T\boldsymbol{x} = 0 \quad \text{or} \quad w(\boldsymbol{x}, y - \boldsymbol{x}^T\boldsymbol{\beta}) = 0\}$$

This fact contradicts the assumption and part (ii) is proved. The technical condition in (ii) is satisfied for example if $w(\boldsymbol{x}, r) > 0$ for all $(\boldsymbol{x}, r)$ and the marginal distribution of $\boldsymbol{x}$ is not concentrated on a proper linear subspace of R^{p+1}.

(iii) Since the (unweighted) least squares functional is Fisher-consistent under our assumptions, the weights defined by (7.5.9) are of the form

$$w = w(\boldsymbol{x}, y - \boldsymbol{x}^T\boldsymbol{\beta}) = w(\boldsymbol{x}, \epsilon).$$

Using the regression structure (7.5.8), and writing F for F_β,

$$\begin{aligned} \gamma_w(F) &= \mathrm{E}_F[w(\boldsymbol{x}, y - \boldsymbol{x}^T\boldsymbol{\beta})y\boldsymbol{x}] \\ &= \mathrm{E}\left[\boldsymbol{x}\mathrm{E}[w(\boldsymbol{x}, y - \boldsymbol{x}^T\boldsymbol{\beta})(y - \boldsymbol{x}^T\boldsymbol{\beta}) \mid \boldsymbol{x}]\right] \\ &\quad + \mathrm{E}\left[\boldsymbol{x}\boldsymbol{x}^T\boldsymbol{\beta}\mathrm{E}[w(\boldsymbol{x}, y - \boldsymbol{x}^T\boldsymbol{\beta}) \mid \boldsymbol{x}]\right]. \end{aligned} \tag{7.5.14}$$

The first term in (7.5.14) is zero because the inner expectation is an integral of an odd function of ϵ and thus:

$$\mathrm{E}_\epsilon[w(\boldsymbol{x}, \epsilon)\epsilon] = 0. \tag{7.5.15}$$

Thus

$$\begin{aligned} \gamma_w(F) &= \mathrm{E}_F[w\boldsymbol{x}\boldsymbol{x}^T]\boldsymbol{\beta} \\ \boldsymbol{T}_w(F) &= \Sigma_w^{-1}(F)\gamma_w(F) = \boldsymbol{\beta}. \end{aligned}$$

(iv) First we prove that $\Sigma_{\hat{w}}(F_n)$ converges in probability to $\Sigma_w(F)$. Expressing this quantity as a sum

$$\Sigma_{\hat{w}}(F_n) = [\Sigma_{\hat{w}}(F_n) - \Sigma_w(F_n)] + \Sigma_w(F_n),$$

we note that the second term converges in probability to $\Sigma_w(F)$ by the Law of Large Numbers. The first term has (i, j) term bounded in absolute value by

$$\frac{1}{n}\sum_k c\|T(F_n) - T(F)\| \, \|\boldsymbol{x}_k\| \, |X_{ki}X_{kj}|$$

which converges to zero in probability by the moment condition on the rows $\boldsymbol{x}^T$ and the weak consistency of the initial estimator.

A similar argument applies to $\gamma_{\hat{w}}(F_n)$, and the consistency of the weighted least squares estimator then follows from the assumed invertibility of $\Sigma_w(F)$ and the continuity of the inverse. □

Remark

Having obtained $\boldsymbol{T}_w(G)$ by (7.5.12), it may then be substituted for $\boldsymbol{T}(G)$ in the definition of the residual (7.5.10), to obtain new weights. By (iii) above, this will yield the same weights as $\boldsymbol{T}(G)$ when $G = F_\beta$. Thus $\boldsymbol{T}_w$ is also a solution of the normal equations $\Sigma_w(G)\boldsymbol{T}_w(G) = \gamma_w(G)$, when the weight function depends on $\boldsymbol{T}_w(G)$ and $G = F_{\beta}$. Of course such an implicit solution need not exist for every G. We will denote the implicitly defined solution to the normal equations by $\boldsymbol{S}_w(G)$, for every G for which it exists, and call it the *implicit weighted least squares functional.* Such an implicitly defined weighted least squares approach requires more care regarding the existence of solutions, as well as the computation and analysis of them. Assuming a unique solution exists, it can usually be found by the method of iteratively reweighted least squares (Section 7.1.3). When the method converges, the iteration is stopped because the last two estimates are almost the same; i.e. they are close to the solution of $\boldsymbol{\Sigma}_{\hat{w}}(F_n)\boldsymbol{S}_{\hat{w}}(F_n) = \boldsymbol{\gamma}_{\hat{w}}(F_n)$. It is of interest to determine conditions under which $\boldsymbol{S}_{\hat{w}}(F_n)$ exists and converges to $\boldsymbol{S}_w(F_{\beta}) = \boldsymbol{\beta}$.

Summary

We have in this section defined a weighted least squares functional which, when applied to the empirical distribution, describes the procedure that first finds an initial estimate, then calculates the weights which depend on the leverages and residuals from the initial estimate, and finally renders an explicit weighted least squares estimate. The corresponding sequence of finite sample estimators converge in probability to the parameter $\boldsymbol{\beta}$ when the regression structure is in place.

7.5.3 The Influence Function for Weighted Least Squares

We have defined the weighted least squares functional $\boldsymbol{T}_w$ explicitly in three steps: first, choose a Fisher-consistent initial estimator, such as the (unweighted) least squares functional $\boldsymbol{T}(G) = \boldsymbol{\Sigma}^{-1}(G)\boldsymbol{\gamma}(G)$; second, for every point $[x^T, y]$ and distribution G calculate the residual $r = y - \boldsymbol{x}^T\boldsymbol{T}(G)$; and third, with weights $w = w(\boldsymbol{x}, r)$ compute the weighted least squares functional $\boldsymbol{T}_w(G) = \boldsymbol{\Sigma}_w^{-1}(G)\boldsymbol{\gamma}_w(G)$.

The influence function for $\boldsymbol{T}_w(G)$ at F in the direction of a point mass $\Delta_{[x^T,y]}$ has the following form:

$$\mathbf{IF}_{T_w,F}(\boldsymbol{x}^T, y) = \boldsymbol{\Sigma}_w^{-1} w r \boldsymbol{x} - \boldsymbol{\Sigma}_w^{-1}\boldsymbol{C}\,\mathbf{IF}_{T,F}(\boldsymbol{x}^T, y) \tag{7.5.16}$$

where $\boldsymbol{C} = \mathrm{E}_F[w_2 r \boldsymbol{x}\boldsymbol{x}^T]$ and $w_2 = (d/dr)w(\boldsymbol{x}, r)$.

Remarks

1. The first term in (7.5.16) is similar to the influence function for ordinary least squares (7.4.12) in that it is a product of the position $\boldsymbol{\Sigma}_w^{-1}\boldsymbol{x}$ and the residual $r = y - x^T\boldsymbol{T}_w(F)$. Now, however, the matrix $\boldsymbol{\Sigma}(F)$ is replaced by the weighted matrix $\boldsymbol{\Sigma}_w(F)$ which may downweight large

leverage points and $\boldsymbol{T}_w$ appears in place of T. Also, there is another factor in the first term, the weight $w = w(\boldsymbol{x},r)$ itself, which is usually chosen to downweight points with a large residual r and with $\boldsymbol{x}$ having a large leverage. Thus a properly chosen weight function will lead to a bounded first term in the influence function (7.5.16).

2. The second term shows the influence that the initial estimator has on the weighted least squares estimator. If the former has unbounded influence function, so will the latter. This illustrates how important the choice of initial estimator is in one-step weighted least squares.

3. To overcome the lack of robustness caused by the nonrobust initial estimator, we may formally substitute the *weighted* least squares influence function for that of the initial estimator in (7.5.16), and solve for the influence function of the *implicitly* defined weighted least squares estimator, that is, of $\boldsymbol{S}_w(F)$, as the solution of

$$\boldsymbol{\Sigma}_w(F)\boldsymbol{S}_w(F) = \boldsymbol{\gamma}_w(F) \tag{7.5.17}$$

where $w = w(\boldsymbol{x},r) = w(\boldsymbol{x},y - \boldsymbol{x}^T \boldsymbol{S}_w(G))$. This leads to

$$\boldsymbol{IF}_{\boldsymbol{S}_w,F}(\boldsymbol{x}^T,y) = \boldsymbol{M}^{-1} wr\boldsymbol{x} \tag{7.5.18}$$

where $\boldsymbol{M} = \boldsymbol{\Sigma}_w + \boldsymbol{C} = \mathrm{E}_F[(w + w_2 r)\boldsymbol{x}\boldsymbol{x}^T]$. This last expression is simpler than (7.5.16) because only one estimator is involved, but more care is required regarding the existence and the convergence of solutions to $\boldsymbol{S}_w(F)$ by the method of iteratively reweighted least squares.

4. Finally we note that $\boldsymbol{M}$ is of the form $\boldsymbol{M} = \mathrm{E}_F[\eta_2 \boldsymbol{x}\boldsymbol{x}^T]$, where η is given by $\eta(\boldsymbol{x},r) = w(\boldsymbol{x},r)r$ and η_2 is its derivative with respect to r; see (7.5.5). If the usual one-term von Mises expansion [as in (7.4.12)] has the remainder term converging to zero fast enough, then the normalized implicit weighted least squares estimate $n^{1/2}[\boldsymbol{S}_w(F_n) - \boldsymbol{\beta}]$ will have the same limiting distribution as $n^{-1/2}\sum_{i=1}^{n} \boldsymbol{IF}_{\boldsymbol{S}_w,F}(\boldsymbol{x}_i^T, y_i)$. By the central limit theorem (Appendix B.13) for i.i.d. vectors, this limit is multivariate normal with mean vector $\boldsymbol{0}$ and covariance matrix $\boldsymbol{V}/n$, where $\boldsymbol{V}$ is given by

$$\boldsymbol{V} = \mathrm{E}[\boldsymbol{IF}_{\boldsymbol{S}_w,F}\boldsymbol{IF}_{\boldsymbol{S}_w,F}^T] = \boldsymbol{M}^{-1}\boldsymbol{Q}\boldsymbol{M}^{-1}$$

where for ϵ arising from r, see line after (7.5.10),

$$\boldsymbol{Q} = \mathrm{E}_F[w^2(\boldsymbol{x},\epsilon)\epsilon^2\boldsymbol{x}\boldsymbol{x}^T] = \mathrm{E}_F[\eta^2\boldsymbol{x}\boldsymbol{x}^T].$$

This formula for the asymptotic covariance is in agreement with the result of Maronna and Yohai (1981).

Derivation of the Influence Function for Weighted Least Squares

The derivation of (7.5.16) will be carried out without careful statement of regularity conditions required for each step. Assume that $F = F_{\beta,\epsilon}$ satisfies the regression structure (7.5.8) and that H is the difference of two distributions (such as $H = \Delta_{[x^T,y]} - F$.) We will first find the Gateaux derivative of $\boldsymbol{T}_w$ at F in the direction of H, and then extract the influence function from it. For any functional $\boldsymbol{T}(G)$, the Gateaux derivative of T (at F, in the direction H) has been defined by $\boldsymbol{T}_F'(H) = (d/dt)\boldsymbol{T}(F + tH)|_{t=0}$.

In the following discussion we will drop the notational dependence on F, but all integrals are with respect to F, and all Gateaux derivatives are at F unless otherwise stated. In this notation the explicit weighted least squares functional satisfies $\Sigma_w \boldsymbol{T}_w = \gamma_w$. Now we will calculate the Gateaux derivative of $\boldsymbol{T}_w$ in the direction of H. Applying the product rule for derivatives to $\Sigma_w \boldsymbol{T}_w = \gamma_w$ yields

$$\Sigma_w'(H)\boldsymbol{T}_w + \Sigma_w \boldsymbol{T}_w'(H) = \gamma_w'(H). \tag{7.5.19}$$

It will be shown in the problems that

$$\begin{aligned} \gamma_w'(H) &= \gamma_{w'(H)} + \gamma_w(H) \\ \Sigma_w'(H) &= \Sigma_{w'(H)} + \Sigma_w(H). \end{aligned} \tag{7.5.20}$$

Substituting (7.5.20) into (7.5.19) and rearranging terms we find

$$\boldsymbol{T}'_w(H) = \Sigma_w^{-1}[\gamma_w(H) - \Sigma_w(H)\boldsymbol{T}_w] + \Sigma_w^{-1}[\gamma_{w'(H)} - \Sigma_{w'(H)}\boldsymbol{T}_w]. \tag{7.5.21}$$

The first term in (7.5.21) is, with (7.5.13), easily seen to be of the form $\mathrm{E}_H[\Sigma_w^{-1} wr\boldsymbol{x}]$, where $r = y - \boldsymbol{x}^T \boldsymbol{T}_w$ is the residual for the weighted estimate. The kernel of this integral is the first term in the influence function (7.5.16).

The second term of (7.5.21) requires more attention. Assuming that F satisfies the regression structure (7.5.8) and that $\boldsymbol{T}_w$ is Fisher-consistent at F, we may add and subtract $\boldsymbol{x}^T\boldsymbol{\beta}$ to y to obtain

$$\begin{aligned} \gamma_{w'(H)} &= \mathrm{E}_F[w'(H)y\boldsymbol{x}] \\ &= \mathrm{E}_F[w'(H)(y - \boldsymbol{x}^T\boldsymbol{\beta})\boldsymbol{x}] + \mathrm{E}_F[w'(H)\boldsymbol{x}\boldsymbol{x}^T\boldsymbol{\beta}] \\ &= \mathrm{E}_F[w'(H)\epsilon\boldsymbol{x}] + \Sigma_{w'(H)}\boldsymbol{T}_w. \end{aligned} \tag{7.5.22}$$

Thus the second term in (7.5.21) reduces to $\Sigma_w^{-1}\mathrm{E}_F[w'(H)\epsilon\boldsymbol{x}]$. We will next find $w'(H)$.

By the chain rule for compositions, the weight function has Gateaux derivative

$$w'(H) = w'(\boldsymbol{x}, r)(H) = w_2(\boldsymbol{x}, r)r'(H). \tag{7.5.23}$$

Thus we will need $r'(H)$. For r given by (7.5.10), the Gateaux derivative is

$$r'(\boldsymbol{x}, y, H) = -\boldsymbol{x}^T\boldsymbol{T}'(H). \tag{7.5.24}$$

The second term of (7.5.21) becomes, after substituting (7.5.22),

$$\Sigma_w^{-1}\{\mathrm{E}[w_2 r'(H)\epsilon \boldsymbol{x}]\}. \tag{7.5.25}$$

Then (7.5.25) reduces [with substitution of (7.5.24)] to

$$-\Sigma_w^{-1}\mathrm{E}[w_2(\boldsymbol{x},r)r\boldsymbol{x}\boldsymbol{x}^T]\boldsymbol{T}'(H) = -\Sigma_w^{-1}\boldsymbol{C}\boldsymbol{T}'(H).$$

Summary

We have derived the influence function (7.5.16) for an explicit weighted least squares functional, and used it to find the influence function (7.5.18) for an implicity defined one. It has been tactily assumed throughout that differentiation and integration can be interchanged when necessary. We clearly require that $w(\boldsymbol{x}, r)$ be an even function of r, and that the errors be symmetrically distributed about 0 and have finite variance.

7.6 COMPUTATION OF ROBUST ESTIMATES

In this section we present methods for obtaining robust regression estimates and their standard errors using fairly simple computational methods. The robustness is due to the fact that cases that have a large leverage or residual are downweighted according to the value of DFITS. There are more efficient robust estimators, but they require more mathematics to justify and more intensive computation to evaluate. Such estimators are developed in detail in Chapter 6 of Hampel et al. (1986), wherein a complete bibliography to the literature can be found. Some of the material in Section 7.6.1 and a general overview are given in Hettmansperger (1987). We will first give some background material and define the Welsch estimator (1980) in Section 7.6.1, and then illustrate the computations of it on several examples in Sections 7.6.2–7.6.5. It is assumed that the reader has absorbed Sections 7.1, 7.2, and 7.5.1, even if the starred sections in Sections 7.3–7.5 have been bypassed.

7.6.1 Background Material and Definitions

We begin by considering M-estimating equations of the form (7.5.3), but with the residual r_i replaced by r_i/σ, where σ is also to be estimated simultaneously. (Recall that to obtain scale equivariance of the M-estimates of $\boldsymbol{\beta}$, we need to estimate the scale factor σ by some scale equivariant estimator.)

Because of the identity $\eta(\boldsymbol{x}, r/\sigma) = w(\boldsymbol{x}, r/\sigma)r/\sigma$, the M-estimating equations are equivalent to the following weighted least squares equations:

$$\sum_i (y_i - x_i^T \hat{\beta}) w_i \boldsymbol{x}_i = 0 \tag{7.6.1}$$

where $w_i = w\left(\boldsymbol{x}_i, (y_i - x_i^T \hat{\beta})/\sigma\right)$. We will restrict attention to functions $w(\boldsymbol{x}, r/\sigma)$ which are of the form

$$w\left(\boldsymbol{x}, \frac{r}{\sigma}\right) = \frac{\sigma v(\boldsymbol{x})}{r} \psi_c \left(\frac{r}{\sigma v(\boldsymbol{x})}\right) \tag{7.6.2}$$

where $v(\boldsymbol{x})$ is a known function, and ψ_c is Huber's ψ function. The weight function w downweights points for which the magnitude of the rescaled residual is greater than $cv(\boldsymbol{x})$. Usually $v(\boldsymbol{x})$ depends on $\boldsymbol{x}$ through its leverage.

The class of estimators $\hat{\beta}_w$ defined by (7.6.1) and (7.6.2) are computed by the method of iteratively reweighted least squares discussed in Section 7.1.2. This class contains the following estimators:

1. Least squares, by letting $c \to \infty$.
2. Huber's original proposal for estimators that are robust to large residuals, by taking $v(\boldsymbol{x}) = 1$; see Huber (1973a), Chapter 7.
3. Schweppe's original proposal, by taking $v(\boldsymbol{x}_i) = (1 - h_{ii})^{1/2}$; see Handschin, Schweppe, Kohlas and Fiechter (1975). When scale is estimated by the ith deleted residual $s_{(i)}$ given by (7.2.6), the weights depend only on the Studentized residuals (7.2.7).
4. Welsch's (1980) proposal, by letting $v(\boldsymbol{x}_i) = (1 - h_{ii})/\sqrt{h_{ii}}$ and estimating σ by the deleted $s_{(i)}$. In this case the weights depend on the combined effect of leverage and Studentized residual through DFITS.

We are mainly interested in Welsch's proposal, since it is intuitively a natural way to guard against aberrant observations without losing too much efficiency. Furthermore this estimator and its covariance matrix may be calculated easily using weighted least squares. To see this, note that $w(\boldsymbol{x}, r/\sigma)$ in (7.6.2) is of the form

$$w\left(\boldsymbol{x}, \frac{r}{\sigma}\right) = \min\left\{\frac{c\sigma v(\boldsymbol{x})}{|r|}, 1\right\}.$$

When Welsch's suggestions are employed,

$$w_i = w\left(\boldsymbol{x}_i, \frac{r_i}{s_{(i)}}\right) = \min\left\{\frac{c}{|d_i|}, 1\right\} \tag{7.6.3}$$

where d_i = DFITS$_i$ is defined by (7.2.10). This estimator clearly only downweights observations that have $|d_i| > c$. Thus Welsch's proposal uses the statistics that were calculated for diagnostic purposes (and discussed earlier in Section 7.2) to find a robust fit. Minitab programs for simple calculation of these robust regression estimates and their standard errors are given in Appendix D, and their use is illustrated below. The estimates of standard errors are motivated by the following considerations.

In Section 7.5.3 the asymptotic covariance matrix of the weighted least squares estimator was found to be $\boldsymbol{V}/n$, with $\boldsymbol{V} = \boldsymbol{M}^{-1}\boldsymbol{Q}\boldsymbol{M}^{-1}$. It may be estimated by the following means. First, for $\eta(\boldsymbol{x}, r/\sigma) = w(\boldsymbol{x}, r/\sigma)r/\sigma$ a natural estimate of $\boldsymbol{Q}$ is given by $\boldsymbol{Q} = \mathrm{E}_F[\eta^2\boldsymbol{x}\boldsymbol{x}^T] \approx \mathrm{E}_{F_n}[\eta^2\boldsymbol{x}\boldsymbol{x}^T]$. Using this approximation and (7.6.3), we obtain

$$\boldsymbol{Q} = \mathrm{E}_F\left[\frac{r^2}{\sigma^2}w^2\left(\boldsymbol{x}, \frac{r}{\sigma}\right)\boldsymbol{x}\boldsymbol{x}^T\right] \approx \frac{1}{n\sigma^2}\sum_{i=1}^{n} r_i^2 w_i^2 \boldsymbol{x}_i \boldsymbol{x}_i^T. \tag{7.6.4}$$

We do not estimate the factor σ^{-2} because it cancels out with other σ's arising in $\boldsymbol{M}^{-1}\boldsymbol{Q}\boldsymbol{M}^{-1}$. We estimate

$$\boldsymbol{M} = \mathrm{E}_F\left[\frac{\partial\eta(x, r/\sigma)}{\partial r}\boldsymbol{x}\boldsymbol{x}^T\right]$$

when $w(\boldsymbol{x}, r/\sigma)$ is defined by (7.6.2), and find that terms cancel so that we are left with

$$\boldsymbol{M} = \mathrm{E}_F\left[\frac{1}{\sigma}\psi_k'\left(\frac{r}{\sigma v(\boldsymbol{x})}\right)\boldsymbol{x}\boldsymbol{x}^T\right] \approx \frac{1}{n\sigma}\sum_{i=1}^{n}\psi_k'(d_i)\boldsymbol{x}_i\boldsymbol{x}_i^T. \tag{7.6.5}$$

Now let $\boldsymbol{D}_1$ be the diagonal matrix with ith diagonal entry $\psi_k'(d_i)$ and let $\boldsymbol{D}_2$ be the diagonal matrix with ith diagonal entry $w_i^2 r_i^2$. Then the approximations (7.6.4) and (7.6.5) may be combined to obtain an estimate of the asymptotic covariance matrix $\boldsymbol{V}/n$ which is conveniently expressed

$$\frac{\boldsymbol{V}}{n} \approx \frac{1}{n}(\boldsymbol{X}^T\boldsymbol{D}_1\boldsymbol{X})^{-1}(\boldsymbol{X}^T\boldsymbol{D}_2\boldsymbol{X})(\boldsymbol{X}^T\boldsymbol{D}_1\boldsymbol{X})^{-1}. \tag{7.6.6}$$

We multiply this estimate by the correction factor $n/(n-p-1)$ in our calculation of the asymptotic estimates of the standard error (see Appendix D.4.3). When all the weights are 1, the weighted least squares procedure reduces to ordinary least squares. Even in this case the estimate (7.6.6) of the covariance matrix (times the correction) will not in general agree with $s^2(\boldsymbol{X}^T\boldsymbol{X})^{-1}$ given in Section 7.1.1. The reason for the discrepancy is that the latter is based on an exact finite sample result while the former is based on asymptotic theory. Nevertheless, we have chosen the correction factor $n/(n-p-1)$ because when the weights do *not* depend on position, then a simpler formula for the asymptotic covariance matrix (7.6.8) suggests this factor, as explained below.

The reader who is familiar with the random row formulation of regression of Sections 7.4 and 7.5 may use a conditioning argument to show that when $\eta(\boldsymbol{x}, r/\sigma) = \psi(r/\sigma)$ and r is the residual of a Fisher-consistent estimator, then the matrices $\boldsymbol{M}$ and $\boldsymbol{Q}$ become

$$\begin{aligned} \boldsymbol{M} &= \mathrm{E}\left[\frac{1}{\sigma}\psi'\left(\frac{r}{\sigma}\right)\right]\mathrm{E}[\boldsymbol{x}\boldsymbol{x}^T] \\ \boldsymbol{Q} &= \mathrm{E}\left[\psi^2\left(\frac{r}{\sigma}\right)\right]\mathrm{E}[\boldsymbol{x}\boldsymbol{x}^T]. \end{aligned} \tag{7.6.7}$$

Hence in this case the asymptotic covariance matrix is of the form

$$\begin{aligned} \frac{\boldsymbol{V}}{n} &= \frac{\mathrm{E}[\sigma^2\psi^2(r/\sigma)]}{\left(\mathrm{E}[\psi'(r/\sigma)]\right)^2}\frac{\left(\mathrm{E}[\boldsymbol{x}\boldsymbol{x}^T]\right)^{-1}}{n} \\ &\approx \frac{n\sum\sigma^2\psi^2(r_i/\sigma)}{\left(\sum\psi'(r_i/\sigma)\right)^2}(\boldsymbol{X}^T\boldsymbol{X})^{-1} \end{aligned} \tag{7.6.8}$$

where σ is estimated by $s_{(i)}$, say. When ψ is the identity function, the M-estimator reduces to ordinary least squares. In this case the estimate of σ drops out of the last expression and the above approximation to $\boldsymbol{V}/n$ becomes

$$\frac{\boldsymbol{V}}{n} \approx \frac{\sum r_i^2}{n}(\boldsymbol{X}^T\boldsymbol{X})^{-1}.$$

If this approximation is multiplied by the correction factor $n/(n-p-1)$ as recommended by Huber (1981), p. 173, we obtain the classical estimate of the covariance matrix for least squares.

7.6.2 Illustration of Computations

We assume in this and the following examples that the statistician has already completed preliminary diagnostic and model building procedures and wants to proceed with the data and the classical normal model for errors, even though outliers and large leverage points may still be present. We will use the Minitab macros RWLS1, RWLS2, and RWLS3 of Appendix D, which in turn calculate the least squares estimates and initial Welsch estimates, further iterations of the Welsch estimates, and finally the estimated standard errors of these estimates, as described above. We will use the notation Welsch ($k = 2$) to indicate that the constant $k = 2$ defines the Welsch weights in (7.6.3) with $c = k\sqrt{(p+1)/n}$. We have found that $k = 1$ leads to a better compromise between the desire for efficiency and robustness.

Example 1: Growth of Prices (Continued from Section 7.1.1). Recall that in this example what appears to be an obvious significant regression is not found to be so by the least squares approach, because the presence of one outlier inflates the estimated standard error of the least squares coefficients.

To calculate the robust weighted least squares estimates using the Minitab macros in Appendix D, the dependent variable must be in column c1 and the dependent variables in columns c2-ck1, where k1 is the number of dependent variables, $= p + 1$, including a constant term. The value of k defining the Welsch weights must be stored in k4. More details regarding the output are stored in the macros themselves. A brief illustration of the commands and some abbreviated output follows:

```
MTB>  let k1=2 # This defines p+1.
MTB>  let k4=2 # k4 * SQRT((p+1)/n) defines the value of c in (7.6.3).
MTB>  exec 'rwls1'
LScoef = least squares estimates = -1.25321 and 0.07536
Wcoef = initial Welsch estimates = -1.30574 and 0.07536
MTB>  exec 'rwls2' 3
Latest set of Welsch estimates = -1.30965 and 0.07536
Latest set of Welsch estimates = -1.30991 and 0.07536
Latest set of Welsch estimates = -1.30993 and 0.07536
MTB>  exec 'rwls3'
Estimated variance-covariance matrix of the coefficients
1.90097 -0.04470
-0.04470 0.00105
Estimates of the standard errors of the coefficients
1.37876 0.03248
```

Note that the coefficient of interest β_1 is estimated to be .07536 by both the least squares approach and the weighted least squares approach. However, the standard error of the least squares estimate was found earlier to be $\approx .064$, while that of the weighted least squares estimate is $\approx .032$. The robust approach reveals the significance of the regression in the presence of the outlier, while the least squares approach does not. Of course it is important to know whether these estimates of the standard error are reliable. A small Monte Carlo study by Sheather and Hettmansperger (1987) showed excellent agreement between the covariance matrix estimates described above and bootstrap estimates for the growth of prices example. However, more research needs to be done into the reliability of these estimates of standard error.

The Minitab macros also print out the weights utilized in each stage of weighted least squares, but this is not shown above. All the weights were 1.0 except for case 4, which was given successive weights .479, .443, .441, and .440 in the four iterations.

For the sake of comparison, we also repeated the above calculations with $k = 1$. This value of k should lead to more robust but less efficient results than the above estimates ($k = 2$).

The fourth iterate of reweighted least squares yielded $y = -1.815 + .087x$ and the final weights were all 1 except $w_4 = .22$ and $w_7 = .55$. Note that two points are downweighted in this case.

The estimated standard errors of the regression coefficients were 1.57 and .038, respectively, which is not very different from those for the $k = 2$ results.

Growth of Prices Data Augmented by an Additional Observation

To illustrate how the iterative Welsch ($k = 2$) estimators behave on an example dominated by a gross outlier at a large leverage point, we now augment the growth of prices data of Table 7.1 by including the value 15.5 of the year 1947. The initial (least squares) and successive Welsch regression equations are

$$\hat{y}_{LS} = -47.49 + 1.18x$$

$$\hat{y}_1 = -4.289 + .148x$$

$$\hat{y}_2 = -3.094 + .119x$$

$$\hat{y}_3 = -3.074 + .119x.$$

These least squares and successive Welsch iterative solutions are plotted in Figure 7.8, along with the modified data. The weights were all 1.0 except for case 8, which received successive weights of .039, .024, and .023.

The estimated standard errors for the least squares estimates are, respectively, 27.8 and .637, while the corresponding estimated standard errors for the Welsch estimates [based on (7.6.5)] are 3.23 and .077. The regression parameter is found by ordinary least squares to have a t-value of $1.18/.637 \approx 1.85$, while the corresponding value for the robust estimate is $.119/.077 \approx 1.55$. □

Example 2: Stack Loss Data (Continued from Section 7.2.5.) In this example and those to follow we table the Welsch robust regression estimates and their standard errors for $k = 1$ and $k = 2$ for comparison with the least squares estimates. The first two examples have been chosen because they have already been analyzed by a variety of robust methods in the literature,

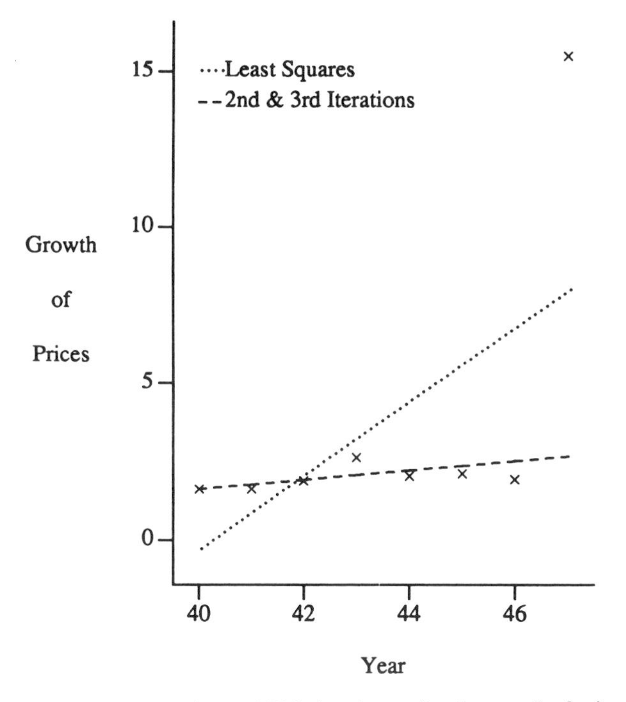

FIGURE 7.8 Least squares and iterated Welsch estimates for the growth of prices over 1940–1947.

and we may therefore see how well the Welsch proposal does relative to them.

The most robust version of the Welsch procedures considered here (the case $k = 1$) appears to downweight precisely those points which are identified by other methods as outliers. And the Welsch ($k = 1$) procedure does compare favorably with a number of other robust regression estimators in terms of the pragmatic criteria of the median absolute deviation and interquartile range of the residuals. These quantities measure how well the fitted model agrees with the bulk of the data. When the residuals are symmetrically distributed, the interquartile range will be twice the median absolute deviation, but in general they will give different robust measures of the spread of the residuals.

We will not show the details of the calculations that were illustrated in the last section, but the student may check them by applying the Minitab macros to the data. The Welsch ($k = 2$) estimates are quite different from the least squares estimates, even though only the case 21 was downweighted with $w_{21} = .34$. The Welsch ($k = 1$) estimates provide an even better fit

Table 7.7 Stack Loss: Welsch Regression Estimates, Their Estimated Standard Errors, and IQR and MAD of the Residuals

	Least Squares	$k = 2$	$k = 1$
$\hat{\beta}_0$	−39.92 (11.9)	−42.13 (5.76)	−40.93 (5.25)
$\hat{\beta}_1$	.7156 (.135)	.8170 (.148)	.8391 (.137)
$\hat{\beta}_2$	1.2953 (.368)	1.0156 (.403)	.7970 (.276)
$\hat{\beta}_3$	−.1521 (.156)	−.1258 (.079)	−.1049 (.067)
MAD	1.87	1.84	1.58
IQR	4.31	3.82	2.92
$\overline{w}$	1	.97	.89

in terms of the mean absolute deviation (MAD) and interquartile range (IQR) of the residuals, and in this case the method downweights four cases: $w_1 = .49$, $w_3 = .59$, $w_4 = .46$, and $w_{21} = .16$ (Table 7.7). □

In a study of trimmed least squares estimation Ruppert and Carroll (1980) compared five robust procedures with regard to the criteria MAD and IQR of the residuals on the stack loss data and the water salinity data (see the next example). For this data set the Welsch ($k = 1$) procedure does about as well as the Huber and trimming estimators, but not as well as the least absolute deviation estimator or the Andrews redescending estimator.

We also calculated the Welsch ($k = 1000$) estimates and found (not surprisingly) that they agreed with the least squares estimates. Of more interest were the estimates of their standard errors, which were 7.13, .177, .496, and .096, respectively. These values do not agree with the classical estimates given in the least squares column of Table 7.7, because they are based on the asymptotic formula (7.6.6). While they are not misleading, they are not as close as we would like.

Example 3: Water Salinity Data. The water salinity data set is described in Appendix C and is another example that is widely discussed in the literature. [See Hettmansperger (1987) and other references below.] The Welsch ($k = 2$) procedure downweights only two cases ($w_{15} = .88$ and $w_{16} = .09$). The more robust procedure with $k = 1$ downweights five cases ($w_5 = .40$, $w_8 = .86$, $w_9 = .57$, $w_{15} = .43$, and $w_{16} = .04$).

The median absolute deviation and the interquartile range of the residuals were about the same for the Welsch ($k = 2$) and ($k = 1$) procedures. These values compare quite favorably with five other robust estimators as reported in Table 4 of Ruppert and Carroll (1980); only the least absolute deviations and Andrew M-procedure are marginally better in terms of these criteria. In a separate study of trimmed-mean and bounded influence regression estimators by de Jongh, de Wet, and Welsh (1988) it was found

Table 7.8 Water Salinity: Welsch regression estimates, their standard errors, and IQR and MAD of the residuals

	Least Squares	$k = 2$	$k = 1$
$\hat{\beta}_0$	9.59 (3.12)	16.89 (3.84)	18.88 (4.61)
$\hat{\beta}_1$	.777 (.086)	.715 (.059)	.725 (.051)
$\hat{\beta}_2$	−.025 (.161)	−.142 (.161)	−.185 (.151)
$\hat{\beta}_3$	−.295 (.107)	−.570 (.144)	−.656 (.181)
MAD	.73	.47	.51
IQR	1.47	1.05	1.07
$\overline{w}$	1	.96	.90

that only two estimators gave (slightly) better fits to the data according to these same criteria than the results from the Welsch procedure shown in Table 7.8.

While here we have not plotted the Studentized residuals or examined in detail the data which have large values of DFITS, it must be emphasized that such diagnostic procedures are routine and an essential part of linear regression analysis. The robust procedures are complementary methods for identifying aberrant cases (those that are downweighted) and for finding a robust fit to the bulk of the data. □

Example 4: Scottish Hill Races Data. The Scottish Hill Races data were originally reported and analyzed by Atkinson in the discussion of the article by Chatterji and Hadi (1986). The dependent variable is the record time in minutes and the independent variables are distance in miles and climb in feet. The least squares line and a table of fitted values and diagnostics are given below in the form of edited Minitab output:

```
The regression equation is
RecTime = - 8.99 + 6.22 Distance + 0.0110 Climb
Predictor        Coef         Stdev        t-ratio          p
Constant       -8.991         4.303          -2.09      0.045
Distance       6.2179        0.6011          10.34      0.000
Climb        0.011048      0.002051           5.39      0.000
s = 14.68     R-sq = 91.9%     R-sq(adj) = 91.4%

Unusual Observations
Obs. Distance     RecTime        Fit   Stdev.Fit     Residual     St.Resid
7        16.0      204.62     173.36        9.52        31.26       2.80RX
11       28.0      192.67     188.31       12.19         4.36       0.53 X
18        3.0       78.65      13.53        3.45        65.12       4.57R
R denotes an obs. with a large st. resid.
X denotes an obs. whose X value gives it large influence.
```

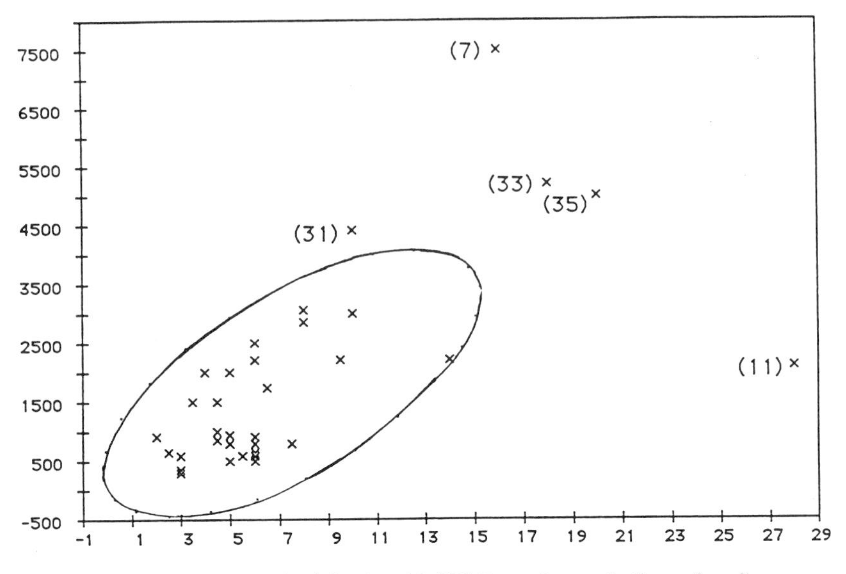

FIGURE 7.9 (Distance, Climb) for Scottish Hill Races data and ellipse of equileverage.

Note that Minitab routinely draws attention to observations 7, 11, and 18, but further diagnostics were requested and are given in Table 7.9. We also plot the explanatory variables and ellipse of equileverage in Figure 7.9 to visually identify the points of high leverage. This is easily done with the Minitab macros in Appendix D.3.

Atkinson notes that normal plots of Studentized residuals (t-residuals) reveal cases 7 and 18 as outliers. If these two observations are deleted, then an additional plot reveals that case 33 is an outlier, the point being that cases 7 and 18 are masking the presence of an outlier in case 33.

Atkinson then applies Rousseeuw's least median of squares regression [Rousseeuw (1984)] which identifies cases 7, 18, 11, 33, and 35 as having the largest absolute residuals and hence being worthy of further attention. When all five of these values are deleted from the fit, Atkinson finds that an analysis of the resulting Studentized residuals shows that cases 11 and 35 agree with the bulk of the data, and so again cases 7, 18, and 33 are anomalous.

By way of comparison, the diagnostics in Table 7.9 pick up cases 7 and 18 as having a large t-residual, cases 7, 11, 33, and 35 as having more than twice the average leverage, and cases 7, 11, and 18 as having large DFITS by criterion (7.2.11). Thus even though case 33 has a large leverage, it

Table 7.9 Scottish Hill Races: Data,[†] Fitted Values, and Diagnostics

Case	Rec Time	Dist	Climb	r_i^*	$\hat{Y}_i$	$r_{(i)}^*$	h_{ii}	DFITS$_i$
1	16.083	2.5	650	.16	13.7	.16	.05	.04
2	48.350	6.0	2500	−.53	55.9	−.52	.05	−.12
3	33.650	6.0	900	−.32	38.3	−.32	.04	−.06
4	45.600	7.5	800	−.06	46.5	−.06	.05	−.01
5	62.267	8.0	3070	−.87	74.7	−.87	.06	−.21
6	73.217	8.0	2866	.06	72.4	.06	.05	.01
7	204.617	16.0	7500	2.80	173.4	3.17	.42	2.70
8	36.367	6.0	800	−.05	37.2	−.05	.04	−.01
9	29.750	5.0	800	−.08	30.9	−.08	.04	−.02
10	39.750	6.0	650	.30	35.5	.29	.05	.06
11	192.667	28.0	2100	.53	188.3	.53	.69	.79
12	43.050	5.0	2000	−.08	44.2	−.08	.04	−.02
13	65.000	9.5	2200	−.65	74.4	−.64	.03	−.12
14	44.133	6.0	500	.72	33.8	.71	.05	.17
15	26.933	4.5	1500	−.60	35.6	−.59	.04	−.12
16	72.250	10.0	3000	−.98	86.3	−.98	.04	−.21
17	98.417	14.0	2200	−.28	102.4	−.28	.08	−.08
18	78.650	3.0	350	4.56	13.5	7.6	.06	1.84
19	17.417	4.5	1000	−.88	30.0	−.87	.04	−.17
20	32.567	5.5	600	.05	31.8	.05	.05	.01
21	15.950	3.0	300	.21	13.0	.21	.06	.05
22	27.900	3.5	1500	−.10	29.3	−.10	.05	−.02
23	47.650	6.0	2200	−.35	52.6	−.34	.04	−.07
24	17.933	2.0	900	.32	13.4	.31	.06	.08
25	18.683	3.0	600	.17	16.3	.16	.05	−.04
26	26.217	4.0	2000	−.82	38.0	−.82	.05	−.20
27	34.433	6.0	800	−.19	37.2	−.19	.04	−.04
28	28.567	5.0	950	−.28	32.6	−.27	.04	.05
29	50.500	6.5	1750	−.02	50.8	−.02	.03	.00
30	20.950	5.0	500	−.47	27.6	−.46	.05	−.10
31	85.583	10.0	4400	−1.18	101.8	−1.19	.12	−.44
32	32.383	6.0	600	−.18	34.9	−.18	.05	−.04
33	170.250	18.0	5200	.74	160.4	.73	.17	.33
34	28.100	4.5	850	−.02	28.4	−.02	.04	.00
35	159.833	20.0	5000	−.82	170.6	−.81	.19	−.39

[†] Reprinted with permission of the Institute of Mathematical Statistics.

Table 7.10 Scottish Hill Races: Welsch Regression Estimates, Their Standard Errors, and IQR and MAD of the Residuals

	Least Squares	$k = 2$	$k = 1$
$\hat{\beta}_0$	−8.99 (4.30)	−9.56 (2.82)	−8.92 (2.10)
$\hat{\beta}_1$	6.22 (0.60)	6.58 (0.17)	6.61 (0.12)
$\hat{\beta}_2$	.011 (.002)	.009 (.002)	.008 (.001)
MAD	4.16	3.80	3.66
IQR	9.98	8.37	7.46
$\overline{w}$	1	.96	.93

would not be picked up by the least squares diagnostics as contributing badly to the fit.

The question arises as to whether the robust Welsch procedure will downweight all five cases or only those that are inconsistent with the bulk of the data.

The Welsch ($k = 2$) approach downweights cases 7 and 18 ($w_7 = .15$ and $w_{18} = .32$); these are two of the three cases that are identified by criterion (7.2.11). However, Welsch's approach with $k = 1$ downweights case 33 as well: $w_7 = .07$, $w_{18} = .16$, and $w_{33} = .42$. It is interesting that this iterative procedure does not downweight cases 11 and 35, in agreement with Atkinson's finding that they are consistent with the bulk of the data.

Finally [as Atkinson (1988) notes], routine checking of the cases identified by the diagnostics above reveals that case 18 is indeed a blunder. The 78 minutes listed for a 3-mile race should have been 18 minutes. This change should be made before reanalyzing the data.

A striking feature of the results in Table 7.10 is the closeness of the regression estimates. However, the standard errors of the Welsch estimators are much smaller than those of the least squares estimates. An obvious reason for this is that the outliers are inflating the least squares standard errors. □

These examples show that the Welsch procedure with $k = 1$ consistently identifies and downweights cases that are likely to affect the way the bulk of the data agree with the resulting fitted model. It is simple to understand and implement, and reasonable estimates of its standard errors are available. The above examples and others indicate that it is reliable in downweighting points which would adversely affect the fit of the model, but not those high leverage points where information is consistent with the bulk of the data. The only question is whether this method is too inefficient for uncontaminated data to be recommended for general usage, and more research on the efficiency of this method (as defined in Section 7.4.5) relative to least squares is required.

7.7 PROBLEMS

Section 7.1

1. For any random vector $\boldsymbol{Y} = [Y_1, \ldots, Y_n]^T$ define its expectation by $\mathrm{E}[\boldsymbol{Y}] = (\mathrm{E}[Y_1], \ldots, \mathrm{E}[Y_n])^T$. Define $\mathrm{Cov}[\boldsymbol{Y}]$ to be the matrix with (i, j)th element $\mathrm{Cov}[Y_i, Y_j]$.

(a) For any matrix $\boldsymbol{M}$ of random variables, define the corresponding matrix $\mathrm{E}[\boldsymbol{M}]$ of elementwise expected values: $(\mathrm{E}[\boldsymbol{M}])_{ij} = \mathrm{E}[\boldsymbol{M}_{ij}]$. Show that $\mathrm{Cov}[\boldsymbol{Y}] = \mathrm{E}[(\boldsymbol{Y} - \mathrm{E}[\boldsymbol{Y}])(\boldsymbol{Y} - \mathrm{E}[\boldsymbol{Y}])^T]$.

(b) Show that if matrix multiplication is well defined, $\mathrm{E}[\boldsymbol{MY}] = \boldsymbol{M}\mathrm{E}[\boldsymbol{Y}]$.

(c) Show further that $\mathrm{Cov}[\boldsymbol{MY}] = \boldsymbol{M}\mathrm{Cov}(\boldsymbol{Y})\boldsymbol{M}^T$.

(d) Verify (7.1.6).

2. Derive the normal equations (7.1.4) and the least squares estimator (7.1.5).

3. Verify consequences 1 and 2 following (7.1.8).

4. In Example 1 of Section 7.1.1 use:

(a) (7.1.5) to find the least squares estimates.

(b) (7.1.7) to estimate σ^2.

(c) Item 4 following (7.1.8) to estimate the standard errors of the regression coefficients.

Section 7.2

5. Verify the basic properties of the hat matrix, Section 7.2.2.

6. Calculate the leverage vector $\boldsymbol{H}_i$ for each design point $\boldsymbol{x}_i^T = [1, 39 + i]$, $i = 1, \ldots, 7$, in Example 1. Verify that the length squared of the leverage vector equals the value shown in Table 7.2.

7. In each of the following cases verify the expression given for the (i, j)th element of the hat matrix.

(a) Simple regression through the origin.

$$\boldsymbol{X} = \begin{bmatrix} x_1 \\ \vdots \\ x_n \end{bmatrix}; \qquad h_{ij} = \frac{x_i x_j}{\sum_{k=1}^{n} x_k^2}.$$

Points have large leverage if they are far from the origin.

(b) Simple linear regression.

$$X^T = \begin{bmatrix} 1 & 1 & \dots & 1 \\ x_1 & x_2 & \dots & x_n \end{bmatrix}; \qquad h_{ij} = \frac{1}{n} + \frac{(x_i - \overline{x})(x_j - \overline{x})}{\sum_{k=1}^{n}(x_k - \overline{x})^2}.$$

Here points $[1, x_i]$ have large leverage if x_i is far from $\overline{x}$.

8. (a) Verify that $r_i^* \xrightarrow{d} N(0,1)$ as $n \to \infty$ for fixed p.

(b) Show that under the classical assumptions for linear regression $(r_i^*)^2/(n-p-1)$ has a Beta distribution.

9. Verify (7.2.1) and (7.2.2):

(a) $\boldsymbol{r} = (\boldsymbol{I} - \boldsymbol{H})\boldsymbol{\epsilon} = (\boldsymbol{I} - \boldsymbol{H})\boldsymbol{Y}$.

(b) $\text{Cov}[\boldsymbol{r}] = (\boldsymbol{I} - \boldsymbol{H})\sigma^2$.

10. Use (7.2.8) to show that $Y_i - \hat{Y}_{(i)} = (Y_i - \hat{Y}_i)/(1 - h_{ii})$ and hence that $\text{Var}[Y_i - \hat{Y}_{(i)}] = \sigma^2/(1 - h_{ii})$.

11. Check the standardized residuals found in Table 7.3.

12. Show that the mean residual sum of squares for the ith deleted sample $s_{(i)}^2 = \Sigma_i(Y_{(i)} - \hat{Y}_{(i)})^2/(n-p-2)$ can be calculated from the mean residual sum of squares s^2 for the full sample, h_{ii}, and the ith residual $Y_i - \hat{Y}_i$; i.e., verify (7.2.7).

13. Verify the details leading from (7.2.6) to the claim that the Studentized residuals $r_{(i)}^*$ have the t-distribution.

14. Verify the effect of deletion on $\hat{\beta}_i$ and $\hat{Y}_i$ shown in (7.2.9).

15. Verify the diagnostics shown in Table 7.4 for the growth of prices data. Which points, if any, warrant further scrutiny?

16. Carry out the details showing the complementary role of the three diagnostics for simple linear regression (Section 7.2.5).

17. Try to determine which points, if any, are causing trouble with the fit in Brownlee's stack loss data.

(a) Find the points of most concern with regard to leverage.

(b) Find the points of significant t-residual.

(c) Find the points suggested by the rule of thumb (7.2.11) for DFITS.

(d) Repeat the entire procedure after dropping the third explanatory variable.

*Section 7.3**

18. For a fixed sample size n, the leverages are nondecreasing in the number of explanatory variables, that is, $h_{ii} = h_{ii}(p)$ is nondecreasing in p. [*Hint*: From (7.3.1) it follows that

$$(\boldsymbol{H})_{ii} = (\boldsymbol{H}_A)_{ii} + (\boldsymbol{H}_B^*)_{ii} \geq (\boldsymbol{H}_A)_{ii}$$

using the fact that the diagonal members of the hat matrix are nonnegative (property 2 of $\boldsymbol{H}$).] Hence increasing the number of variables from q to $p+1$ cannot decrease the leverage of the ith point. Geometrically, this says that in a higher dimensional design space, there is more room for a point to apply leverage.

19. Multiple regression with $p = 2$ explanatory variables. Let $z_{ij} = x_{ij} - \overline{x}_{\cdot j}$, $j = 1, 2$ and $i = 1, \ldots, n$. Then

$$Z^T Z = \begin{bmatrix} S_{11} & S_{12} \\ S_{12} & S_{22} \end{bmatrix}$$

where $S_{11} = \sum_{i=1}^n z_{i1}^2$, $S_{12} = \sum_{i=1}^n z_{i1} z_{i2}$, and $S_{22} = \sum_{i=1}^n z_{i2}^2$. The two eigenvalues of $Z^T Z$ are

$$\lambda_+ = \frac{S_{11} + S_{22}}{2} + \frac{1}{2}\sqrt{(S_{11} - S_{22})^2 + 4S_{12}^2}$$

$$\lambda_- = \frac{S_{11} + S_{22}}{2} - \frac{1}{2}\sqrt{(S_{11} - S_{22})^2 + 4S_{12}^2}.$$

If $D = S_{11} - S_{22} = 0$, then the eigenvector corresponding to λ_+ is

$$\boldsymbol{U}_+ = \begin{bmatrix} \frac{1}{\sqrt{2}} \\ \frac{1}{\sqrt{2}} \end{bmatrix}.$$

If $D \neq 0$ and $\gamma = S_{12}/D$,

$$\boldsymbol{U}_+ = \begin{bmatrix} u \\ \sqrt{1 - u^2} \end{bmatrix}$$

where

$$u^2 = \frac{2\gamma^2}{[1 + \operatorname{sgn}(D)\sqrt{1 + 4\gamma^2} + 4\gamma^2]}.$$

20. **(a)** Check the details supplied for the proof of Theorem 7.1.
(b) Supply the missing details in parts (iii) and (v).

21. Verify that simplicial designs, as defined in Section 7.3.5, are:
(a) In fact scalar designs.
(b) Nothing more than the vertices of a translated simplex in R^{p+1}.

*Section 7.4**

22. In simple linear regression the leverages are given by expression (7.4.2). For each of the following examples, determine whether the least squares estimates have an asymptotic normal distribution by appealing to Theorem 7.2, after showing that:
(a) $x_i = i$, $i = 1,\ldots,n$ implies $\max_i h_{ii} \sim 4/n$.
(b) $x_i = 2^i$, $i = 1,\ldots,n$, implies $\max_i h_{ii} \sim \frac{3}{4}$.

23. Given any sequence $v_1, v_2, \ldots$ define for each n a matrix $\boldsymbol{X} = \boldsymbol{X}^n$ (the dependence on n is so clearly understood that the superscript notation may be suppressed) with

$$\boldsymbol{X}^T = \begin{bmatrix} 1 & 1 & \cdots & 1 \\ v_1 & v_2 & \cdots & v_n \end{bmatrix}.$$

Unit scaling of $\boldsymbol{X}$ simply means dividing each column of $\boldsymbol{X}$ by its length so that the resulting columns have length 1. Call

$$\boldsymbol{X}^T = \begin{bmatrix} \dfrac{1}{\sqrt{n}} & \dfrac{1}{\sqrt{n}} & \cdots & \dfrac{1}{\sqrt{n}} \\ x_1^{(n)} & x_2^{(n)} & \cdots & x_n^{(n)} \end{bmatrix}$$

where $x_i^{(n)} = v_i / \|\boldsymbol{v}\|^{(n)}$, with $\|\boldsymbol{v}\|^{(n)} = \{\sum_{i=1}^n v_i^2\}^{1/2}$. Then

$$\boldsymbol{X}^T\boldsymbol{X} = \begin{bmatrix} 1 & \dfrac{\sum_{i=1}^n v_i}{\sqrt{n}\|\boldsymbol{v}\|^{(n)}} \\ 0 & 1 \end{bmatrix}$$

may well have a limit as $n \to \infty$.
(a) Give an example of a sequence $\{v_i\}$ such that $(1/n)(\boldsymbol{X}^T\boldsymbol{X}) \to \boldsymbol{\Sigma}$, a positive definite matrix.
(b) Give an example of $\{v_i\}$ where the convergence in (a) fails.

24. Give the details of (7.4.6).

25. Prove the identities in (7.4.7) and (7.4.8).

26. Show that when the independent variables have a Poisson distribution in simple linear regression (see Section 7.4.2), then $\max_i h_{ii} \xrightarrow{\text{a.s.}} 0$.

27. Let $T'_F(H) = \lim_{t \to 0}[(T(F+tH) - T(F))/t]$ be the Gateaux derivative of T at F in the derivation H for any functional $T(F)$.

(a) Show that if $T(F) = \int k(x)\,dF(x) = \mathrm{E}_F[k(x)]$ for some kernel function k, then $T'_F(H) = \mathrm{E}_H[k(x)]$.

(b) Let $\boldsymbol{\gamma}(F) = \mathrm{E}_F[\boldsymbol{x}y]$ and $\boldsymbol{\Sigma}(F) = \mathrm{E}_F[\boldsymbol{x}\boldsymbol{x}^T]$. Use (a) to find $\boldsymbol{\gamma}'_F(H)$ and $\boldsymbol{\Sigma}'_F(H)$.

(c) Define $\boldsymbol{T}(F)$ by $\boldsymbol{\Sigma}(F)\boldsymbol{T}(F) = \boldsymbol{\gamma}(F)$. Apply the product rule for derivatives to the left-hand side of this expression to effortlessly find (7.4.11).

28. Show directly that $\mathrm{E}_F[\boldsymbol{IF}(\boldsymbol{x}^T, y; F)] = 0$, when $\boldsymbol{IF}$ is given by (7.4.11).

29. (a) State the conditions (7.4.16) and (7.4.17) in the context of simple linear regression (Example 1 of Section 7.4.2).

(b) Are the conditions of (a) satisfied if the ϵ_i's have a symmetric distribution with finite variance and the x_i's have a Poisson distribution?

(c) Are the conditions of (a) satisfied if the Poisson distribution is replaced by one with only two moments existing?

30. Use the assumptions of the corollary to Theorem 7.3 to verify the condition (7.4.17).

*Section 7.5**

31. Verify the claim after (7.5.7) that if the marginal distribution of $\boldsymbol{x}$ is nonsingular in R^p, that is, if $\mathrm{Cov}_G[\boldsymbol{x}]$ is positive definite, then so is $\boldsymbol{\Sigma}(G) = \mathrm{E}_G[\boldsymbol{x}\boldsymbol{x}^T]$, and hence $\boldsymbol{\Sigma}(G)$ may be inverted.

32. Suppose that $\boldsymbol{x}$ has a bivariate normal distribution with mean $\left[\begin{smallmatrix}0\\0\end{smallmatrix}\right]$ and covariance matrix

$$\Sigma = \begin{bmatrix} \sigma_1^2 & \sigma_{12} \\ \sigma_{12} & \sigma_2^2 \end{bmatrix}.$$

Find the "leverage" of $\boldsymbol{x}$ defined by $q(\boldsymbol{x}) = (\boldsymbol{x} - \boldsymbol{\mu})^T \Sigma^{-1} (\boldsymbol{x} - \boldsymbol{\mu})$.

33. Suppose that $G = F_{\beta,\sigma}$ satisfies the regression structure (7.5.8), and that γ, Σ are defined by (7.5.6). Verify that $\gamma(G) = \Sigma(G)\boldsymbol{\beta}$.

34. Find the asymptotic covariance function associated with the explicit weighted least squares estimator by using (7.5.16), that is, show that

$$\mathrm{E}[\boldsymbol{IF}_{T_w,F}\boldsymbol{IF}_{T_w,F}^T] = \boldsymbol{Q} - 2\mathrm{E}[\eta\epsilon\boldsymbol{x}\boldsymbol{x}^T]\Sigma^{-1}\boldsymbol{C} + \sigma^2\boldsymbol{C}\Sigma^{-1}\boldsymbol{C}.$$

Section 7.6

35. **(a)** Show that the estimators defined by (7.5.3) and (7.5.5) may be rewritten in weighted least squares form (7.6.1).

(b) Show that when Welsch's suggestions are employed (proposal 4 of Section 7.6.1) that the weights are given by (7.6.3).

36. **(a)** Substitute F_n for F in the definition of $\boldsymbol{Q}$ after (7.5.18) to obtain the approximation in (7.6.4).

(b) After evaluation of $\boldsymbol{M}$ given in (7.5.18) substitue F_n for F to obtain the approximation after (7.6.4).

37. Find the least squares estimates and standard errors for the growth of prices example given in Section 7.1. Use the results to find the iterated Welsch estimates and their standard errors, with or without the aid of the Minitab programs given in Appendix D.4.

38. Continuing Problem 37, modify the growth of prices data of Table 7.1 by replacing 2.64 of the year 1943 by 2.0 and 1.94 of the year 1946 by 3.00. Using the Minitab macros in Appendix D.4, show that the successive Welsch regression equations are

$$\hat{y}_{\mathrm{LS}} = -6.077 + .189x$$

$$\hat{y}_1 = -3.628 + .130x$$

$$\hat{y}_2 = -3.344 + .123x$$

$$\hat{y}_3 = -3.324 + .123x.$$

Plot these least squares and successive Welsch iterative solutions, along with the modified data. Note that the weights are all 1.0 except for case 7, which receives successive weights of .144, .087, and .083. Also show that the estimated standard errors for the least squares estimates are, respectively, 1.98 and .05, while the corresponding estimated standard errors for the Welsch estimates [based on (7.6.5)] are .89 and .02.

The regression parameter is found by ordinary least squares to have a t-value of $.189/.05 \approx 4$, while the corresponding value for the robust estimate is $.123/.02 \approx 6$.

39. Find the iterated Welsch ($k = 1$) estimates and their standard errors for the stack loss data of Section 7.2, using only the first two explanatory variables. Compare them with the results for ordinary least squares.

40. Find the iterated Welsch ($k = 1$) estimates and their standard errors for the water salinity data, and compare the t-values for least squares regression with those for the robust regression estimates. Would different conclusions be drawn using the two methods?

41. Find the iterated Welsch ($k = 1$) estimates and their standard errors for the Scottish Hill Races data after correcting the case 18 record time. It should be 18.65 minutes. Compare your results with the ordinary least squares analysis.

42. Consider simple linear regression through the origin based on a sample of size n. In this problem we illustrate the effects of one additional datum $(\boldsymbol{x}_{n+1}^T, Y_{n+1}) = [x \; y]$ on the three diagnostic measures: t-residuals, leverage, and DFITS. Then we will see how Welsch's robust regression procedure accomodates the additional observation. Let

$$\boldsymbol{Y}^T = [y_1, y_2, \ldots, y_n, y]$$
$$\boldsymbol{X}^T = [x_1, x_2, \ldots, x_n, x]$$

and assume $\boldsymbol{Y} = \boldsymbol{X}\boldsymbol{\beta} + \boldsymbol{\epsilon}$, where $\boldsymbol{\epsilon} \sim N(0, \sigma^2 I_{n+1})$. Denote the leverage h_{n+1} of x by h.

(a) Show that regardless of the value of y, the leverage h at x satisfies $0 \le h$ and $h \to 1$ as $x \to \infty$. Hence h effectively takes on all values from 0 to 1 as x varies.

(b) Show that the t-residual of (x, y) is

$$r_{(n+1)}^* = \frac{[y - \hat{\beta}_n x]}{s_n}(1 - h)^{1/2}.$$

where $\hat{\beta}_n$ and s_n are the classical estimates of β and σ based on $x_1, \ldots, x_n$. Hence for fixed x the t-residual $r_{(n+1)}^*$ takes on all real values as y varies.

(c) Show that

$$\text{DFITS}_{n+1} = h^{1/2}\left[\frac{y - \hat{\beta}_n x}{s_n}\right].$$

Plot contours of equal DFITS in the (x, y) plane.

(d) Explain in general terms how to calculate the Welsch robust regression estimates. Then in particular concentrate on the simple linear regression setting discussed above and assume that the $(n+1)$st point (x, y) is the only point with a large value of DFITS. Show explicitly how it would be downweighted by Welsch's method. Then express the first iterated estimate of Welsch's procedure in terms of the initial estimate and other functions of the data.

7.8 COMPLEMENTS

7.8.1 Other Approaches to Robust Regression

Bounded influence estimators are not known to be optimal in any of the partial orderings of estimators determined by the relations in Section 7.4.5. Some optimality results have been obtained which are extensions of Hampel's bounded influence result for location (Theorem 4.3). The idea is to place a bound on the influence function (7.5.18), or a standardized version of it, and subject to this bound to choose the M-estimator that minimizes the trace of the asymptotic covariance matrix. The form of the solution depends on the norm used to determine the bound. See Krasker and Welsch (1982) and Hampel et al. (1986), Chapter 6, for a thorough discussion and further references.

Welsh (1987) has proposed an extension of the trimmed mean to linear regression, which is relatively simple to compute and which does not require the concomitant estimate of a scale estimate. However, it is highly dependent on the choice of initial estimator to avoid a low breakdown point. The preferred choice of initial estimator of Welsh and some of the discussants of his paper is the least absolute deviations estimator.

The bounded influence regression methods explained in Section 7.6 can be computed easily using standard computer packages. However, Maronna, Bustos, and Yohai (1979) have shown that the breakdown point of these estimators cannot exceed $1/p$, where p is the number of regression parameters. Thus certain configurations of mutiple outliers can break down these estimators, and the proportion of outliers required is very small in higher dimensions. Thus alternative procedures are desired for such dimensions, and some progress has been made.

A number of estimators with breakdown point equal to 0.5 have been proposed in the last few years. The first of these was the repeated median estimator of Siegel (1982). Rousseeuw (1984) proposed the least median of squares estimator, which is just the value of β which minimizes the median of the squared residuals. Rousseeuw and Yohai (1984) proposed a class of estimators based on the minimization of a robust estimate M-estimator of scale (S-estimators). All these high breakdown point estimators are highly inefficient for regression models with normal errors, but they provide very robust starting values to be followed by more efficient reweighted procedures.

Recently, Yohai (1987) has presented a new class of estimators (MM-estimators), which have both high breakdown point and high efficiency under normal errors. MM-estimators are defined by a three-stage procedure.

Rousseeuw and Leroy (1987) have written an excellent introduction to robust regression and outlier detection, which is centered on Rousseeuw's least median of squares method for robust regression [Rousseeuw (1984)]. Their book argues the case for using robust regression methods to identify outliers, as well as to obtain robust estimates.

There has been lively discussion in the literature of the relative merits of least squares regression diagnostics and robust regression in detecting outliers. Carroll and Ruppert (1988) make the point that robust estimators often provide regression diagnostics as by-products and these can be quite useful, but they are not substitutes for specifically designed diagnostics. Unfortunately, in data sets with several influential points some of these points can exert such a strong influence on the least squares fit that other influential points are masked and, hence, are not detected by least squares based diagnostics. McKean, Sheather, and Hettmansperger (1988) have developed analogues of a number of the popular least squares diagnostics based on a robust fitting procedure rather than least squares. The resulting diagnostics appear to be more powerful than the least squares based methods.

7.8.2 Bootstrap Methods in Regression

The computation of standard errors for robust regression estimates can be based on asymptotic formulas for the covariance matrices or alternatively estimated from resampling of the observations. See Wu (1986), with 18 discussants, for an introduction to the bootstrap methods in regression.

APPENDIX A

Linear Algebra Results

Much of the material collected here is for quick reference within Chapter 7 on regression analysis. There are also applications to other sections of the book whenever multivariate distributions arise, such as the joint limiting behavior of two order statistics or two moments in Chapter 4.

A.1 MATRIX RESULTS

There are at least three thorough introductions to the matrix theory that is useful in statistics, Appendix 1 of Anderson (1957), Chapter 1 of Rao (1973) and Graybill (1982). We restate some definitions and results here to help fix the notation. The reader may find the problems in Section A.3 helpful.

A.1.1 Definitions and Basic Results

A *matrix* of real numbers of dimension $m \times n$ is a real array of m rows and n columns:

$$A = [a_{ij}] = \begin{bmatrix} a_{11} & \dots & a_{1n} \\ \vdots & & \vdots \\ a_{m1} & \dots & a_{mn} \end{bmatrix}.$$

Matrix addition is defined by $\boldsymbol{A} + \boldsymbol{B} = [a_{ij} + b_{ij}]$. If $\boldsymbol{A}$ is of dimension $m \times n$ and $\boldsymbol{B}$ is of dimension $n \times p$ then we may define the product $\boldsymbol{AB} = \boldsymbol{C} = [c_{ij}]$, where $c_{ij} = \sum_k a_{ik} b_{kj}$. The transpose of $\boldsymbol{A}$ is denoted by $\boldsymbol{A}^T = [a_{ji}]$.

For square matrices (of dimension $m \times m$), define two special cases, the *zero matrix* $\mathbf{0}_m = [0]$ and the *identity matrix* $\boldsymbol{I}_m = [\delta_{ij}]$, where δ_{ij} is the Kronecker delta ($= 1$ if $i = j$ and 0 otherwise). A square matrix $\boldsymbol{A}$ is *symmetric* if $A = A^T$. For a square matrix A there may exist C such that $AC = CA = I$; C is called the inverse of A and is denoted by A^{-1}. When A^{-1} exists, we say A is *nonsingular*. A square matrix A is *orthogonal* if $AA^T = I$. A symmetric matrix is *idempotent* if $AA = A$. The *determinant* of A is defined in the above references, and is denoted by det(A) or $|A|$. The *trace* of A is defined by $\text{tr}(A) = \Sigma_i a_{ii}$.

Some elementary properties of matrices are collected in the following result.

Theorem A.1: Elementary Properties of Matrices

(i) $(A + B)^T = A^T + B^T$ and $(AB)^T = B^T A^T$.

(ii) For a square matrix A with $\det(A) \neq 0$, there exists a unique inverse A^{-1}.

(iii) When the inverses exist, $(AB)^{-1} = B^{-1}A^{-1}$.

(iv) $\det(A^T) = \det(A)$ and $\det(AB) = \det(A)\det(B)$.

(v) $\text{tr}(A + B) = \text{tr}(A) + \text{tr}(B)$ and $\text{tr}(AB) = \text{tr}(BA)$.

There are two important results on inverses of matrices which we state as separate theorems.

Theorem A.2: Inverse of Partitioned Matrices

When A,D are symmetric matrices such that the inverses that occur in the expressions exist, then

$$\begin{bmatrix} A & B \\ B^T & D \end{bmatrix}^{-1} = \begin{bmatrix} A^{-1} + MN^{-1}M^T & -MN^{-1} \\ -N^{-1}M^T & N^{-1} \end{bmatrix}$$

where $M = A^{-1}B$ and $N = D - B^T A^{-1} B$.

Theorem A.3: Updating Formula

Assuming inverses exist,

$$(A + B^T C)^{-1} = A^{-1} - A^{-1}B^T(I + CA^{-1}B^T)^{-1}CA^{-1}.$$

The following property is important in statistics because it, together with symmetry, characterizes covariance matrices (see Section A.1.2). A square matrix A is *nonnegative definite* if for all vectors $b \neq 0$, $b^T A b \geq 0$. It is *positive definite* if $b^T A b > 0$ for all $b \neq 0$.

Theorem A.4 If A is nonnegative definite, then A is *positive definite* if and only if A is nonsingular.

Associated with each $p \times p$ nonnegative definite matrix A is a *quadratic form*

$$q_A(\boldsymbol{x}) = \boldsymbol{x}^T A \boldsymbol{x} = \Sigma_{i,j} x_i x_j a_{ij}, \qquad \boldsymbol{x} \in \mathsf{R}^p.$$

In addition, a family of ellipsoids, indexed by $k > 0$, is determined by the quadratic form of A:

$$\mathcal{E}_A(k) = \{\boldsymbol{x} : q_A(\boldsymbol{x}) \leq k\}, \qquad k > 0.$$

The volume of $\mathcal{E}_A(k)$ in R^p is given by

$$\frac{(\pi k)^{p/2} \det(A)^{1/2}}{\Gamma(p/2+1)}.$$

Let C be a $p \times p$ matrix. An *eigenvalue* of C is any real λ such that there is a nonzero vector U (called an *eigenvector* associated with λ) such that $CU = \lambda U$.

Theorem A.5: Spectral Decomposition Theorem

If the matrix C is square and symmetric, there exists an orthogonal matrix U such that $U^T C U = \Lambda$, where $\Lambda = [\lambda_j]$ is a diagonal matrix. The jth column of U is a unit length eigenvector associated with the eigenvalue λ_j.

A.1.2 Random Vectors and Matrices

A random matrix $Y = [Y_{ij}]$ is a matrix whose elements are random variables, and $\mathrm{E}[Y]$ is used to denote the matrix of expectations $[\mathrm{E}[Y_{ij}]]$, assuming that they exist. For a random vector Y we also define the covariance matrix by $\mathrm{Cov}[Y] = \mathrm{E}[(Y - \mathrm{E}[Y])(Y - \mathrm{E}[Y])^T]$, assuming that second moments of the elements of Y exist.

Theorem A.6: Characterization of Covariance Matrices

Every covariance matrix is symmetric and nonnegative definite. Conversely, given a symmetric, nonnegative definite matrix A, there exists a random vector X with $\mathrm{Cov}[X] = A$.

Theorem A.7: Mean and Covariance Matrix of Linear Transformation

If $Y = AX + b$ where X is a random vector, A is a constant matrix, and b is a constant vector, then

$$\mathrm{E}[Y] = A\mathrm{E}[X] + b$$

and

$$\text{Cov}[Y] = A\text{Cov}[X]A^T.$$

The most important multivariate distribution is the multivariate normal. We say $\boldsymbol{X}$ is *multivariate normal* if every nonzero linear combination of the components of $\boldsymbol{X}$ has a univariate normal distribution. We write $\boldsymbol{X} \sim N(\boldsymbol{\mu}, \Sigma)$, where $\boldsymbol{\mu} = \text{E}[\boldsymbol{X}]$ and $\Sigma = \text{Cov}[\boldsymbol{X}]$.

The distribution $N(\boldsymbol{\mu}, \Sigma)$ is completely determined by $\boldsymbol{\mu}, \Sigma$. It has density function

$$f_X(\boldsymbol{x}) = \frac{1}{(2\pi)^{1/2}\det(\Sigma)^{1/2}} \exp\left\{-\frac{1}{2}(\boldsymbol{x}-\boldsymbol{\mu})^T \Sigma^{-1}(\boldsymbol{x}-\boldsymbol{\mu})\right\}.$$

Theorem A.8: Distribution of Quadratic Forms

If $\boldsymbol{Y} \sim N(\boldsymbol{0}, \sigma^2 \boldsymbol{I}_n)$ and $\boldsymbol{A}$ is idempotent of rank p, then $\boldsymbol{Y}^T\boldsymbol{A}\boldsymbol{Y} \sim \sigma^2\chi_p^2$.

Theorem A.9: Independence of Quadratic Forms

If $\boldsymbol{A}, \boldsymbol{B}$ have orthogonal columns ($\boldsymbol{A}^T\boldsymbol{B} = \boldsymbol{0}$), and $\boldsymbol{Y} \sim N(\boldsymbol{0}, \sigma^2\boldsymbol{I}_n)$, then the quadratic forms $\boldsymbol{Y}^T\boldsymbol{A}\boldsymbol{Y}$ and $\boldsymbol{Y}^T\boldsymbol{B}\boldsymbol{Y}$ are independent.

A.2 VECTOR SPACE RESULTS

A.2.1 Inner Product Spaces

For a detailed development of inner product spaces, see Ash (1972) or Brockwell and Davis (1987, Chapter 2). Much of this material (omitting proofs) is taken from the latter reference.

Let R^n denote the set of all column vectors of n real numbers,

$$\boldsymbol{x} = [x_1, \ldots, x_n]^T = \begin{bmatrix} x_1 \\ \vdots \\ x_n \end{bmatrix}.$$

R^n is the n-fold Cartesian product of the real numbers R. The set $\mathcal{V} = \mathsf{R}^n$ is a vector space under the operations of vector addition and scalar multiplication by real numbers. This means that for all $\boldsymbol{x}, \boldsymbol{y} \in \mathcal{V}$, $c \in \mathsf{R}$,

$$\begin{array}{ll} \text{(i)} & c\boldsymbol{x} \in \mathcal{V} \\ \text{(ii)} & \boldsymbol{x} + \boldsymbol{y} \in \mathcal{V} \\ \text{(iii)} & c(\boldsymbol{x}+\boldsymbol{y}) = c\boldsymbol{x} + c\boldsymbol{y}. \end{array} \tag{A.2.1}$$

Elementary concepts are length, inner product, angle, and orthogonality. The *inner product* of $\boldsymbol{x},\boldsymbol{y} \in \mathcal{V}$ is

$$\langle \boldsymbol{x},\boldsymbol{y} \rangle = \sum_{i=1}^{n} x_i y_i = \boldsymbol{x}^T \boldsymbol{y}$$

the *length* of $\boldsymbol{x}$ is $\|\boldsymbol{x}\| = \langle \boldsymbol{x},\boldsymbol{x} \rangle^{1/2} = \{\sum_{i=1}^{n} x_i^2\}^{1/2}$; the *angle* between nonzero $\boldsymbol{x},\boldsymbol{y}$ is the angle $\theta \in [0,\pi]$ such that

$$\cos\theta = \frac{\langle \boldsymbol{x},\boldsymbol{y} \rangle}{\|\boldsymbol{x}\|\,\|\boldsymbol{y}\|}.$$

Two vectors are *orthogonal* if $\langle \boldsymbol{x},\boldsymbol{y} \rangle = 0$.

Theorem A.10 The Cauchy-Schwartz inequality is

$$|\langle \boldsymbol{x},\boldsymbol{y} \rangle| \le \|\boldsymbol{x}\|\,\|\boldsymbol{y}\|$$

and the triangle inequality is

$$\|\boldsymbol{x} + \boldsymbol{y}\| \le \|\boldsymbol{x}\| + \|\boldsymbol{y}\|.$$

The vector space $\mathcal{V}$ is an *inner product space* in that for all $\boldsymbol{x},\boldsymbol{y} \in \mathcal{V}$ and $c \in R$,

$$\begin{aligned}
&\text{(i)} && \langle \boldsymbol{x},\boldsymbol{y} \rangle = \langle \boldsymbol{y},\boldsymbol{x} \rangle \\
&\text{(ii)} && \langle \boldsymbol{x}+\boldsymbol{y},\boldsymbol{z} \rangle = \langle \boldsymbol{x},\boldsymbol{z} \rangle + \langle \boldsymbol{y},\boldsymbol{z} \rangle \\
&\text{(iii)} && \langle c\boldsymbol{x},\boldsymbol{y} \rangle = c\langle \boldsymbol{x},\boldsymbol{y} \rangle \\
&\text{(iv)} && \langle \boldsymbol{x},\boldsymbol{x} \rangle \ge 0 \text{ with equality if and only if } \boldsymbol{x} = \boldsymbol{0}.
\end{aligned} \tag{A.2.2}$$

Statisticians use properties of inner product spaces in many contexts (for example, linear regression, asymptotic approximations, and prediction theory).

The space $\mathcal{V} = \mathrm{R}^n$ is also a *normed* space, with norm $\|x\| = \langle \boldsymbol{x},\boldsymbol{x} \rangle^{1/2}$. This means that for all $\boldsymbol{x},\boldsymbol{y} \in \mathcal{V}$ and $c \in \mathrm{R}$,

$$\begin{aligned}
&\text{(i)} && \|\boldsymbol{x}+\boldsymbol{y}\| \le \|\boldsymbol{x}\| + \|\boldsymbol{y}\| \\
&\text{(ii)} && \|c\boldsymbol{x}\| = |c|\,\|\boldsymbol{x}\| \\
&\text{(iii)} && \|\boldsymbol{x}\| \ge 0 \text{ with equality if and only if } \boldsymbol{x} = \boldsymbol{0}.
\end{aligned} \tag{A.2.3}$$

A sequence of vectors $\{\boldsymbol{x}_m\}$, $\boldsymbol{x}_m \in \mathcal{V}$, is said to *converge* in *norm* to $\boldsymbol{x} \in \mathcal{V}$ if

$$\|\boldsymbol{x}_m - \boldsymbol{x}\| \longrightarrow 0 \qquad \text{as} \quad n \to \infty.$$

A sequence of vectors $\{x_m\}$ is called a *Cauchy* sequence if

$$\|x_l - x_m\| \longrightarrow 0 \qquad \text{as} \quad l, m \to \infty.$$

An inner product space (such as R^n) in which every Cauchy sequence converges to some member of the space is said to be a *complete inner product space*, or a *Hilbert space*. We are mainly interested here in the notion of the projection of a vector in a Hilbert space onto certain subspaces.

Let $\mathcal{M}$ be a linear subspace of a Hilbert space $\mathcal{V}$ [that is, $\mathcal{M}$ itself satisfies conditions (A.2.1) with $\mathcal{M}$ replacing $\mathcal{V}$]. $\mathcal{M}$ is called a *closed subspace* of $\mathcal{V}$ if every Cauchy sequence in $\mathcal{M}$ converges to a member of $\mathcal{M}$. Define the *orthogonal complement* of a subset $\mathcal{M}$ of $\mathcal{V}$ by

$$\mathcal{M}^{\perp} = \{y \in \mathcal{V} : \langle x, y \rangle = 0 \quad \forall\, x \in \mathcal{M}\}.$$

Theorem A.11: The Projection Theorem

If $\mathcal{M}$ is a closed subspace of the Hilbert space $\mathcal{V}$, and $y \in \mathcal{V}$, then there exists a unique element $\hat{y}$ of $\mathcal{M}$ which is closest to x in that

$$\|y - \hat{y}\| = \inf_{x \in \mathcal{M}} \|x - y\|. \tag{A.2.4}$$

Furthermore, $\hat{y} \in \mathcal{M}$ satisfies (A.2.4) if and only if $\hat{y} \in \mathcal{M}$ and $y - \hat{y} \in \mathcal{M}^{\perp}$. Thus given an element $y \in H$, there is a unique element $\hat{y} \in \mathcal{M}$ which satisfies

$$\langle y - \hat{y}, x \rangle = 0 \qquad \text{for all} \quad x \in \mathcal{M}. \tag{A.2.5}$$

These equations are called the *prediction equations*.

Example 1:* *Linear Regression. Let $\mathcal{V} = R^n$, and let $y = X\beta + \epsilon$ be a linear regression model (see Chapter 7). Let $\mathcal{M}$ be the vector subspace of $\mathcal{V}$ which consists of all the linear combinations of the columns of X, which we denote by $X_0, X_1, X_2, \ldots, X_p$. By the projection theorem, each $y \in R^n$ there is a unique element $\hat{y}$ of $\mathcal{M}$ which is closest to y. Since $\hat{y} \in \mathcal{M}$, it may be written as a linear combination of the columns of X:

$$\hat{y} = \sum_{i=0}^{p} b_i X_i = XB$$

for some $B = [b_0, b_1, \ldots, b_p]^T$. The prediction equations (A.2.5) yields

$$\langle \hat{y}, X_j \rangle = \langle y, X_j \rangle, \qquad j = 0, \ldots, p$$

which may be summarized in the matrix form

$$X^T \hat{y} = X^T y$$

or the *normal equations*

$$(X^T X)B = X^T y. \tag{A.2.6}$$

The existence of a solution B to (A.2.6) is guaranteed by the projection theorem. Of course if $X^T X$ is nonsingular, there exists a unique solution to (A.2.6) given by $B = (X^T X)^{-1} X^T y$. □

A second important application of the projection theorem is given in Appendix B.3.

A.2.2 Derivatives and Approximations in R^p

Suppose that $g : R \to R$ has partial derivatives

$$\frac{\partial g}{\partial x_1}, \ldots, \frac{\partial g}{\partial x_n},$$

all existing at $x = x_0$. Also suppose that

$$g(x_0; t) = \sum_{i=1}^{p} \left. \frac{\partial g}{\partial x_i} \right|_{x=x_0} \cdot t_i, \qquad t \in R^p$$

satisfies the following property: for every $\epsilon > 0$, there exists a neighborhood $N_\epsilon(x_0)$ of x_0 such that

$$|g(x) - g(x_0) - g(x_0; x - x_0)| \leq \epsilon \|x - x_0\| \qquad \forall\, x \in N_\epsilon(x_0).$$

Then we say that the *differential* $g(x_0; t)$ of g exists at x_0.

Theorem A.12: Apostol (1961)

(i) If g has a differential at x_0, then g is continuous at x_0.

(ii) If all the partial derivatives $\partial g / \partial x_i$ exist in a neighborhood of x_0 and are continuous at x_0, then g has a differential at x_0.

A.3 PROBLEMS

1. Verify the elementary results (i), (iii), and (v) of Theorem A.1.

2. Verify Theorem A.2 by multiplication of the partitioned matrix by its putative inverse.

3. Verify Theorem A.3 by multiplication of the updating matrix by its putative inverse.

4. Prove Theorem A.6. (*Hint*: For the converse use the fact that the multivariate normal distribution is completely determined by its mean vector and covariance matrix.)

APPENDIX B

Asymptotic Results

B.1 DEFINITIONS AND UNIVARIATE CONVERGENCE THEOREMS

B.1.1 Modes of Convergence

1. Let $\{X_n\}_{n\geq 1}$ be a sequence of random variables (not necessarily defined on the same sample space). Let $X_n \sim F_n$. We say that X_n converges to a random variable $X \sim F$ *in distribution* if $F_n(x) \to F(x)$ as $n \to \infty$ for all continuity points x of the limit distribution F.

   If $\{X_n\}_{n\geq 1}$ and X are defined on the *same* sample space, then we may define convergence in probability, convergence with probability 1, and convergence in the rth mean as follows.

2. The sequence $\{X_n\}$ converges *in probability* to X, written $X_n \xrightarrow{p} X$, if for every $\epsilon > 0$,

$$P\{|X_n - X| > \epsilon\} \to 0 \qquad \text{as} \quad n \to \infty.$$

   If a sequence $\{X\}$ converges in probability to a constant 0, we sometimes write $X_n = o_p(1)$. Convergence in probability of X_n to X is also denoted by $X_n = X + o_p(1)$.

3. The sequence $\{X_n\}$ converges to X *with probability* 1 if

$$P\{w : X_n(\omega) \to X(\omega) \text{ as } n \to \infty\} = 1.$$

   This property is also called *almost sure convergence* and is denoted by $X_n \xrightarrow{\text{w.p.1}} X$ or $X_n \xrightarrow{\text{a.s.}} X$.

4. The sequence $\{X_n\}$ converges to X *in the rth mean* (written $X_n \xrightarrow{r} X$) if

$$\mathrm{E}|X_n - X|^r \to 0 \qquad \text{as} \quad n \to \infty.$$

The above definitions are often employed in statistics to describe the limiting behavior of a sequence of estimators $\hat{\theta}_n$ of a parameter θ. If $\hat{\theta}_n \xrightarrow{\text{w.p.1}} \theta$ for all θ, we say that $\hat{\theta}_n$ is *strongly consistent* for θ. If $\hat{\theta}_n \xrightarrow{p} \theta$, we say that $\hat{\theta}_n$ is *weakly consistent* for θ. *Consistency* used by itself means *weak* consistency.

Often $\sqrt{n}(\hat{\theta}_n - \theta) \xrightarrow{d} vZ$, where Z is a standard normal random variable and v is a positive constant. Such convergence in distribution to a normal distribution is called *asymptotic normality*, and it is discussed in Section B.1.3 below. Statisticians prefer to prove the stronger result, mean square convergence: $\sqrt{n}(\hat{\theta}_n - \theta) \xrightarrow{2} vZ$, which implies asymptotic normality and also that the mean and variance of the sequence $\sqrt{n}(\hat{\theta}_n - \theta)$ converge to the mean and variance of the limiting distribution.

B.1.2 Univariate Convergence Results

Proofs or references to proofs of Theorems B.1 to B.9 and B.11 to B.17 may be found in Serfling (1980). The book also contains more specialized results on the asymptotic theory of robust statistics.

The first theorem states that certain modes of convergence are preserved by continuous functions.

Theorem B.1: Continuity Theorem

Let $X_1, X_2, \ldots$ and X be random variables on the same probability space. Suppose that g is a vector-valued function on $\mathbb{R}^k$ and that g is continuous on a set that has probability 1 under the limit distribution F_X. Then

(i) $X_n \xrightarrow{\text{w.p.1}} X$ implies $g(X_n) \xrightarrow{\text{w.p.1}} g(X)$.

(ii) $X_n \xrightarrow{p} X$ implies $g(X_n) \xrightarrow{p} g(X)$.

(iii) $X_n \xrightarrow{d} X$ implies $g(X_n) \xrightarrow{d} g(X)$.

The next result gives the relationships between the different modes of convergence.

Theorem B.2

$$X_n \xrightarrow{\text{w.p.1}} X \qquad \text{implies} \quad X_n \xrightarrow{p} X, \quad \text{which in turn implies} \quad X_n \xrightarrow{d} X.$$

Also,

$$X_n \xrightarrow{r} X, \qquad \text{for some} \quad r > 0 \quad \text{implies} \quad X_n \xrightarrow{p} X.$$

The next result shows that convergence in distribution must be uniform convergence, if the limit is continuous.

Theorem B.3: Pólya's Theorem

If $F_n \xrightarrow{d} F$ and F is continuous, then

$$\lim_{n\to\infty} \sup_x |F_n(x) - F(x)| = 0.$$

Slutsky's lemma reduces many convergence arguments to a study of simpler sequences of random variables.

Theorem B.4: Slutsky's Lemma

Let $X_n \xrightarrow{d} X$ and $Y_n \xrightarrow{p} c$, where c is a finite constant. Then

(i) $X_n + Y_n \xrightarrow{d} X + c$

(ii) $X_n Y_n \xrightarrow{d} cX$

(iii) $X_n/Y_n \xrightarrow{d} X/c$ if $c \neq 0$.

Theorem B.5: Law of Large Numbers

If $X_1, \ldots, X_n$ are i.i.d. with finite mean $\mu = \mathrm{E}[X_1]$, then

$$\overline{X}_n = \frac{1}{n}\sum_{i=1}^{n} X_i \xrightarrow{\text{w.p.1}} \mu.$$

B.1.3 Asymptotic Normality

A sequence of random variables $\{X_n\}$ is said to be *asymptotically normal* if there exist constants (a_n, b_n), $b_n > 0$, such that

$$\frac{X_n - a_n}{b_n} \xrightarrow{d} Z$$

where $Z \sim \Phi$, that is, where the limit is the standard normal distribution. In this case we write $X_n \sim N(a_n, b_n^2)$, or preferably $X_n \sim AN(a_n, b_n^2)$, to emphasize the asymptotic nature of the statement. The sequence $\{a_n\}$ is called the *asymptotic mean* and often denoted by $\{\mu_n\}$. Frequently $\mu_n = \mathrm{E}[X_n]$.

However, the asymptotic mean is clearly not unique, and furthermore the μ_n need not equal the mean of X_n. In fact examples will show that the mean of X_n need not exist for asymptotic normality to hold. Similar statements regarding the sequence $\{b_n^2\}$ hold: it is called the *asymptotic variance*. Often $b_n^2 = \sigma_n^2 = \mathrm{Var}[X_n]$, but the latter need not exist for asymptotic normality to hold.

Theorem B.6: Lindeberg-Levy Central Limit Theorem

Let $\{X_i\}$ be i.i.d. with mean μ and finite variance σ^2. Then

$$\sqrt{n}\left(\frac{1}{n}\sum_{i=1}^{n} X_i - \mu\right) \xrightarrow{d} N(0,\sigma^2), \qquad \text{that is,} \quad \frac{1}{n}\sum_{i=1}^{n} X_i \text{ is } AN\left(\mu, \frac{\sigma^2}{n}\right).$$

While Theorem B.6 is very commonly employed, there are many situations where the random terms of a sum are not identically distributed, but the contribution of each term to the sum still becomes negligible as the number of terms increases. The next result uses the "Lindeberg condition" to make this precise.

First consider a sequence of row sums of a *double array* of random variables $\{X_{nj},\ 1 \le j \le k_n,\ n = 1,2,\ldots\}$, where $k_n \to \infty$ as $n \to \infty$. It will be assumed that for each fixed n the k_n random variables in the nth row are independent, but not necessarily identically distributed, or even defined on the same space as variables in other rows. This includes the i.i.d. case of Theorem B.6.

Theorem B.7: Lindeberg Central Limit Theorem

Given a double array of random variables with independent terms in each row as described above, it is also assumed that the sum of variables in the nth row, namely $S_n = \sum_{j=1}^{k_n} X_{nj}$, has finite mean A_n and finite variance B_n^2. Let $\mu_{nj} = \mathrm{E}[X_{nj}]$. Further assume the Lindeberg condition: for every $\epsilon > 0$,

$$\frac{\sum_{j=1}^{k_n} \mathrm{E}(X_{nj} - \mu_{nj})^2 I\{|X_{nj} - \mu_{nj}| > \epsilon B_n\}}{B_n^2} \to 0 \qquad \text{as} \quad n \to \infty.$$

Then S_n is asymptotically normal with parameters A_n, B_n^2.

Earlier in the continuity theorem it was seen that continuous functions preserve certain modes of convergence. Differentiable functions also preserve asymptotic normality in the following way:

Theorem B.8

If X_n is $AN(\mu,\sigma_n^2)$, where $\sigma_n^2 \to 0$, and if g is differentiable at μ, then $g(X_n)$ is $AN(g(\mu),[g'(\mu)]^2\sigma_n^2)$.

Theorem B.9: Asymptotic Normality of Order Statistics

Let $U_{(1)},\ldots,U_{(n)}$ be the order statistics of n i.i.d. observations from the uniform (0,1) distribution. Let $U_{(np)}$ denote the $[np]$th order statistic, where $[r]$ is the greatest integer less than or equal to r. Then

$$U_{(np)} \sim AN\left(p, \frac{p(1-p)}{n}\right).$$

Corollary. Let $X_{(1)},\ldots,X_{(n)}$ denote the order statistics of n i.i.d. observations from a distribution F with density f which is continuous and positive at $x_p = F^{-1}(p)$. Then $X_{(np)} \sim AN(x_p, p(1-p)/nf^2(x_p))$.

Under certain smoothness criteria on a function g, the convergence of $X_n \to c$ will yield a Taylor series for $g(X_n)$ about $g(c)$.

Theorem B.10

Assume that the function g has $k+1$ derivatives at a point c and that the sequence X_n converges to c at a rate faster than some sequence of positive constants $b_n \to 0$. That is, $X_n = c + o_p(b_n)$, which is defined by

$$\frac{X_n - c}{b_n} \xrightarrow{p} 0 \qquad \text{as} \quad n \to \infty.$$

Then

$$g(X_n) = g(c) + g'(c)(X_n - c) + \cdots + \frac{g^{(k)}(c)}{k!}(X_n - c)^k + o_p(b_n^{k+1}).$$

This last result is proved in Brockwell and Davis (1987), Section 6.2.

B.2 MULTIVARIATE CONVERGENCE THEOREMS

The definitions of convergence in Section B.1.1 Theorems B.1 and B.2 extend to random vectors $X = (X_1,\ldots,X_n)^T$ taking values in $\mathbb{R}^n$, provided the absolute value signs are replaced by the Euclidean norm:

$$\|X\| = (X^T X)^{1/2} = \left(\sum_{i=1}^{n} X_i^2\right)^{1/2}.$$

In particular we state the continuity theorem for vector-valued functions of random vectors.

Theorem B.11: Continuity Theorem

Let $X_1, X_2, \ldots$ and X be random vectors on the same probability space with range in R^k. Suppose that g is a vector-valued function on R^k and that g is continuous on a set that has probability 1 under the limit distribution F_X. Then

(i) $X_n \xrightarrow{\text{w.p.1}} X$ implies $g(X_n) \xrightarrow{\text{w.p.1}} g(X)$.

(ii) $X_n \xrightarrow{p} X$ implies $g(X_n) \xrightarrow{p} g(X)$.

(iii) $X_n \xrightarrow{d} X$ implies $g(X_n) \xrightarrow{d} g(X)$.

Corollary: Linear Transformations and Quadratic Forms Preserve Convergence.

(i) Let $g(X_n) = AX_n$, where A is an $m \times k$ matrix. Then since matrix multiplication is a continuous function,

$$X_n \to X \qquad \text{implies} \quad AX_n \to AX$$

in the given mode of convergence.

(ii) Similarly, if A is a $k \times k$ matrix, then

$$X_n \to X \qquad \text{implies} \quad X_n^T AX_n \to X^T AX.$$

The next result reduces convergence in distribution for random *vectors* to convergence in distribution for random *variables.*

Theorem B.12: Cramer-Wold Device

If $X_1, X_2, \ldots$ and X are random vectors with values in R^k, then $X_n \xrightarrow{d} X$ if and only if $u^T X_n \xrightarrow{d} u^T X$ for every $u \in \mathsf{R}^k$.

In view of the Cramer-Wold device and the definition of multivariate normal distribution given in Appendix A, we now may define *asymptotic* multivariate normality. A sequence of random vectors $\{X_n\}$ taking values in R^k is said to be *asymptotically (multivariate) normal* if there exists a sequence of vectors $\{\mu_n\}$ and symmetric matrices $\{\Sigma_n\}$ such that:

1. The Σ_n matrices have no zero diagonal elements.
2. For every vector $u \in \mathsf{R}^k$ for which $u^T \Sigma_n u > 0$ for all sufficiently large n, the sequence $\{u^T X_n\} \sim AN(u^T \mu_n, u^T \Sigma_n u)$.

The sequence $\{\boldsymbol{\mu}_n\}$ is called the *asymptotic mean vector*, and $\{\Sigma_n\}$ is called the *asymptotic covariance matrix*, whether or not moments of the components of $\boldsymbol{X}_n$ exist. It is often the case that $\{\Sigma_n\}$ can be replaced by $n^{-1}\Sigma$, for some fixed Σ, and then Σ itself is sometimes also referred to as the asymptotic covariance matrix.

Theorem B.13: Central Limit Theorem for Random Vectors

Let $\{\boldsymbol{X}_i\}$ be i.i.d. random vectors with finite mean vector $\boldsymbol{\mu}$ and covariance matrix Σ. Then

$$\sqrt{n}\left(\frac{1}{n}\sum_{i=1}^{n}\boldsymbol{X}_i - \boldsymbol{\mu}\right) \xrightarrow{d} N(\mathbf{0}, \Sigma), \qquad \text{that is,}$$

$$\frac{1}{n}\sum_{i=1}^{n}\boldsymbol{X}_i \quad \text{is} \quad AN\left(\boldsymbol{\mu}, \frac{1}{n}\Sigma\right).$$

The following two theorems show that under certain differentiability conditions on a function g, asymptotic normality of $\{\boldsymbol{X}_n\}$ implies asymptotic normality of $\{g(\boldsymbol{X}_n)\}$.

Theorem B.14

Assume that $\boldsymbol{X}_n = [X_{n1}, \ldots, X_{np}]^T$ is $AN(\boldsymbol{\mu}, b_n^2\Sigma)$, with $b_n \to 0$ and Σ a covariance matrix. Also assume that $\boldsymbol{g} : \mathrm{R}^p \to \mathrm{R}^q$ is a vector-valued function for which each component $g_i(\boldsymbol{x})$ has a nonzero differential $g_i(\boldsymbol{\mu} : \boldsymbol{t})$, $\boldsymbol{t} = (t_1, \ldots, t_q)^T$ at $\boldsymbol{x} = \boldsymbol{\mu}$. Define the $p \times q$ matrix

$$D = \left[\left.\frac{\partial g_i}{\partial x_j}\right|_{\boldsymbol{x}=\boldsymbol{\mu}}\right].$$

Then $g(\boldsymbol{X}_n)$ is $AN(g(\boldsymbol{\mu}),\ b_n^2 D^T \Sigma D)$.

Corollary. If $b_n = n^{-1/2}$ and $\Sigma = [\sigma_{ij}]$ in Theorem B.14, then

$$g(\boldsymbol{X}_n) \quad \text{is} \quad AN\left(g(\boldsymbol{\mu}), \frac{1}{n}\sum_{i=1}^{k}\sum_{j=1}^{k}\sigma_{ij}\left.\frac{\partial g}{\partial x_i}\right|_{\boldsymbol{x}=\boldsymbol{\mu}}\left.\frac{\partial g}{\partial x_j}\right|_{\boldsymbol{x}=\boldsymbol{\mu}}\right). \qquad \square$$

Theorem B.15

Let $\boldsymbol{X}_n$ be $AN(\boldsymbol{\mu}, n^{-1}\Sigma)$. Let $g(\boldsymbol{x})$ be a real-valued function having continuous partial derivatives of order $m > 1$ in a neighborhood of $\boldsymbol{x} = \boldsymbol{\mu}$; assume also that all partials of order j, $1 \le j \le m$, vanish at $\boldsymbol{x} = \boldsymbol{\mu}$. Then

$$n^{m/2}[g(\boldsymbol{X}_n) - g(\boldsymbol{\mu})] \xrightarrow{d} \frac{1}{m!} \sum_{i_1=1}^{p} \cdots \sum_{i_m=1}^{p} \left. \frac{\partial^m g}{\partial x_{i_1} \cdots \partial x_{i_m}} \right|_{x=\mu} \prod_{j=1}^{m} Z_{ij}$$

where $\mathbf{Z} = (Z_1, \ldots, Z_p)^T \sim N(\mathbf{0}, \boldsymbol{\Sigma})$.

Let $\alpha_k = \mathrm{E}[X^k]$, $k = 1, 2, \ldots$, define the kth moments of X about 0. Let $a_k = (1/n) \sum_{i=1}^{n} X_i^k$ denote the kth sample moments. The following result may be derived from the above theorems.

Theorem B.16: Consistency and Asymptotic Normality of Sample Moments

(i) If α_k exists, the $a_k \xrightarrow{\text{w.p.1}} \alpha_k$.

(ii) If $\alpha_{2k} < \infty$, then $n^{1/2}[a_1 - \alpha_1, a_2 - \alpha_2, \ldots, a_k - \alpha_k] \xrightarrow{d} N(\mathbf{0}, \boldsymbol{\Sigma})$

where $\boldsymbol{\Sigma} = [\sigma_{ij}]$ and $\sigma_{ii} = \alpha_{ij} - \alpha_i \alpha_j$.

Corollary: Asymptotic Normality of Mean and Standard Deviation. If $\mathrm{E}_F[X^4] < \infty$, then the joint limiting distribution of the sample mean and sample standard deviation is given by

$$n^{1/2} \begin{bmatrix} \overline{x}_n - \mu \\ s_n - \sigma \end{bmatrix} \xrightarrow{d} N(\mathbf{0}, \boldsymbol{\Sigma})$$

where

$$\boldsymbol{\Sigma} = \sigma^2 \begin{bmatrix} 1 & \dfrac{\gamma_3}{2} \\ \dfrac{\gamma_3}{2} & \dfrac{(\gamma_4 - 1)}{4} \end{bmatrix}$$

and $\mu = \mathrm{E}_F[X]$, $\sigma^2 = \mathrm{Var}_F[X]$, and $\gamma_k = \mathrm{E}_F[(X - \mu)/\sigma]^k$, $k = 3, 4$.

Theorem B.17: Asymptotic Normality of Order Statistics

Let $X_1, \ldots, X_n$ be i.i.d. with distribution F, which has a positive continuous density f. Let $x_\alpha = F^{-1}(\alpha)$ denote the α quantile and let $F_n^{-1}(\alpha)$ be the sample α quantile. Then for $0 < \alpha < \beta < 1$,

$$n^{1/2} \begin{bmatrix} F_n^{-1}(\alpha) - F^{-1}(\alpha) \\ F_n^{-1}(\beta) - F^{-1}(\beta) \end{bmatrix} \xrightarrow{d} N(\mathbf{0}, \boldsymbol{\Sigma})$$

where

$$\Sigma = \begin{bmatrix} \frac{\alpha(1-\alpha)}{f^2(x_\alpha)} & \frac{\alpha(1-\beta)}{f(x_\alpha)f(x_\beta)} \\ \frac{\alpha(1-\beta)}{f(x_\alpha)f(x_\beta)} & \frac{\beta(1-\beta)}{f^2(x_\beta)} \end{bmatrix}.$$

B.3 U-STATISTICS AND THE PROJECTION METHOD

In this section $X_1,\ldots,X_n$ are i.i.d. random variables with distribution F. (They may also be random vectors, but our notation will not indicate this.) A U-statistic of order k is of the form

$$U_k = \frac{1}{\binom{n}{k}} \sum_{i_1} \cdots \sum_{i_k} \phi_k(X_{i_1},\ldots,X_{i_k})$$

where ϕ_k is a symmetric function of its k arguments and the summation is extended over all $\binom{n}{k}$ choices of distinct ordered subscripts $1 \leq i_1 < i_2 < \cdots < i_k \leq n$. The U-statistics are unbiased estimators of their kernel means $\mathrm{E}_F[\phi_k(X_1,\ldots,X_k)]$, which we always assume to exist. We are mainly interested in the case $k \geq 2$, since for $k = 1$ standard results for sums of i.i.d. random variables apply. Some important examples follow.

Example 1: The Sample Variance. Let $\phi_2(x_1,x_2) = (x_2 - x_1)^2/2$. Then it is not hard to show that the kernel mean is the variance of F, namely, $\mathrm{Var}_F[X]$. The U-statistic of order $k = 2$ defined by this kernel is

$$U_2 = \sum_{1 \leq i < j \leq n} \phi_2(X_i,X_j)$$

which can be shown to equal the sample variance s^2.

Example 2: The Wilcoxon Signed Rank Test Statistic. Let $\phi_2(x_1,x_2) = \mathrm{sgn}(x_1 - x_2)$. Then the Wilcoxon signed rank statistic can be shown (see Chapter 5, Problems) to be a linear function of the U-statistic defined by ϕ_2.

Example 3: Kendall's τ Statistic. Let $\mathbf{Z}^T = (X,Y) \sim F$, where F is a continuous bivariate distribution, and define

$$\phi_2(\mathbf{Z}_1,\mathbf{Z}_2) = \mathrm{sgn}\{(X_2 - X_1)(Y_2 - Y_1)\}.$$

Then $\mathrm{E}_F[U_2] = \tau(F)$, and $\tau(F_n)$ is Kendall's τ statistic.

The asymptotic normality of U-statistics has been proved by Hoeffding (1948) using the projection method to be discussed later in Section B.4 and is formally stated in the next theorem.

Theorem B.18: Asymptotic Normality of U-Statistics

Let $\psi = \phi_k - \mathrm{E}[\phi_k]$ and define

$$\psi_1(x) = \mathrm{E}_F[\psi(X_1,\ldots,X_k) \mid X_1 = x].$$

Then if $\mathrm{E}[\phi_k^2] < \infty$, the U-statistic

$$n^{1/2}(U_k - \mathrm{E}[U_k]) \xrightarrow{d} N(0, k^2 \mathrm{Var}_F[\psi_1(X)].$$

Hoeffding proved the theorem by projecting the kth-order U-statistic onto the space of linear functions of the form $\sum_{i=1}^n a_i(X_i)$ and showing that the error involved was asymptotically negligible. Hajek (1968) later formalized this projection method in Hilbert space terms, as follows. Given $X_1,\ldots,X_n$ independent random variables, define

$$\mathcal{H} = \{T_n = T_n(X_1,\ldots,X_n) : \mathrm{E}[T_n^2] < \infty\}.$$

Then (after identification of almost surely equal functions), it is shown that $\mathcal{H}$ is a Hilbert space with inner product $\mathrm{E}[TU]$. The linear subspace

$$\mathcal{L} = \left\{ L_n = \sum_{i=1}^n a_i(X_i) : \mathrm{E}[a_i(X_i)]^2 < \infty \right\}$$

is closed in the topology generated by the inner product. Hence by the classical projection theorem (see Appendix A.2, Theorem A.11) there is a unique member $\hat{T}_n$ of L which is closest to T_n in that it minimizes $\mathrm{E}[T_n - L_n]^2$ as L_n ranges over L. Hajek showed that if $\mathrm{E}[T_n] = 0$ and $\mathrm{E}[T_n^2] < \infty$, then this *projection* $\hat{T}_n$ of T_n onto L was given by $\sum_{i=1}^n a_i(X_i)$, where $a_i(x) = \mathrm{E}[T_n \mid X_i = x]$. Moreover, he showed $\mathrm{E}[T_n - \hat{T}_n]^2 = \mathrm{Var}[T_n] - \mathrm{Var}[\hat{T}_n]$, so that only a comparison of variances was needed to assess the accuracy of the approximation. Two advantages of using this method are:

1. Precise directions are given for finding the projection.
2. The projection is very often a good approximation in that $\mathrm{Var}[T_n] - \mathrm{Var}[\hat{T}_n] \to 0$ as $n \to \infty$. In such cases T_n and $\hat{T}_n$ share the same limiting distribution if any.

A proof of Theorem B.18 for the case of $k = 2$ is given in Section B.5, and numerous applications of the theorem are in the exercises.

B.4 ASYMPTOTIC NORMALITY VIA DIFFERENTIATION

In Section 3.4 it was shown that continuity of the functional $T(F)$ guaranteed consistency and robustness of the associated estimator sequence $T(F_n)$. It turns out that differentiability of $T(F)$ leads to additional desirable properties such as asymptotic normality. In fact many statistics $T(F_n)$ are well approximated by one or two terms of their von Mises expansion, a Taylor series for $T(F_n)$ about $T(F)$ in "powers" of $F_n - F$. We will informally discuss these expansions in Section B.4.1 before looking closely in Sections B.4.2 and B.4.3 at the problem of defining the derivative of a functional T at F. This amounts to a *linearization*: we choose a continuous linear functional $G \to \int \Omega \, dG$, call it T_F', such that when G is close to F, $T(G) - T(F)$ is close to $T_F'(G-F)$. As a by-product we obtain the asymptotic equivalence of $n^{1/2}[T(F_n) - T(F)]$ and

$$n^{1/2}[T_F'(F_n - F)] = n^{1/2} \sum_i [\Omega(X_i) - \mathrm{E}[\Omega(X_i)]]$$

from which asymptotic normality follows. This procedure is called the Δ-method.

Another powerful and general method for proving asymptotic normality is the projection method formalized by Hajek (1968). It is compared with the Δ-method in Section B.4.3.

B.4.1 von Mises Expansions and U-Statistics

A von Mises expansion for $T(F_n)$ is a series of the form

$$\begin{aligned} T(F_n) = T(F) &+ \int \Omega_1(x_1) \, d(F_n - F)(x_1) \\ &+ \int \int \Omega_2(x_1, x_2) \, d(F_n - F)(x_1) \, d(F_n - F)(x_2) + \cdots \\ &+ \int \cdots \int \Omega_k(x_1 \ldots x_k) \, d(F_n - F)(x_1) \cdots d(F_n - F)(x_k). \end{aligned} \tag{B.4.1}$$

Such an expansion may often be derived formally from the Taylor series for $a(t) = T(F + tH)$ about $t = 0$. To obtain a k-term expansion, find the first k derivatives of a,

$$a^{(k)}(0) = \left. \frac{d^k}{dt^k} a(t) \right|_{t=0}$$

and express them as

$$a^{(k)}(0) = \int \Omega_k(x_1,\ldots,x_k)dH(x_1)\cdots dH(x_k) \tag{B.4.2}$$

for some "kernel" functions Ω_k. Then evaluate the Taylor series for $a(t)$ about 0,

$$a(t) = a(0) + ta'(0) = \frac{t^2a''(0)}{2} + \cdots + \frac{t_k a^{(k)}(0)}{k}$$

at $t = n^{-1/2}$ and $H = n^{1/2}(F_n - F)$ to obtain (B.4.1).

Example 1: M-Estimators of Location. Let ψ implicitly define the location functional $T(G)$ through

$$\int \psi(x - T(G))dG(x) = 0. \tag{B.4.3}$$

Proceeding heuristically, we will obtain the first two terms of the expansion (B.4.1). Assume for simplicity of exposition that ψ is odd and that F is symmetric about 0, so $T(F) = 0$. Let $a(t) = T(F + tH)$, and (B.4.3) becomes

$$\int \psi(x - a(t))dF(x) + t\int \psi(x - a(t))dH(x) = 0.$$

Differentiating this equation twice yields

$$a'(0) = \int \Omega_1(x)dH(x)$$

where

$$\Omega_1(x) = \frac{\psi(x)}{\int \psi'\,dF},$$

and

$$a''(0) = \int\int \Omega_2(x,y)dH(x)dH(y)$$

where

$$\Omega_2(x,y) = \frac{-2\psi(x)\psi'(y)}{(\int \psi'\,dF)^2}.$$

Thus

$$\begin{aligned} T(F_n) = T(F) &+ \frac{1}{[\int \psi'\,dF]}\frac{\sum_{i=1}^n \psi(X_i)}{n} \\ &- \frac{1}{[n\int \psi'\,dF]^2}\Sigma_i\Sigma_j[\psi'(X_j) - \mathrm{E}_F[\psi']]\psi(X_i) \end{aligned}$$

or

$$n^{1/2}[T(F_n)-T(F)] = n^{-1/2}\frac{\Sigma_i\psi(X_i)}{(\int\psi'\,dF)}\left[2-\frac{\Sigma_j\psi'(X_j)}{n(\int\psi'\,dF)}\right]. \qquad \text{(B.4.4)}$$

The first term in (B.4.4) suggests the limiting distribution $N(0,\mathrm{E}_F[\Omega_1^2(X)])$ while the second term provides a correction term to bias the first for finite n.

Example 2: L-Estimators. Since $T(F)=\int_0^1 F^{-1}(q)J(q)dq$ we consider first the quantiles $T_q(F)=F^{-1}(q)$. Defining $a_q(t)=T_q(F+tH)$, we may find a_q by differentiating $q=F(a-q(t))+tH(a_q(t))$. This leads to

$$a_q'(0)=\int\Omega_q(x)\,dH(x)$$

where

$$\Omega_q(x)=\frac{-1}{f(T_q(F))}I_{(-\infty,T_q(F)]}(x).$$

For the L-estimator itself we have $a(t)=T(F+tH)$, so that

$$\begin{aligned}
a'(0) &= \int_0^1 a_q'(0)J(q)\,dq=\int_0^1\frac{-H(F^{-1}(q))}{f(F^{-1}(q))}J(q)\,dq\\
&= -H(x)J(F(x))\,dx=\int\left[\int\Omega_q(x)J(q)\,dq\right]dH(x)\\
&= \int\Omega(x)\,dH(x).
\end{aligned}$$

The one-term von Mises expansion is then

$$\begin{aligned}
T(F_n) &= T(F)+\int\Omega(x)\,d(F_n-F)(x)\\
&= T(F)+\frac{1}{n}\sum_{i=1}^{n}\left[\int_x^{\infty}J(F(y))\,dy-\int_{-\infty}^{\infty}F(y)J(F(y))\,dy\right].
\end{aligned}$$

In the above examples we have not worried about the domain of T, the validity of our calculations, or the accuracy of a k-term expansion as an approximation to $T(F_n)-T(F)$. Von Mises in (1947) was the first author to seriously deal with these problems, and it is a testament to their difficulty that no one else ventured into the area until Fillipova (1962) and Kallianpur (1963). They continued to use slight modifications of von Mises' messy conditions, and again the theory was largely ignored. When Hampel

(1974) showed that the influence function (kernel of the lead term in the von Mises expansion) could be a valuable heuristic tool, statisticians again became interested in derivatives of functionals. Reeds (1976) has greatly clarified the situation, and many of his ideas form the basis of Sections B.4.1 to B.4.3. More recently Fernholz (1983) has further developed the theory and applications of the Hadamard (compact) derivative. Clarke (1983, 1986) has shown that M-estimators with bounded influence functions are Fréchet differentiable under weak smoothness conditions.

About the same time that von Mises was developing his Taylor series for functionals, Hoeffding was studying the properties of U-statistics. It is an interesting fact that the kth term in a von Mises expansion is a kth-order U-statistic, plus cross product terms which are asymptotically negligible. Thus the U-statistics play the role of polynomials in Taylor series, and it is not surprising that their theory is far more tractable.

B.4.2 Theory of the Δ-Method

The one-dimensional Δ-method is an application of the well known result, which is a special case of Theorem B.8: if Z_n are random variables such that $n^{1/2}(Z_n - c)$ converges in distribution, and if g is differentiable at the point c, then

$$n^{1/2}[g(Z_n) - g(c)] = g'(c)n^{1/2}(Z_n - c) + o_p(1).$$

In particular, if $n^{1/2}(Z_n - c) \to N(0, \sigma^2)$, then

$$n^{1/2}[g(Z_n) - g(c)] \to N(0, [g(c)]^2\sigma^2).$$

The terminology *Δ-method* arises from the fact that the existence of the derivative at a allows exploitation of the one term Taylor expansion:

$$g(c + \Delta) - g(c) = \Delta g'(c) + o(\Delta). \qquad \text{(B.4.5)}$$

Our goal is to formulate the above idea for functionals on the class of cumulative distribution functions F, replacing a by F, Z_n by F_n, and g by T. It will be necessary to extend F to the linear space of distributions corresponding to finite signed measures

$$\mathcal{M} = \{aF + bG : a, b \text{ real};\ F, G \in \mathcal{F}\}.$$

$\mathcal{M}$ is a normed linear space with respect to $\|H\| = \sup_x |H(x) - H(-\infty)|$ or $\|H\|$ = total variation of H. Most statistical functionals $T : \mathcal{M} \to \mathcal{N}$, where $\mathcal{N}$ is a finite dimensional Euclidean space R^q, but in the following definition $\mathcal{M}, \mathcal{N}$ can be arbitrary normed linear spaces.

Define *derivatives* for T at $F \in \mathcal{M}$ as follows. First, for any continuous linear map $L : \mathcal{M} \to \mathcal{N}$ and $H \in \mathcal{M}$ define an associated *remainder*

$$R(F+tH) = \begin{cases} T(F+tH) - T(F) - L(tH), & t \neq 0 \\ 0, & t = 0. \end{cases} \tag{B.4.6}$$

Suppose

$$\frac{R(F+tH)}{t} \to 0 \text{ in the norm on } \mathcal{N} \text{ as } t \to 0. \tag{B.4.7}$$

(i) T is *Gateaux differentiable* at F with derivative $T'_F = L$ if (B.4.7) holds for all $H \in \mathcal{M}$.

(ii) T is *Hadamard differentiable* (or compact differentiable) if (B.4.7) holds *uniformly* for H lying in an arbitrary compact subset of $\mathcal{M}$; and T_F is called the *compact derivative* of T at F.

(iii) T is *Fréchet differentiable* if (B.4.7) holds uniformly for H lying in an arbitrary bounded subset of M; and T'_F is called the *Fréchet derivative* of T at F.

These derivatives are successively stronger. Moreover, it can be shown that T is Fréchet differentiable at F if and only if

$$\|T(G) - T(F) - T'_F(G-F)\| = o(\|G-F\|) \tag{B.4.8}$$

for some continuous linear functional T'_F. Taking $G = F_n$ in (B.4.8) we see that $n^{1/2}[T(F_n) - T(F)]$ and $n^{1/2}T_F(F_n - F)$ differ by $o_p(n^{1/2}\|F_n - F\|)$, which is $o_p(1)$ in view of a Kolmogorov limit theorem. A proof that certain L-estimators are asymptotically normal is obtained via this Fréchet derivative method in the next subsection. Unfortunately, Fréchet derivatives of statistical functionals rarely exist. On the other hand, the weak Gateaux derivative often exists, even when $n^{1/2}[T(F_n) - T(F)]$ is not normally distributed in the limit. These facts suggest looking for an intermediate strength derivative to prove asymptotic normality. Reeds (1976) suggests that the *compact* derivative is appropiate because of the following result:

Theorem B.19

Let $\mathcal{M}$,$\mathcal{N}$ be arbitrary normed linear spaces with $T : \mathcal{M} \to \mathcal{N}$. Assume T is compactly differentiable at F; and that F_n are random quantities in $\mathcal{M}$ such that $n^{1/2}(F_n - F)$ is *tight*. Then the remainder (B.4.6) satisfies

$$n^{1/2}R(F_n) \equiv n^{1/2}[T(F_n) - T(F) - L(F_n - F)] = o_p(1). \tag{B.4.9}$$

Note that despite our function space notation Theorem B.19 actually generalizes the Δ-method to arbitrary normed linear spaces. In applications, the norm on $\mathcal{M}$ must be chosen strong enough so that a compact derivative exists, but not so strong that tightness of $n^{1/2}(F_n - F)$ fails.

Proof. Given $\epsilon > 0$, "tightness" of $n^{1/2}(F_n - F)$ means there exists a compact set $K \subset M$ such that for all n, $P\{n^{1/2}(F_n - F) \in K\} \geq 1 - \epsilon$.

By assumption, there exists a continuous linear functional L, approximating T, such that the associated remainder $n^{1/2}R(F + n^{-1/2}H) = o_p(1)$ as $n \to \infty$ uniformly for H in any compact set. Hence given a neighborhood N of $0 \in n$, $n^{1/2}R(F + n^{-1/2}H) \in N$ for all $H \in K$, and all sufficiently large n. It follows that $P\{n^{1/2}R(F_n) \in N\} \geq 1 - \epsilon$ for all sufficiently large n. □

A final remark: When invoking Theorem B.19, one knows only that the limiting distribution of $n^{1/2}L(F_n - F)$, if any, is the same as that of $n^{1/2}[T(F_n) - T(F)]$. However, if the linear functional L can be expressed as $L(G) = \int \Omega(x)\,dG(x)$, $G \in M$, then $n^{1/2}[T(F_n) - T(F)] \to N(0, \mathrm{E}_F[\Omega^2(X)])$, as mentioned before. This is the case whenever L is *weakly* continuous. The following example shows that such an integral representation of continuous linear functionals is not always possible.

Example 3. $T(G)$ = mass in discrete part of G. Then T is continuous in the total variation metric on M. But $n^{1/2}[T(F_n) - T(F)] \equiv n^{1/2}$ if F is continuous.

B.4.3 Some Applications of the Δ-Method

Now we consider two examples which illustrate the Δ-method. The first functional considered is the Fréchet differentiable, while the second is only compactly differentiable.

Theorem B.20: *L*-Estimators (Boos, 1979)

Let $T(F)$ be a robust L-estimator. That is, $T(F) = \int_0^1 F^{-1}(t)J(t)\,dt$, where $J(t) = 0$ for $0 < t \leq \alpha$ and $\beta \leq t < 1$. Assume also that J is bounded and continuous a.e. Lesbesgue and a.e. F^{-1}. Then T is Fréchet differentiable at F with derivative $T_F'(H) = -\int H(x)J(F(x))\,dx$, $H \in M$.

Proof. Let $K(t) = \int_0^t J(s)\,ds$. Then an integration by parts shows that $T(G) - T(F) = \int [K(F(x)) - K(G(x))]\,dx$. The integrand divided by $F - G$ is a difference qoutient which approaches $J(F(x))$ at all continuity points of J. Since J is continuous a.e. F^{-1}, the set of x where J is discontinuous at

$F(x)$ has Lebesgue measure 0, and hereafter we omit this null set from our range of integration. We need to show that [see (B.4.8)]

$$\left| \int \left[\frac{K(F(x)) - K(G(x))}{F(x) - G(x)} - J(F(x)) \right] [F(x) - G(x)] \, dx \right|$$

is of smaller order than $\delta = \|G - F\|$. Clearly it is less than or equal to

$$\delta \int \left| \frac{K(F(x)) - K(G(x))}{F(x) - G(x)} - J(F(x)) \right| dx \tag{B.4.10}$$

so the claim is established if we can justify applying the dominated convergence theorem. The integrand in (B.4.10) is 0 outside the finite interval $[a_\delta, b_\delta] = [F^{-1}(\alpha - \delta), F^{-1}(\beta + \delta)]$, and within this interval the integrand is bounded by $2\|J\|$. Thus T is Fréchet differentiable at F. □

Corollary to Theorem B.20. Under the assumption of Theorem B.20, the estimator sequence $T(F_n)$ satisfies $n^{1/2}[T(F_n) - T(F)] \to N(0, \sigma^2)$, where

$$\sigma^2 = \int \int J(F(x)) J(F(y)) [F(\min(xy)) - F(x)F(y)] \, dx \, dy.$$

The above L-estimators require a very smooth J function, which excludes finite linear combinations of order statistics. In fact, *quantiles are not Fréchet differentiable* with respect to the sup norm on M. □

Example 4.

Let F be the uniform distribution on (0,1), and for small Δ define

$$G(x) = \begin{cases} \frac{1}{2}, & \frac{1}{2} - \Delta \leq x < \frac{1}{2} \\ F(x), & \text{otherwise.} \end{cases}$$

Then if $T(G) \equiv$ median of G, the Fréchet derivative (if it exists) must be consistent with the influence function:

$$T_F'(G) = \int \Omega_{1/2}(x) \, dG(x) = \frac{-G(F^{-1}(\frac{1}{2}))}{f(F^{-1}(\frac{1}{2}))} = \frac{1}{2} - G\left(\frac{1}{2}\right).$$

However, the remainder

$$R(G_\Delta) = T(G_\Delta) - T(F) - T_F'(G - F) = -\Delta = -\|G - F\|$$

so $R(G_\Delta)$ is not $o(\|G - F\|)$, as required by (B.4.8) for Fréchet differentiabilty. Quantiles are, however, *compactly* differentiable [see Reeds (1976)].

B.4.4 The Projection Method versus the von Mises Method

To gain some insight into the relative merits of the projection method with the von Mises argument, let us prove by both methods Hoeffding's result (Theorem B.18 stated earlier) for the case $k = 2$.

We have $X_1, \ldots, X_n$ i.i.d. F and $U_n = \binom{n}{2}^{-1} \Sigma\Sigma_{i<j} \Omega(X_i, X_j)$, where $\Omega(x,y) = \Omega(y,x)$ for all x, y. Hoeffding's theorem assumes $\mathrm{E}[\Omega^2(X_1, X_2)] < \infty$ and concludes that $T_n = n^{1/2}[U_n - \mathrm{E}[U_n]] \to N(0, 4\mathrm{Var}[\psi_1](X_1))$, where

$$\psi_1(x) \equiv \mathrm{E}[\Omega(X_1, X_2) \mid X_2 = x] - \mathrm{E}_\Omega[(X_1, X_2)].$$

Proof by Projection. We need to find $\hat{T}_n = \sum_{k=1}^n \mathrm{E}[T_n \mid X_k]$. But

$$\mathrm{E}[T_n \mid X_k = x] = n^{1/2} \binom{n}{2}^{-1} \Sigma\Sigma_{i<j} \mathrm{E}\left[\Omega(X_i, X_j) - \mathrm{E}[\Omega \mid X_k = x]\right]$$
$$= 2n^{-1/2} \psi_k(x).$$

Now $\hat{T}_n = 2n^{-1/2} \sum_{i=1}^n \psi_1(X_i)$ has the limitimg distribution claimed for T_n; we need only to show $\mathrm{Var}[T_n] - \mathrm{Var}[\hat{T}_n] \to 0$ as $n \to \infty$. $\mathrm{Var}[\hat{T}_n] = 4\mathrm{Var}[\psi_1(X_1)]$ is obvious, but $\mathrm{Var}[T_n]$ requires a lengthy if elementary calculation. [See, e.g., Example 6 of A.1 of Lehmann (1975).] Some intermediate values are

$$\mathrm{Var}[T_n] = \frac{4}{n(n-1)^2} \mathrm{Var}[\Sigma\Sigma_{i<j} \Omega(X_i, X_j)]$$
$$= \frac{4}{n(n-1)^2} \left[\frac{n(n-1)}{2} \mathrm{Var}[\Omega(X_1, X_2)] + n(n-1)(n-2)\mathrm{Var}[\psi_1(x)] \right]$$
$$= 4\frac{n-2}{n-1} \mathrm{Var}[\psi_1(X_1)] + o(1).$$

To summarize, the proof by projection is probabilistic, long, and elementary. □

Proof by the von Mises Method. Let $T(G) = \int \int \Omega(x,y)\, dG(x)\, dG(y)$. Then

$$T(F_n) - T(F) = \frac{2}{n^2} \Sigma\Sigma_{i<j} \Omega(X_i, X_j) + \frac{1}{n^2} \Sigma_i \Omega(X_i, X_i) - \mathrm{E}[\Omega(X_1, X_2)]$$

and

$$n^{1/2}[T(F_n) - T(F)] = n^{1/2}[U_n - \mathrm{E}[U_n]] + o_p(1).$$

Expanding $T(F + tH)$, we obtain

$$T(F + tH) = T(F) + t \int \left[\int \Omega(x,y)\, dF(x) \right] dH(y) + t^2 T(H).$$

So our candidate for compact derivative of T is

$$T'_F(H) = \int \left[\int \Omega(x,y)\,dF(x) \right] dH(y).$$

The remainder in a one-term expansion is $R(F + tH) = t^2 T(H)$; we need to show that $tT(H) \to 0$ as $t \to 0$ uniformly for H lying in compact sets, or equivalently that T is bounded on compact sets. □

At this point we need to confront the problem of how to choose a norm topology on the domain of T for which $n^{1/2}[T(F_n) - T(F)]$ is tight and for which the above compactness condition holds. Then by Theorem B.19, $n^{1/2}[T(F_n) - T(F)]$ and $n^{1/2}T'_F(F_n - F)$ have the same limiting distribution, namely, normal with parameters 0 and $4\,\mathrm{Var}[\psi(X_i)]$.

If we take the domain of T to be all finite signed measures M endowed with the sup norm, then $n^{1/2}(F_n - F)$ is tight. The functional T is continuous (and hence bound on compacts) if Ω is continuous and bounded or if Ω is of bounded variation. Possibly weaker conditions on Ω will imply that T is bounded on compacts, but we are not aware of them. To summarize, the proof by the von Mises method begins nicely in that the expansion of $T(F + tH)$ clearly indicates the derivative T'_F and what is required to show that the remainder goes to zero. However, the subsequent choice of the norm for the domain of T may be a difficult one; in our case the obvious choice led to unnecessarily stringent requirements on Ω.

Now we can see salient differences in the two methods of proof. In the projection method we need to calculate $\hat{T} = \Sigma_i \mathrm{E}[T \mid X_i]$ and $\mathrm{Var}[T]$, $\mathrm{Var}[\hat{T}]$. If we are lucky, the projection will yield a good approximation, and the desired result is obtained. The method is probabilistic, straightfoward, but often tedious and opaque. The von Mises method is analytic and clear in outline, but requiring imagination and knowledge of functional analysis to complete. The von Mises method required stronger assumptions to obtain the same result, but this can be blamed perhaps on our inadequate analysis. Perhaps a different norm on the domain of T would allow for weaker conditions on Ω.

In fact since the von Mises method is based on normed space arguments while the projection method is based on Hilbert space arguments, it is reasonable to conjecture that the former method will apply whenever the latter does; a proof remains to be seen.

B.5 PROBLEMS

These exercises are intended to illustrate the results in Appendix B by means of commonly encountered examples and applications. More applications are found in the text proper.

Section B.1

1. The Markov inequality states that if Y is a positive random variable and a is a positive constant, then $P\{Y > a\} \leq \mathrm{E}[Y]/a$.
   (a) Show that convergenge in mean square implies convergence in probability; that is, $\mathrm{E}[X_n - X]^2 \to 0$ implies $X_n - X = o_p(1)$. *Hint*: start with the definition of convergence in probability and apply the Markov inequality to $Y = (X_n - X)^2$.
   (b) Show by example that convergence in probability does not imply convergence in mean square.

2. Show that $X_n \xrightarrow{p} X$ implies $X_n \xrightarrow{d} X$, but that the converse is false.

3. Use the continuity theorem to show that if $X_n \xrightarrow{d} X$, where $X \sim \Phi$, the standard normal distribution, then $X_n^2 \xrightarrow{d} Y$, where $Y \sim \chi_1^2$, the chi-squared distribution with one degree of freedom.

4. Let X_n have the continuous uniform distribution on the interval $(0, 1/n)$.
   (a) Show that $X_n \xrightarrow{d} 0$, and identify the limit distribution function.
   (b) Demonstrate that the conclusion of Pólya's theorem fails to hold, and explain why.

5. Let $X_n \sim (1 - 1/n)\Phi + (1/n)C$, where Φ and C are the standard normal and Cauchy distributions, respectively. Show that $X_n \xrightarrow{d} Z \sim \Phi$ so that $X_n \sim AN(0, 1)$, even though $\mathrm{E}[X_n]$ fails to exist.

6. Show that for i.i.d. random variables with finite variance the Lindeberg condition holds and hence that Theorem B.7 implies Theorem B.6.

7. Let $X_1, \ldots, X_n$ be i.i.d. uniform (0,1) variables, and let $\overline{X}_n$ and $X_{(n)}$ represent the mean and maximum of these n variables.
   (a) Show that $\overline{X}_n \sim AN(\mu_n, \sigma_n^2)$, for some choice of parameters.

(b) Show that $X_{(n)} \xrightarrow{p} c$, for some $c \neq 0$.

(c) Show that $\sqrt{n}((2\overline{X}_{n-1})/X_{(n)}) \xrightarrow{d} H$, for some H, and identify the limit distribution H.

8. Suppose that $X_n \sim AN(a_n, b_n^2)$. Let $d_n > 0$ be another sequence of constants with $b_n/d_n \to 1$. Find the most general conditions on c_n for which we may conclude that $X_n \sim AN(c_n, d_n^2)$.

9. Let $X_n \sim B(n, p)$, the binomial distribution.

(a) Show that $X_n \sim AN(np, np(1-p))$.

(b) Show that

$$\sup_x \left| p\{X_n \leq x\} - \Phi\left(\frac{x - np}{\sqrt{np(1-p)}}\right)\right| \to 0 \qquad \text{as} \quad n \to \infty.$$

10. In this problem we use Taylor's theorem for random variables to find an expansion for a function of the sample mean $\overline{X}_n$ of n i.i.d. variables having three finite central moments α_k, $k = 1, 2, 3$. Let $\mu_k = \mathrm{E}[X - \alpha_1]^k$ denote the kth central moment.

(a) Find an expression for $g(\overline{X}_n)$ in terms of the moments and use it to prove that $g(\overline{X}_n)$ is asymptotically normal under certain conditions on g.

(b) Apply your result in part (a) to obtain the limiting distribution of the reciprocal of the sample mean which is based on i.i.d. observations from a continuous uniform (0,1) distribution.

(c) Does the expansion in part (a) suggest an expansion for $\mathrm{E}[g(\overline{X}_n])$ in terms of $g(\mu)$ and the moments?

11. Let F have a density f which is continuous and positive at the quantile $x_p = F^{-1}(p)$.

(a) Show that $g(u) = F^{-1}(u)$ is differentiable at $u = p$ with derivative $g'(p) = 1/f(x_p)$.

(b) Verify the corollary to Theorem B.9.

12. Assume that $\sqrt{n}(X_n - a) \xrightarrow{2} Y$ (mean square convergence), where Y has mean 0 and variance V, and a is a constant.

(a) Show that $\sqrt{n}(\mathrm{E}[X_n] - a) \to 0$.

(b) Show that $n\,\mathrm{Var}[X_n] \to V$.

(c) Use parts (a) and (b) to verify the remarks at the end of Section B.1 regarding $\sqrt{n}(\hat{\theta}_n - \theta)$.

Section B.2

13. Let $X_1, \ldots, X_n$ be i.i.d. with finite mean μ and variance $\sigma^2 > 0$.

(a) Show by means of the law of large numbers and Slutsky's lemma that the sample variance converges:

$$s_n^2 = \frac{1}{n-1} \sum_{i=1}^{n} [X_i - \overline{X}_n]^2 \xrightarrow{p} \sigma^2.$$

(b) Use other results stated above to demonstrate that the "Studentized" mean is asymptotically normal. That is, if $\overline{X}_n$ is centered at its mean and divided by an estimate of its standard deviation, it converges to a normal distribution. In fact

$$\frac{\overline{X}_n - \mu}{s_n/\sqrt{n}} \xrightarrow{d} Z \sim \Phi.$$

14. Use the continuity theorem and the law of large numbers to prove Theorem B.16.

15. Prove the corollary to Theorem B.16.

16. Use the corollary to Theorem B.16 or other means to show that the sample coefficient of variation $s/\overline{x}$ is asymptotically normal, and find the asymptotic mean and standard deviation.

Section B.3

17. Verify the asymptotic normality of the sample variance (Example 1 of Section B.3) using the projection theorem, Theorem B.18.

18. Derive the asymptotic normality of the Wilcoxon signed rank statistic (Example 2 of Section B.3) using the projection theorem.

19. Derive the asymptotic normality of Kendall's τ statistic (Example 3 of Section B.3) using the projection theorem.

20. Verify the details of Hajek's projection method by confirming that the classical projection theorem applies to the situation described after Theorem B.18.

(a) The vector space $\mathcal{H}$ of square integrable functions of the observations is indeed a Hilbert space with inner product $\langle T,U\rangle = \mathrm{E}[TU]$.

(b) The space $\mathcal{L}$ of homogeneous, linear, and square integrable functions of the observations is a *closed* subspace of $\mathcal{H}$.

(c) The projection of T_n onto the subspace $\mathcal{L}$ is of the form $\sum_i a_i(X_i)$, where $a_i(x) = \mathrm{E}[T_n \mid X_i = x]$.

Section B.4

21. (a) Explain why the three notions of derivative in (B.4.7) are successively stronger: (iii) implies (ii) implies (i). Also, if $\mathcal{M} = R^p$ for some p, (ii) implies (iii); if $\mathcal{M} = R$, (i) implies (iii).

(b) The quantile example in the notes shows that (B.4.2) (iii) is strictly stronger than (ii). Show that (ii) is strictly stronger than (i) by consideration of $g : R^2 \to R$ defined by $g(x,x^2) = 1$ for $x \neq 0$; $g(x,y) = 0$ otherwise.

22. Let $g : R^p \to R^q$, so g is of the form $(g_1,\ldots,g_q)$, and each $g_i : R^p \to R$. Let $A_{ij}(\boldsymbol{x})$ denote the partial derivative of the ith component function of the jth variable x_j at the point $\boldsymbol{x} \in R^p$. Let $A_{\boldsymbol{x}} = [A_{ij}(\boldsymbol{x})]$ denote this Jacobian matrix.

(a) If the Fréchet derivative g' of g exists at $\boldsymbol{a}$, it coincides with the linear transformation $\boldsymbol{y} \to A_{\boldsymbol{a}}\boldsymbol{y}$. Conversely, if the partial derivatives of g exist and are continuous at $\boldsymbol{a}$, then the Fréchet derivative exists and is given by $\boldsymbol{y} \to A_{\boldsymbol{a}}\boldsymbol{y}$. [Bartle (1964), Chapter VII].

(b) If $\boldsymbol{Z}_n \in R^p$ satisfy $n^{1/2}(\boldsymbol{Z}_n - \boldsymbol{a}) \to$ normal (o,Σ), and if the partial derivatives of g exist and are continuous at $\boldsymbol{a}$, then $n^{1/2}[g(\boldsymbol{Z}_n) - g(\boldsymbol{a})] \to$ normal $(0, A\Sigma A^T)$, where A is the Jacobian matrix of g, evaluated at $\boldsymbol{a}$.

23. Verify the equivalence of the two definitions of Fréchet derivative (B.4.7) (iii) and (B.4.8).

24. (a) Let $\mathcal{M}$ be the space of distributions corresponding to all finite signed measures, and let $\mathcal{N}$ be the real line. If $L : \mathcal{M} \to \mathcal{N}$ is a *weakly* continuous functional, then there exists a continuous bounded function Ω such that $L(G) = \int \Omega(x)\,dG(x)$. [Let $\omega(x) \equiv L(\delta_x)$ where δ_x is the point mass on x.]

(b) Extend part (a) to the case where $\mathcal{N} = R^k$.

25. Show that the chain rule holds for compact derivatives: if $\mathcal{M}$, $\mathcal{N}$, $\mathcal{O}$ are normed spaces and $T : \mathcal{M} \to \mathcal{N}$ has compact derivative T'_F at F, and

$U : \mathcal{N} \to O$ has compact derivative U'_G at $G = T(F)$, then $U \circ T$ has compact derivative $(U \circ T) = U'_G \circ T'_F$ at F.

26. **(a)** Show that the Hodges-Lehmann estimator $\text{med}\{X_i + X_j\}/2$ is asymptotically equivalent to the estimator $(U \circ T) n^{1/2}(F_n - F)$, where $T(F) = G =$ distribution of $(X_1 + X_2)/2$ when X_1, X_2 i.i.d. F and $U(G) = G^{-1}(\frac{1}{2})$.

(b) Show that $n^{1/2}(U \circ T)'(F_n - F)$ is asymptotically normal with parameters 0 and $\frac{1}{12}[\int f^2(x)\,dx]^{-2}$.

(c) The limiting distribution of the Hodges-Lehmann estimator can be derived by the von Mises method using parts (a), (b) above, Problem 25, and Theorem B.19 provided only that T is shown to be compactly differentiable for a norm topology on its domain in which $n^{1/2}(F_n - F)$ is tight. Can you complete the proof?

APPENDIX C

Index of All Data Sets and Listing of Larger Data Files

The following data sets are analyzed in the text and/or problems. The data are found in the first-named section given below, unless they have been relegated to this appendix because of their size. The sources of the data are referenced with them.

Topic and Sections
Fingerprint ridge counts: 1.2, 6.1, 6.5, 6.6, C.1
Melbourne's rainfall: 2.1, C.2
EMT6 cell lifetimes: 2.1, 2.4, 3.5, 4.1, 4.3
Effects of sleep-inducing drugs: 4.1, 5.1
Velocity of light: 4.7, C.3
Percentage of shrimp: 4.7
PPM of DDT in kale: 4.7
Crossed vs. self-fertilized plants: 4.7, C.4
Atomic weight of carbon: 6.1, 6.5, 6.6
Cloth mill data: 6.7
Growth of prices in Taiwan: 7.1–7.3, 7.6
Stack loss data: 7.2, 7.3, 7.6
Water salinity in Pamilco Sound: 7.6, C.5
Scottish hill races: 7.6

C.1 FINGERPRINT RIDGE COUNTS

The following columns give the fingerprint ridge count totals for the left and right hands of two distinct groups, an Australian aboriginal tribe from Western Arnhem Land, and a first-year statistics class at La Trobe University. It is expected that the former group be much more homogeneous than the latter, which is composed of unrelated persons from a large number of ethnic backgrounds. The first nine rows give the left and right hand totals, respectively, of the tribal members; and the last nine rows give the corresponding results for the statistics students. The authors thank Dr. Neville G. White of the Department of Genetics and Human Variation, La Trobe University, for permission to use the aboriginal ridge count data which he collected.

Ridge Counts of Tribal Members

	left	right	left	right	left	right
1	74	92	71	89	58	66
2	113	116	76	83	91	91
3	69	73	64	72	57	58
4	68	73	62	66	73	71
5	61	75	100	110	61	58
6	70	83	77	78	71	85
7	99	105	62	75	74	76
8	46	52	87	87	78	96
9	74	78	81	79	74	80

Ridge Counts of University Students

	left	right	left	right	left	right
1	90	91	64	60	45	48
2	75	68	87	89	73	72
3	60	53	90	85	67	62
4	54	62	77	71	112	131
5	39	37	34	38	87	95
6	58	48	34	38	89	87
7	35	35	91	99	66	73
8	7	4	37	41	67	67
9	53	48	66	73	27	42

C.2 MELBOURNE'S RAINFALL

The following columns show Melbourne's rainfall, in tenths of millimeters, for the months of June, July, and August, respectively for the years 1981, 1982, and 1983.

	1981			1982			1983		
Day	June	July	August	June	July	August	June	July	August
1	64	0	0	30	0	0	4	40	0
2	22	0	28	4	0	0	2	24	40
3	0	0	42	0	0	12	0	0	0
4	26	82	80	0	0	2	0	0	0
5	0	144	0	0	0	0	0	0	0
6	0	6	24	0	0	0	0	98	0
7	0	8	0	0	10	0	0	126	0
8	0	0	0	0	0	0	40	0	0
9	0	118	80	0	0	0	4	0	36
10	14	0	28	0	0	0	32	0	0
11	40	0	0	0	0	162	2	0	12
12	0	36	24	0	6	46	0	4	0
13	16	2	4	0	6	8	74	0	0
14	112	6	0	0	10	0	0	0	0
15	48	0	22	0	0	46	4	36	0
16	32	0	0	14	0	0	4	0	0
17	0	0	0	0	0	2	0	12	302
18	0	0	6	0	0	0	0	2	8
19	4	0	28	0	74	0	0	0	0
20	0	0	2	64	0	0	0	72	0
21	0	60	120	80	0	0	8	0	16
22	0	40	20	84	0	6	0	0	0
23	110	0	14	6	0	0	0	4	56
24	0	10	0	14	0	0	6	0	0
25	22	10	0	0	0	0	16	0	0
26	32	0	0	0	0	0	26	0	14
27	2	2	48	0	0	0	0	10	56
28	8	0	34	8	0	0	0	148	6
29	8	128	164	0	16	0	2	18	0
30	0	0	10	0	4	0	100	40	4
31		10	36		0	4		2	30

Source: Bureau of Meterology, Victoria.

C.3 NEWCOMB'S MEASUREMENTS OF THE PASSAGE TIME OF LIGHT

The following 66 measurements are from Stigler (1977), Table 5, and were the Third Series of Measurements recorded by Newcomb in 1882. The given values $\times 10^{-3} + 24$ give the time in millionths of a second for light to traverse a known distance. The "true value" has subsequently been shown to be 33.02.

28	−44	29	30	24	28
37	32	36	27	26	28
29	26	27	22	23	20
25	25	36	23	31	32
24	27	33	16	24	29
36	21	28	26	27	27
32	25	28	24	40	21
31	32	28	26	30	27
26	24	32	29	34	-2
25	19	36	29	30	22
28	33	39	25	16	23

C.4 DARWIN'S DATA ON GROWTH RATES OF PLANTS

The data were originally gathered by Charles Darwin in 1878. He was interested in whether seedlings from cross-fertilized flowers were in any way superior to seedlings from self-fertilized flowers. The data which are shown are the heights in inches. More explanation regarding the experimental difficulties encountered by Darwin are quoted in Andrews and Herzberg (1985), wherein these data are reproduced, and a further discussion is found.

	Cross-	Self-	Diff.
1	23.500	17.375	6.125
2	12.000	20.375	−8.375
3	21.000	20.000	1.000
4	22.000	20.000	2.000
5	19.125	18.375	0.750
6	21.500	18.625	2.875
7	22.125	18.625	3.500
8	20.375	15.250	5.125

	Cross-	Self-	Diff.
9	18.250	16.500	1.750
10	21.625	18.000	3.625
11	23.250	16.250	7.000
12	21.000	18.000	3.000
13	22.125	12.750	9.375
14	23.000	15.500	7.500
15	12.000	18.000	−6.000

C.5 WATER SALINITY IN PAMLICO SOUND

The data are taken from Ruppert and Carroll (1980). The first column gives a measure of the water salinity of a river discharge in North Carolina's Pamlico Sound. The predictor variables in columns 2–4 are, respectively, the salinity two weeks earlier, a linear time trend (the biweekly period in March–May), and the amount of river discharge. This data set is also analyzed in Rousseeuw and Leroy (1987), Hettmansperger (1987) and Welsh (1987).

OBS	SALINITY	SALLAG	TREND	H20FL0W
1	7.6	8.2	4	23.005
2	7.7	7.6	5	23.873
3	4.3	4.6	0	26.417
4	5.9	4.3	1	24.868
5	5.0	5.9	2	29.895
6	6.5	5.0	3	24.200
7	8.3	6.5	4	23.215
8	8.2	8.3	5	21.862
9	13.2	10.1	0	22.274
10	12.6	13.2	1	23.830
11	10.4	12.6	2	25.144
12	10.8	10.4	3	22.430
13	13.1	10.8	4	21.785
14	12.3	13.1	5	22.380
15	10.4	13.3	0	23.927
16	10.5	10.4	1	33.443
17	7.7	10.5	2	24.859
18	9.5	7.7	3	22.686
19	12.0	10.0	0	21.789
20	12.6	12.0	1	22.041
21	13.6	12.1	4	21.033

OBS	SALINITY	SALLAG	TREND	H20FL0W
22	14.1	13.6	5	21.005
23	13.5	15.0	0	25.865
24	11.5	13.5	1	26.290
25	12.0	11.5	2	22.932
26	13.0	12.0	3	21.313
27	14.1	13.0	4	20.769
28	15.1	14.1	5	21.393

APPENDIX D

Minitab Macros

The following macros are presented without further comment, since it is assumed that the reader is familiar with the basic commands of the Minitab software package. The sections of the book where these macros are discussed are given in parentheses.

The authors are pleased to acknowledge the cooperation of Minitab, and its permission to include material on instructions and output from the software package in the text.*

D.1 BOOTSTRAP MACROS

D.1.1 Example of Parametric Bootstrap (2.4)

```
# 'boot1.mtb'
#
# INITIALIZE THE ITERATION BY SETTING k1=0.
# LET k2=THE ESTIMATE OF THE EXPONENTIAL PARAMETER.
# LET k3=THE SAMPLE SIZE n.
# COLUMN c4 IS FILLED WITH B BOOTSTRAP ESTIMATES IF THIS STORED
# PROGRAM IS EXECUTED B TIMES
noecho
let k1=k1+1
rand k3 c2;
expo k2.
# THE EXPONENTIAL DISTRIBUTION AND ITS PARAMETER ESTIMATE k2
```

*Further information on latest versions of the package are available from Minitab Data Analysis Software, 3081 Enterprise Drive, State College, PA 16801 USA. Telephone: 814/238-3280. Telex: 881612. Minitab is a registered trademark.

```
# MAY BE REPLACED BY ANY OTHER DISTRIBUTION ON MINITAB'S MENU.
let c4(k1)=mean(c2)
# THE MEAN MAY BE REPLACED BY ANY OTHER DESIRED ESTIMATOR.
end
```

D.1.2 Nonparametric Bootstrap (2.4)

```
# 'boot2.mtb'
#
# INITIALIZE BY SETTING k1=0 and k3=SAMPLE SIZE n.
# THIS MACRO PROCEEDS AS IN 'boot1.mtb' BUT THE EMPIRICAL
# DISTRIBUTION OF THE DATA MUST BE STORED IN COLUMNS c1 c2.
# COLUMN c4 IS FILLED WITH B BOOTSTRAP ESTIMATES IF THIS STORED
# PROGRAM IS EXECUTED B TIMES
noecho
let k1=k1+1
rand k3 c3;
discrete c1 c2.
let c4(k1)=mean(c3)
# THE MEAN MAY BE REPLACED BY ANY OTHER DESIRED ESTIMATOR.
end
```

D.2 ONE- AND TWO-SAMPLE PROCEDURES

D.2.1 One-Sample Trimmed Mean (4.3.2)

```
# 'tmean.mtb'
#
# THIS MINITAB MACRO CALCULATES A TRIMMED MEAN, AN ESTIMATE OF ITS
# ASYMPTOTIC STANDARD ERROR AND THE ASSOCIATED APPROXIMATE 95%
# CONFIDENCE LIMITS FOR THE DATA CONTAINED IN c1. THE CONFIDENCE
# LIMITS ARE BASED ON THE T-DISTRIBUTION WITH DEGREES OF FREEDOM
# AS GIVEN BY TUKEY AND MClAUGHLIN (1963). THE USER MUST SET
# k1 = NUMBER OF DATA POINTS TRIMMED OF EACH END OF THE SAMPLE.
let k2 = count(c1)
sort c1 c2
copy c2 c2;
omit 1:k1.
let k3 = k2 - 2*k1 + 1
let k4 = k2 - k1
copy c2 c3;
```

```
omit k3:k4.
let k10 = mean(c3)
let k5 = c3(1)
let k6 = c3(k3-1)
set c4
k1(k5)
set c5
k1(k6)
stack c3 c4 c5 c6
let k11 = stdev(c6)
let k12 = 1. - 2.*(k1/k2)
let k13 = k11/(k12*sqrt(k2))
let k14 = k2 - 2*k1 - 1
invcdf 0.975 k15;
t k14.
let k16 = k10 - k15*k13
let k17 = k10 + k15*k13
# TRIMMED MEAN
print k10
# FRACTION OF DATA REMAINING AFTER TRIMMING
print k12
# ESTIMATE OF ASYMPTOTIC STANDARD ERROR
print k13
# APPROXIMATE 95% CONFIDENCE LIMITS
print k16 k17
end
```

D.2.2 Two-Sample Trimmed Means (6.5)

```
# 'tmean2.mtb'
#
# THIS MINITAB MACRO CALCULATES AN ESTIMATE OF DELTA = MU1 - MU2
# BASED ON TRIMMED MEANS AS WELL AS AN APPROXIMATE 95% CONFIDENCE
# INTERVAL FOR DELTA. tHIS CONFIDENCE INTERVAL IS BASED ON A
# TWO-SAMPLE TRIMMED T-STATISTIC AND USES AN ADAPTATION OF
# WELCH'S DEGREES OF FREEDOM FORMULA, DUE TO YUEN (1974).
# THE DATA FROM THE 1ST SAMPLE IS ASSUMED TO BE IN c11, WHILE THAT
# FROM THE 2ND SAMPLE IS ASSUMED TO BE IN c21. THE USER MUST SET
# k11 = NUMBER OF DATA POINTS TRIMMED OF EACH END OF THE 1ST SAMPLE
# k21 = NUMBER OF DATA POINTS TRIMMED OF EACH END OF THE 2ND SAMPLE.
let k12 = count(c11)
sort c11 c12
```

```
copy c12 c12;
omit 1:k11.
let k1 = k12 - 2*k11 + 1
let k2 = k12 - k11
copy c12 c13;
omit k1:k2.
let k10 = mean(c13)
let k3 = c13(1)
let k4 = c13(k1-1)
set c14
k11(k3)
set c15
k11(k4)
stack c13 c14 c15 c16
let k5 = stdev(c16)
let k6 = 1. - 2.*(k11/k12)
let k7 = k5/(k6*sqrt(k12))
let k8 = k12 - 2*k11 - 1
let k22 = count(c21)
sort c21 c22
copy c22 c22;
omit 1:k21.
let k31 = k22 - 2*k21 + 1
let k32 = k22 - k21
copy c22 c23;
omit k31:k32.
let k20 = mean(c23)
let k33 = c23(1)
let k34 = c23(k31-1)
set c24
k21(k33)
set c25
k21(k34)
stack c23 c24 c25 c26
let k35 = stdev(c26)
let k36 = 1. - 2.*(k21/k22)
let k37 = k35/(k36*sqrt(k22))
let k39 = k7**2 + k37**2
let k40 = sqrt(k39)
let k41 = k10 - k20
let k42 = k41/k40
let k43 = (k7**4)/(k12 - 2*k11 - 1) + (k37**4)/(k22 - 2*k21 - 1)
```

```
let k44 = (k39**2)/k43
cdf k42 k38;
t k44.
let k38 = 1 - k38
invcdf 0.975 k45;
t k44.
let k46 = (k10 - k20) - k45*k40
let k47 = (k10 - k20) + k45*k40
# TRIMMED MEAN OF 1ST SAMPLE
print k10
# FRACTION OF DATA REMAINING IN 1ST SAMPLE AFTER TRIMMING
print k6
# ESTIMATE OF ASYMPTOTIC STANDARD ERROR OF TRIMMED MEAN 1
print k7
#
# TRIMMED MEAN OF 2ND SAMPLE
print k20
# FRACTION OF DATA REMAINING IN 2ND SAMPLE AFTER TRIMMING
print k36
# ESTIMATE OF ASYMPTOTIC STANDARD ERROR OF TRIMMED MEAN 2
print k37
#
# TRIMMED MEAN 1 - TRIMMED MEAN 2
print k41
# ESTIMATE OF ASYMPTOTIC STANDARD ERROR OF DIFFERENCE OF TRIMMED MEANS
print k40
# APPROXIMATE 95% CONFIDENCE LIMITS FOR MU1 - MU2
print k46, k47
#
# YUEN'S 2-SAMPLE TRIMMED WELCH T-STATISTIC FOR TESTING H: MU1 = MU2
print k42
# DEGREES OF FREEDOM FOR YUEN'S 2-SAMPLE TRIMMED WELCH STATISTIC
print k44
end
```

D.2.3 One-Step Huber M-Estimate (4.3.3)

```
# 'huber1.mtb'
#
# THIS MINITAB MACRO CALCULATES A ONE-STEP M-ESTIMATE, BASED ON
# HUBER'S PSI FUNCTION. THE MEDIAN IS USED AS A STARTING VALUE
# AND THE MEDIAN ABSOLUTE DEVIATION IS USED TO ESTIMATE SCALE.
```

```
# THE DATA IS ASSUMED TO BE IN c1. THE USER MUST SET
# k1 = VALUE OF k IN HUBER'S PSI FUNCTION.
let k11 = median(c1)
let c2 = abso(c1 - k11)
invcdf 0.75 k12;
normal 0 1.
let k13 = median(c2)/k12
let c11 = (c1 - k11)/k13
let c12 = (sign(k1 - abso(c11)) + 1)/2
let k2 = -1*k1
code (-100:k2) k2 c11 c13
code (k1:100) k1 c13 c13
let k21 = k11 + k13*sum(c13)/sum(c12)
# ONE-STEP HUBER M-ESTIMATE
print k21
end
```

D.2.4 Iterated Huber M-Estimate (4.3.3), (4.7)

```
# 'huber2.mtb'
#
# THIS MINITAB MACRO ITERATIVELY CALCULATES AN M-ESTIMATE, BASED ON
# HUBER'S PSI FUNCTION. THE MACRO SHOULD BE EXECUTED UNTIL THE
# ESTIMATE HAS BEEN FOUND TO THE DESIRED ACCURACY.
# THIS MACRO ASSUMES THAT THE MACRO 'huber1' HAS ALREADY BEEN RUN.
let c11 = (c1 - k21)/k13
let c12 = (sign(k1 - abso(c11)) + 1)/2
code (-100:k2) k2 c11 c13
code (k1:100) k1 c13 c13
let k21 = k21 + k13*sum(c13)/sum(c12)
# LATEST HUBER M-ESTIMATE
print k21
end
```

D.2.5 Confidence Interval Based on Huber M-Estimate (4.7)

```
# 'huber3.mtb'
#
# THIS MINITAB MACRO CALCULATES AN ESTIMATE OF THE ASYMPTOTIC
# STANDARD ERROR AND THE ASSOCIATED APPROXIMATE 95% CONFIDENCE
# LIMITS FOR AN M-ESTIMATE, BASED ON HUBER'S PSI FUNCTION.
# THIS MACRO ASSUMES THAT BOTH THE MACROS 'huber1' AND 'huber2'
```

```
# HAVE ALREADY BEEN RUN.
#
let c11 = (c1 - k21)/k13
let c12 = (sign(k1 - abso(c11)) + 1)/2
code (-100:k2) k2 c11 c13
code (k1:100) k1 c13 c13
let c14 = c13**2
let k14 = k13**2
let k31 = k14*sum(c14)/(sum(c12)**2)
let k32 = sqrt(k31)
# ESTIMATE OF ASYMPTOTIC STANDARD ERROR OF HUBER'S M-ESTIMATE
print k32
let k41 = k21 - 1.96*k32
let k42 = k21 + 1.96*k32
# APPROXIMATE 95% CONFIDENCE LIMITS
print k41 k42
end
```

D.3 DESIGN MATRIX MANIPULATIONS (7.3), (7.6)

D.3.1 Matrix of Centered Explanatory Variables

The first two macros listed below may be used separately to center a design matrix, or to find the corresponding hat matrix. If the output of the first macro is copied into matrix m1, the second macro will find the hat matrix of the centered variables. The third macro will plot a set of points which lie on the origin-centered unit sphere in $\mathbb{R}^p$ onto the ellipsoid whose eigen-structure is given in c10 and m10. For example, to obtain n points on the ellipse shown in Figure 7.4, the first two columns of explanatory variables in Table 7.5 are centered using center.mtb. Then columns c11 and c12 are copied into m1, and hat.mtb is executed. Third, k2 is set equal to n and n points on the unit disk are selected, say with (7.3.8), and stored in m11. Then ellipsoid.mtb is executed and m15 is copied onto columns c21 and c22, say. A multiplot of (c11,c12) and (c21,c22) yields Figure 7.4.

```
# 'center.mtb'
#
# THIS MINITAB MACRO CENTERS THE FIRST k1 COLUMNS c1-ck1.
# SET k1=THE NUMBER OF COLUMNS TO BE CENTERED.
# SET k2=0 AND EXECUTE THIS PROGRAM k1 TIMES.
# THE OUTPUT IS THE COLUMNS OF CENTERED VARIABLES IN c11-c(10+k1).
#
```

```
let k2=k2+1
let k3=k2+10
mean ck2 k4
let ck3=ck2-k4
end
```

D.3.2 Hat Matrix and Inverse of Second-Moment Matrix

```
# 'hat.mtb'
#
# THIS MINITAB MACRO ASSUMES THAT A DESIGN MATRIX IS IN m1, AND STORES
# THE SECOND MOMENT MATRIX IN m3, and the HAT MATRIX IN m6.
# THE EIGENVALUES AND EIGENVECTORS OF m3 ARE STORED IN c10, m10.
#
tran m1 m2
mult m2 m1 m3
inve m3 m4
mult m1 m4 m5
mult m5 m2 m6
eigen m3 c10 m10
prin m6
end
```

D.3.3 Map of Unit Sphere to the Ellipsoid of Average Leverage

```
# 'ellipsoi.mtb'
#
# THIS MINITAB MACRO CARRIES OUT THE TRANSFORMATION (7.3.9). IT
# REQUIRES E-VALUES IN c10 AND E-VECTORS IN m10, WHICH MAY HAVE
# ARISEN FROM 'hat.mtb'. SET k1=p, THE DIMENSION OF THE
# DESIGN SPACE AND k2=n, THE NUMBER OF POINTS.
# THE OTHER INPUT IS A SET OF POINTS m11
# WHICH LIE ON THE ORIGIN-CENTERED UNIT SPHERE IN p DIM.
# THE OUTPUT LIES ON THE ELLIPSOID DEFINED BY THE E-STRUCTURE AND
# IS STORED IN m15.
#
let c20=sqrt(k1*c10/k2)
diag c20 m12
mult m11 m12 m13
tran m10 m14
mult m13 m14 m15
end
```

D.4 ROBUST REGRESSION ESTIMATES (7.6)

D.4.1 First Iteration of Welsch Regression Estimates

```
# 'rwls1.mtb'
#
# THIS MINITAB MACRO CALCULATES THE FIRST ITERATION OF WELSCH
# BOUNDED INFLUENCE REGRESSION ESTIMATES, USING ORDINARY
# LEAST SQUARES AS STARTING VALUES.
# IT ASSUMES THAT y IS IN c1 AND THE x VARIABLES ARE IN c2-ck1
# [UP TO 9 x VARIABLES ARE ALLOWED]
# IT ALSO ASSUMES THAT
# k1 = THE NUMBER OF x VARIABLES + 1 = p + 1
# k4 = CONSTANT IN THE WELSCH WEIGHTS
brief 0
name c11 'dfits' c12 'LSres' c13 'LScoef' c14 'hi'
name c15 'wel' c16 'w' c17 'Adiff' c19 'Wcoef' c29 'Wres'
let k3=k1-1
let k2=count(c1)
regress c1 k3 c2-ck1;
dfits 'dfits';
residuals 'LSres';
coefficients 'LScoef';
hi 'hi';
let k4 = k4*sqrt(k1/k2)
let 'wel' = k4/abso('dfits')
rmin 1 'wel' into 'w'
regr c1 k3 c2-ck1;
weights 'w';
residuals 'Wres';
coefficients 'Wcoef'.
let 'adiff' = abso('LScoef' - 'Wcoef')
# LEAST SQUARES ESTIMATES
print 'LScoef'
# INITIAL SET OF WELSCH WEIGHTS
print 'w'
# INITIAL WELSCH ESTIMATES
print 'Wcoef'
# ABSOLUTE DIFFERENCE BETWEEN LS AND INTIAL WELSCH ESTIMATES
print 'adiff'
let k5 = max('adiff')
# MAX ABSOLUTE DIFFERENCE BETWEEN LS AND INITIAL WELSCH ESTIMATES
```

```
print k5
let 'dfits' = 'dfits'*('Wres'/'LSres')
let k10 = 20
let k20 = 30
end
```

D.4.2 Iterative Welsch Regression Estimates

```
# 'rwls2.mtb'
#
# THIS MINITAB MACRO ITERATIVELY CALCULATES WELSCH BOUNDED
# INFLUENCE REGRESSION ESTIMATES. THE MACRO SHOULD BE EXECUTED
# UNTIL THE ESTIMATES HAVE BEEN FOUND TO THE DESIRED ACCURACY,
# AS MEASURED BY k5. (UP TO 10 ITERATIONS ARE ALLOWED.)
# THIS MACRO ASSUMES THAT THE MACRO 'rwls1' HAS ALREADY BEEN RUN.
brief 0
noecho
let 'wel'= k4/abso('dfits')
rmin 1 'wel' into 'w'
regress c1 k3 c2-ck1 c18 c44;
weights 'w';
coefficients ck10;
residuals ck20.
let k9 = k10 - 1
let k19 = k20 - 1
# LATEST SET OF WELSCH WEIGHTS
print 'w'
# LATEST SET OF WELSCH ESTIMATES
print ck10
let 'adiff' = abso(ck10 - ck9)
# ABSOLUTE DIFFERENCE BETWEEN LAST 2 SETS OF WELSCH ESTIMATES
print 'adiff'
let k5 = max('adiff')
# MAX ABSOLUTE DIFFERENCE BETWEEN LAST 2 SETS OF WELSCH ESTIMATES
print k5
let 'dfits' = 'dfits'*(ck20/ck19)
let k10 = k10 + 1
let k20 = k20 + 1
# End of iteration
#
#
end
```

D.4.3 Estimates of Regression Covariance Matrix

```
# 'rwls3.mtb'
#
# THIS MINITAB MACRO CALCULATES AN ESTIMATE OF THE VARIANCE-
# COVARIANCE MATRIX OF THE WELSCH BOUNDED INFLUENCE REGRESSION
# ESTIMATES.
# THIS MACRO ASSUMES THAT BOTH THE MACROS 'rwls1' AND 'rwls2'
# HAVE BEEN RUN.
# THE RESIDUALS AND FITTED VALUES FROM THE FINAL ITERATION
# OF THE WELSCH BOUNDED INFLUENCE REGRESSION ARE STORED IN
# 'FinlWres' AND 'FinlWFV', RESPECTIVELY.
brief 0
name c40 'diff' c41 'ind' c42 'wt' c43 'FinlWres' c44 'FinlWFV'
let 'diff' = k4 - abso('dfits')
let 'ind' = 0.5*(sign('diff')+1)
regress c1 k3 c2-ck1;
weights 'ind';
xpxinv m1.
let k21 = k20 - 1
let 'FinlWres' = ck21
let 'wt' = ('w'**2)*('FinlWres'**2)
regress c1 k3 c2-ck1;
weights 'wt';
xpxinv m2.
inverse m2 into m3
multiply m1 by m3 into m4
multiply m4 by m1 into m5
let k30 = 1/(k2-k1)
multiply k30 by m5 into m6
# ESTIMATED VARIANCE-COVARIANCE MATRIX OF THE WELSCH COEFFICIENTS
print m6
diagonal of m6 into c50
let c50 = sqrt(c50)
# ESTIMATES OF THE STANDARD ERRORS OF THE WELSCH COEFFICIENTS
print c50
end
```

References

Anderson, T. W. (1957). *An Introduction to Multivariate Statistical Analysis.* New York: Wiley. [A.1]*

Andrews, D. F. (1974). "A Robust Method for Multiple Linear Regression," *Technometrics* **16**, 523–531. [4.9, 7.2]

Andrews, D. F., Bickel, P. J., Hampel, F. R., Huber, P. J., Rogers, W. H., and Tukey, J. W. (1972). *Robust Estimates of Location: Survey and Advances.* Princeton, NJ: Princeton Univ. Press. [1.2, 1.3, 4.3, 4.9]

Andrews, D. F., and Herzberg, A. M. (1985). *Data: A Collection of Problems from Many Fields for the Student and Research Worker.* New York: Springer. [4.7, C.4]

Apostol, T. M. (1961). *Calculus (Volume 2).* New York: Blaisdell. [A.2]

Ash, R. B. (1972). *Real Analysis and Probability.* New York: Academic Press. [1.3, A.2, B.1]

Atkinson, A. (1985). *Plots, Transformations, and Regression.* Oxford: Clarendon Press. [7]

Atkinson, A. (1988). "Transformations Unmasked," *Technometrics* **30**, 311–317. [7.6]

Bahadur, R. R. (1966). "A Note on Quantiles in Large Samples," *Ann. Math. Stat.* **37**, 577–580. [3.2]

Bartle, R. (1964). *The Elements of Real Analysis.* New York: Wiley. [B.5]

Belsley, D. A., Kuh, E., and Welsch, R. E. (1980). *Regression Diagnostics.* New York: Wiley. [7.2, 7.3]

Beran, R. J. (1977a). "Robust Location Estimates," *Ann. Stat.* **5**, 431–444. [1.6]

*Number in brackets refers to the section of our book in which the work is discussed.

Beran, R. J. (1977b). "Minimum Hellinger Distance Estimates for Parametric Models," *Ann. Stat.* **5**, 445–463. [1.6]

Beran, R. J. (1982). "Estimated Sampling Distributions: The Bootstrap and Competitiors," *Ann. Stat.* **10**, 212–225. [2.6]

Beran, R. J. (1987). "Prepivoting to Reduce Level Error of Confidence Sets," *Biometrika* **74**, 457–468. [2.6]

Best, D. J., and Rayner, J. C. W. (1987). "Welch's Approximate Solution for the Behren's-Fisher Problem," *Technometrics* **29**, 205–210. [6.8]

Bickel, P. J. (1978). "Some Recent Developments in Robust Statistics," Presented at the 4th Australian Statistical Conf. [1.6]

Bickel, P. J., and Doksum, K. A. (1977). *Mathematical Statistics: Basic Ideas and Selected Topics.* San Francisco: Holden-Day. [1.4]

Bickel, P. J., and Lehmann, E. L. (1975). "Descriptive Statistics for Nonparametric Models, I & II," *Ann. Stat.* **3**, 1038–1069. [3, 3.4, 4.3]

Bickel, P. J., and Lehmann, E. L. (1976). "Descriptive Statistics for Nonparametric Models, III," *Ann. Stat.* **4**, 1139–1158. [3.6, 4.4]

Bloomfield, P., and Steiger, W. L. (1983). *Least Absolute Deviations: Theory, Applications and Algorithms.* Boston: Birkhauser. [7.5]

Blyth, C. R. (1972). "Subjective vs. Objective Methods in Statistics," *Am. Stat.* June, 20–22. [1.1]

Blyth, C. R. (1980). "Expected Absolute Error of the Usual Estimator of the Binomial Parameter," *Am. Stat.* **34**, 155–157. [3.7]

Blyth, C. R. (1986). "Approximate Binomial Confidence Limits," *J. Am. Stat. Assoc.* **81**, 843–855. [3.7]

Boente, G., Fraiman, R., and Yohai, V. J. (1987). "Qualitative Robustness for Stochastic Processes," *Ann. Stat.* **15**, 1293–1312. [3.7]

Boos, D. (1979). *A Differential for L-Statistics. Ann. Stat.* **7**, 955–959. B.4

Box, G. E. P. (1953). "Non-Normality and Tests on Variances," *Biometrika* **40**, 318–335. [5]

Box, G. E. P. (1979). "Robustness in the Strategy of Scientific Model Building," in *Robustness in Statistics* R. L. Launer and G. N. Wilkinson, (Eds.). New York: Academic Press. [5.1]

Box, G. E. P. (1983). "An Apology for Ecumenism in Statistics," in *Scientific Inference, Data Analysis, and Robustness* G. E. P. Box, T. Leonard, and C.-F. Wu, (Eds.). New York: Academic Press. [1]

Box, G. E. P., and Draper, N. R. (1975). "Robust Designs," *Biometrika* **62**, 347–352. [7.3]

Box, G. E. P., Leonard, T., and Wu, C.-F. (1983). *Scientific Inference, Data Analysis, and Robustness.* New York: Academic Press. [1.6]

Brockwell, P. J., and Davis, R. A. (1987). *Time Series: Theory and Methods.* New York: Springer. [A.2, B.1]

Brown, B. M. (1982). "Robustness against Inequalities of Variances," *Austr. J. Stat.* **24**, 283–295. [5.5, 6, 6.4]

Brownlee, K. A. (1965). *Statistical Theory and Methodology in Science and Engineering.* 2nd ed., New York: Wiley. [6.1, 6.2, 7.2]

Carroll, R. J., and Ruppert, D. (1988). *Transformations and Weighting in Regression.* London: Chapman and Hall. [7, 7.8]

Chaterjee, S., and Hadi, A. S (1986). "Influential Observations, High Leverage Points and Outliers in Linear Regression," *Stat. Sci.* **1**, 379–393. [7.6]

Clarke, B. R. (1983). "Uniqueness and Fréchet Differentiability of Functional Solutions to Maximum Likelihood Type Equations," *Ann. Stat.* **11**, 1196–1205. [B.4]

Clarke, B. R. (1986). "Nonsmooth Analysis and Fréchet Differentiability of M-Functionals," *Prob. Th. Rel. Fields* **73**, 197–209. [B.4]

Cook, R. D., and Weisburg, S. (1982). *Residuals and Influence in Regression.* London: Chapman and Hall. [3.7, 7, 7.2, 7.3]

Cowan, R., and Staudte, R. G. (1987). "The Bifurcating Autoregression Model in Cell Lineage Studies," *Biometrics* **42**, 769–783. [2.1]

Cramer, H. (1928). "On the Composition of Elementary Errors," *Skand. Aktuatids* **11**, 13–74, 141–180. [1.2]

Cressie, N. (1980). "Relaxing Assumptions in the One Sample t-Test," *Austr. J. Stat.* **22**, 143–153. [5.4, 5.7]

Cressie, N., Sheffield, L. J., and Whitford, H. J. (1984). "Use of the One Sample t-Test in the Real World," *J. Chronic Diseases* **37**, 107–114. [5.7]

Cressie, N., and Whitford, H. J. (1986). "How to Use the Two Sample t-test," *Biometrics* **2**, 131–148. [6.8]

Cushny, A. R., and Peebles, A. R. (1904). "The Action of Optical Isomers II. Hyoscines," *J. Physiol.* **32**, 501–510. [4.1, 4.2, 4.3, 5.1, 5.6]

Daniel, C., and Wood, F. (1980). *Fitting Equations to Data,* 2nd ed. New York: Wiley. [7.2]

David, F. N., and Johnson, N. L. (1948). "The Probability Integral Transformation when Parametres are Estimated from the Sample," *Biometrika* **35**, 182–190. [4.3, 4.4]

David, H. A. (1981). *Order Statistics.* New York: Wiley. [3.1, 3.5]

David, H. T., and David, H. A. (Eds.) (1984). *Statistics: An Appraisal.* Ames: Iowa State Univ. [1.2]

de Jongh, P. J., de Wet, T., and Welsh, A. H. (1988). "Mallows-Type Bounded-Influence-Regression Trimmed Means," *J. Am. Stat. Assoc.* **83**, 805–810. [7.6]

Diciccio, T. J., and Romano, J. P. (1988). "A Review of Bootstrap Confidence Intervals (with Discussion)," *J. R. Stat. Soc. B* **50**, 338–354. [2.6]

Dixon, W. J. (1953). "Processing Data for Outliers," *Biometrics* **9**, 74–89. [4.7]

Dolby, G. R. (1982). "The Role of Statistics in the Methodology of the Life Sciences," *Biometrics* **38**, 1069–1083. [1.1]

Dollinger, M. B., and Staudte, R. G. (1990a). "The Construction of Equileverage Designs for Multiple Linear Regression," *Austr. J. Stat.*, To appear. [7.3]

Dollinger, M. B., and Staudte, R. G. (1990b). "Influence Functions of Iteratively Reweighted Least Squares Estimators," *Submitted for publication.*

Donoho, D. L., and Huber, P. J. (1983). "The Notion of Breakdown Point," in *A Festschrift for Erich Lehmann* P. J. Bickel, K. Doksum, and J. L. Hodges, Jr., (Eds.). Wadsworth. [2.6]

Donoho, D. L., and Liu, R. C. (1988). "The Automatic Robustness of Minimum Distance Functionals," *Ann. Math. Stat.* **16**, 552–586. [1.6, 3.2]

Draper, D. (1988). "Rank-Based Robust Analysis of Linear Models: I Exposition and Review," *Stat. Sci.* **3**, 239–257. [4.3]

Draper, N. R., and Smith, H. (1966, 1981). *Applied Regression Analysis.* 2nd ed. New York: Wiley. [7.2]

Edwards, W., Lindman, L., and Savage, L. J. (1963). "Bayesian Statistical Inference for Psychological Research," *Psych. Rev.* **70**, 193–242; [1.6] also reprinted in Kadane (1984).

Efron, B. (1978). "Controversies in the Foundations of Statistics," *Amer. Math. Month.* **85**, 231–246. [1.2]

Efron, B. (1982). *The Jacknife, the Bootstrap, and Other Resampling Plans.* SIAM. [2.4, 2.6, 3.5, 3.7]

Efron, B. (1987). "Better Bootstrap Confidence Intervals," *J. Am. Stat. Assoc.* **82**, 171–200. [2.6]

Efron, B., and Tibshirani, R. (1986). "Bootstrap Methods for Standard Errors, Confidence Intervals, and Other Measures of Statistical Accuracy," *Stat. Sci.* **1**, 54–75. [2.6]

Eicker, F. (1963). "Asymptotic Normality and Consistency of Least Squares Estimators for Families of Linear Regressions," *Ann. Math. Stat.* **34**, 447–456. [7.4]

Epstein, B., and Sobel, B. (1953). "Life Testing," *J. Am. Stat. Assoc.* **48**, 486–502. [2.3]

Ferguson, T. S. (1967). *Mathematic Statistics: A Decision Theoretical Approach.* New York: Academic Press. [1.2, 2.3, 7.5]

Fernholz, L. T. (1983). "Von Mises Calculus for Statistical Functionals," *Lecture Notes in Statistics* **19**, New York: Springer. [3.7, B.4]

Field, C. A., and Ronchetti, E. (1985). "A Tail Area Influence Function and Its Applications to Testing," *Comm. Stat. Ser. C* **4**, 19–41. [5.3]

Fillipova, A. A. (1962). "Mises' Theorem on the Asymptotic Behaviour of Functionals of Empirical Distribution Functions and Its Statistical Applications," *Theory Prob. Appl.* **7**, 24–57. [B.4]

Finsterwalder, C. E. (1976). "Collaborative Study of an Extension of the Mills et al. Method for the Determination of Pesticide Residues in Foods," *J. Off. Anal. Chem.* **59**, 169–171. [4.7]

Fisher, R. A. (1920). "A Mathematical Examination of the Methods of Determining the Accuracy of an Observation by the Mean Error, and by the Mean Square Error," *Notices R. Astron. Soc.* **80**, 758–770. [1.2]

Fisher, R. A. (1921). "On Mathematical Foundations of Theoretical Statistics," *Phil. Trans. R. Soc. (London) Ser. A* **222**, 309–368. [1.2]

Freedman, D., and Diaconis, P. (1981). "On the Histogram as Density Estimator: L_2 Theory," *Z. Wahrsche. verw. Geb.* **57**, 453–476. [4.3]

Gani, J. (1982). *The Making of Statisticians.* New York: Springer. [1.2]

Gastwirth, J. L, and Rubin, H. (1975a). "The Asymptotic Distribution Theory of the Empiric cdf for Mixing Processes," *Ann. Stat.* **3**, 809–824. [5.4]

Gastwirth, J. L, and Rubin, H. (1975b). "The Behaviour of Robust Estimators on Dependent Data," *Ann. Stat.* **3**, 1070–1100. [5.4, 5.7]

Gauss, C. F. (1821). *Theoria Combinationis Observationum Erroribus Minimis Obnoxiae, Werke 4, Sect. 35.* Gottingen. [7.2]

Gill, R. D. (1989). "Non- and Semi-Parametric Maximum Likelihood Estimators and the von Mises Method (Part 1)," *Scand. J. Stat.* **16**, 97–128. [3.7]

Graybill, F. A. (1983). *Matrices with Applications in Statistics.* California: Wadsworth (2nd ed.). [A1]

Grubbs, F. E. (1969). "Procedures for Detecting Outlying Observations in Samples," *Technometrics* **11**, 1–21. [5.3]

Hajek, J. (1968). "Asymptotic Normality of Simple Linear Rank Statistics under Alternatives," *Ann. Math. Stat.* **2**, 235–346. [B.3, B.4]

Hall, P. J. (1983). "Inverting an Edgeworth Expansion," *Ann. Stat.* **11**, 569–575. [5.7]

Hall, P. J. (1986). "On the Number of Bootstrap Simulations Required to Construct a Confidence Interval," *Ann. Stat.* **14**, 1453–1462. [2.6]

Hall, P. J. (1988). "Theoretical Comparison of Bootstrap Confidence Intervals (with Discussion)," *Ann. Stat.* **16**, 927–985. [2.6]

Hall, P. J., and Selinger, B. (1986). "Statistical Significance: Balancing Evidence against Doubt," *Austr. J. Stat.* **28**, 354–370. [1.6]

Hall, P. J., and Sheather, S. J. (1988). "On the Distribution of a Studentized Quantile," *J. R. Stat. Soc. B* **50**, 380–391. [4.6]

Hall, W. J., Wijsman, R. A., and Ghosh, J. K. (1965). "The Relationship between Sufficiency and Invariance with Applications in Sequential Statistics," *Ann. Math. Stat.* **36**, 565–590. [1.2]

Hampel, F. R. (1968). *Contributions to the Theory of Robust Estimation.* Ph. D. dissertation, Univ. California, Berkeley. [1.2, 2.3, 2.6, 3.1, 3.2, 3.7]

Hampel, F. R. (1971). "A General Qualitative Definition of Robustness," *Ann. Math. Stat.* **42**, 1887–1896. [1.2, 1.3, 3.2]

Hampel, F. R. (1973). "Robust Estimation: A Condensed Partial Survey," *Z. Wahrsch. verw. Geb.* **27**, 87–104. [1.2]

Hampel, F. R. (1974). "The Influence Curve and Its Role in Robust Estimation," *J. Am. Stat. Assoc.* **69**, 383–393. [3.3, B.4]

Hampel, F. R., Ronchetti, E. M., Rousseeuw, P. J., and Stahel, W. J. (1986). *Robust Statistics, the Approach Based on Influence Functions.* New York: Wiley. [2.6, 3.2, 3.4, 3.7, 4.3, 4.5, 5.3, 6.4, 7.5, 7.6]

Handschin, E., Schweppe, F. C., Kohlas, J., and Fiechter, A. (1975). "Bad Data Analysis for Power System State Estimation," *IEEE Trans. Power App. Sys.* **PAS-94**, 329–337. [7.6]

Heathcote, C. R. (1981). "Bounded Influence Curve Estimation and Consistency," in *Interactive Stat.* D. McNeil, (Eds.). Amsterdam: North-Holland. 103–110. [1.6]

Heathcote, C. R., and Silvapulle, P. M. J. (1981). "Minimum Mean Squared Error of Estimation of Location and Scale Parameters Under Misspecification of the Model," *Biometrika* **68**, 501–514. [1.6]

Hettmansperger, T. P. (1984a). *Statistical Inference Based on Ranks.* New York: Wiley. [4.3, 5.4]

Hettmansperger, T. P. (1984b). "Two-Sample Inference Based on One-Sample Sign Statistics," *Appl. Stat.* **33**, 45–51. [6.6]

Hettmansperger, T. P. (1987). "Why Not Try a Robust Regression?," *Austr. J. Stat.* **29**, 1–18. [7.5, 7.6, C.5]

Hettmansperger, T. P., and Sheather, S. J. (1986). "Confidence Intervals Based on Interpolated Order Statistics," *Stat. Prob. Lett.* **4**, 75–79. [4, 4.6]

Hinkley, D. (1977). "Jackknifing in Unbalanced Situations," *Technometrics* **19**, 285–292. [7.4]

Hinkley, D. (1988). "Bootstrap Methods (with Discussion)," *J. R. Stat. Soc. B* **50**, 321–337. [2.6]

Hoaglin, D. C., and Welsch, R. (1978). "The Hat Matrix in Regression and ANOVA," *Am. Stat.* **32**, 17–22. [7.2]

Hodges, J. L. Jr. (1967). "Efficiency in Normal Samples and Tolerance of Extreme Values for Some Estimates of Location," *Proc. Fifth Symp. on Math. Stat. and Prob., Univ. of Cal. Press, Berkeley* **1**, 163–186. [2.3, 2.6]

Hodges, J. L. Jr., and Lehmann, E. H. (1956). "The Efficiency of Some Nonparametric Competitors of the t-Test," *Ann. Math. Stat.* **27**, 324–335. [1.2, 5.1]

Hoeffding, W. (1948). "A Class of Statistics with Asymptotically Normal Distribution," *Ann. Math. Stat.* **19**, 293–325. [B.3, B.4]

Hogg, R. V. (1974). "Adaptive Robust Procedures: A Partial Review and Some Suggestions for Future Applications and Theory (plus Comments and Rejoinder)," *J. Am. Stat. Assoc.* **69**, 909–927. [1.6, 4.3, 4.9]

Hogg, R. V. (1979). "An Introduction to Robust Estimation," in *Robustness in Statistics,* R. L. Launer and G. N. Wilkinson, (Eds.). New York: Academic Press, 1–17. [1.6]

Hogg, R. V., and Craig, A. T. (1978). *Introduction to Mathematical Statistics.* 4th ed. Toronto: Macmillan. [1.4]

Huber, P. J. (1964). "Robust Estimation of a Location Parameter," *Ann. Math. Stat.* **35**, 73–101. [1.2, 1.3, 2.4, 4.3, 4.5]

Huber, P. J. (1973a). "Robust Regression: Asymptotics, Conjectures, and Monte Carlo," *Ann. Stat.* **1**, 799–821. [7.4, 7.5]

Huber, P. J. (1973b). "Robustness and Designs," in *A Survey of Statistical Design and Linear Models* J. N. Srivastava, (Eds.). Amsterdam: North-Holland. [7.3]

Huber, P. J. (1981). *Robust Statistics.* New York: Wiley. [1.6, 3.2, 4.3, 4.4, 4.5, 7.4, 7.6]

Huber, P. J. (1983). "Minimax Aspects of Bounded Influence Regression (with Comments and Rejoinder)," *J. Am. Stat. Assoc.* **78**, 68–70. [7.3]

Huber, P. J. (1984). "Finite Sample Breakdown of M- and P-Estimators," *Ann. Stat.* **12**, 119–126. [4.6]

Huber-Carol, C. (1970). *Etude Asymptotique de Tests Robustes.* Ph.D. Thesis, Eidgen. Tech. Hoch., Zurich. [1.6]

Johnson, N. L., and Kotz, S. (1970). *Continuous Univariate Distributions—1: Distributions in Statistics.* New York: Wiley. [5.2]

Kadane, J. B. (Ed.) (1984). *Robustness of Bayesian Analysis.* Amsterdam: Elsevier North-Holland. [1, 1.6]

Kallianpur, G. (1963). "Von Mises Functionals and Maximum Likelihood Estimation," in *Contributions to Statistics,* C. R. Rao et al., (Eds.). Calcutta: Statistical Publ. Soc., 137–146. [B.4]

King, F. J., and Ryan, J. J. (1976). "Collaborative Study of the Determination of the Amount of Shrimp in Shrimp Cocktail," *J. Off. Anal. Chem.* **59**, 644–649. [4.7]

Kolmogorov, A. N. (1933a). *Grundbegriffe der Wahrscheinlichkeitsrechnung.* Berlin: Julius Springer. [1.2]

Kolmogorov, A. N. (1933b). "Sulla Determinazione Empericas de Una Legge di Distribuzione," *Giorn. Inst. Ital. Attuari* **4**, 83–91. [1.2]

Koopmans, L. H. (1987). *Introduction to Contemporary Statistical Methods.* 2nd ed., Boston: Duxbury. [6.7]

Krasker, W. S., and Welsch, R. E. (1982). "Efficient Bounded Influence Regression Estimation," *J. Am. Stat. Assoc.* **77**, 595–604. [7.8]

Lehmann, E. L. (1975). *Nonparametrics: Statistical Methods Based on Ranks.* San Franciso: Holden-Day. [4.3, 4.6, 5.1, 5.2, 6.6, 8.4]

Leonard, T. (1983). "Some Philosophies of Inference and Modelling," in *Scientific Inference, Data Analysis, and Robustness* G. E. P. Box, T. Leonard, and C.-F. Wu, (Eds.). New York: Academic Press. [1.6]

Likes, J. (1966). "Distribution of Dixon's Statistics in the Case of an Exponential Population," *Metrika* **11**, 46–54. [2.3, 3.6]

Maritz, J. S. (1981). *Distribution-Free Statistical Methods.* London: Chapman and Hall. [Preface]

Maritz, J. S., and Jarrett, R. (1978). "A Note on Estimating the Variance of the Sample Median," *J. Am. Stat. Assoc.* **73**, 194–196. [3.5]

Maritz, J. S., Wu, M., and Staudte, R. G. (1977). "A Location Estimator Based on a U-Statistic," *Ann. Math. Stat.* **5**, 779–786. [4.3]

Maronna, R. A., Bustos, O. H., and Yohai, V. J. (1979). "Bias and Efficiency of General M-Estimators for Regression with Random Carriers," in *Smoothing Techniques for Curve Estimation* T. Gasser and M. Rosenblatt, (Eds.). New York: Springer. [7.8]

Maronna, R. A., and Yohai, V. J. (1981). "Asymptotic Behaiviour of General M-Estimates for Regression and Scale with Random Carriers.," *Z. Wahrsch. verw. Geb.* **58**, 7–20. [7.5]

Martin, R. D., and Yohai, V. J. (1986). "Influence functions for time series (with discussion)," *Ann. Stat.* **14**, 781–855. [1.4]

McKean, J. W., Sheather, S .J., and Hettmansperger, T. P. (1988). "Robust Diagnostics for Rank-Based Inference," Working Paper 88-018, Univ. of New South Wales, Kensington, NSW. [7.8]

Miller, R. G. (1974). "The Jackknife—A Review," *Biometrika* **61**, 1–15. [3.7]

Montgomery, D. C., and Peck, E. A. (1982). *Linear Regression Analysis.* New York: Wiley. [1.4, 7]

Neyman, J., and Pearson, E. S. (1933). "On the Problem of the Most Efficient Tests of Statistical Hypotheses," *Phil. Trans. R. Soc. (London) Ser. A* **231**, 289–337. [1.2]

Noether, G. E. (1955). "On a Theorem of Pitman," *Ann. Math. Stat.* **26**, 64–68. [5.2]

Owen, D. B. (1976). *On the History of Statistics and Probability, Proc. Symp. on the American Mathematical Heritage.* (to celebrate the bicentennial of the United States of America, held at Southern Methodist University, May 27–29, 1974). New York: Marcel Dekker. [1.2, 5.1]

Padmanabhan, A. R. (1985). *Confidence Intervals Based on Robust Estimators for Short-Tailed and Long-Tailed Distributions and an Application.* Statistics Research Rep., Dept. of Mathematics, Monash Univ., Clayton, Vic. [5.2]

Parr, W. C., and Schucany, W. R. (1980). "Minimum Distance and Robust Estimation," *J. Am. Stat. Assoc.* **75**, 616–624. [1.6]

Pearson, K. (1900). "On a Criterion that a Given System of Deviations from the Probable in the Case of a Correlated System of Variables is Such that it can be Reasonably Supposed to have Arisen in Random Sampling," *Phil. Mag. Ser. 5* **50**, 157–175. [1.1]

Pitman, E. J. G. (1949). *Lecture Notes on Nonparametric Statistics.* New York: Columbia Univ. Press. [1.2, 5.1]

Quenouille, M. H. (1949). "Approximate Tests of Correlation in Time Series," *J. R. Stat. Soc. B* **11**, 18–84. [3.7]

Rao, C. R. (1973). *Linear Statistical Inference and Its Applications.* 2nd ed. New York: Wiley. [A.1]

Reeds, J. A. (1976). *On the Definitions of von Mises Functionals.* Ph.D. thesis, Harvard Univ., Cambridge, MA. [B.4]

Rey, W. J. J. (1983). *Introduction to Robust and Quasi-Robust Statistical Methods.* Berlin: Springer. [4.9]

Rieder, H. (1978). "A Robust Asymptotic Testing Model," *Ann. Stat.* **6**, 1080–1094. [1.6]

Rieder, H. (1980). "Estimates Derived from Robust Tests," *Ann. Stat.* **8**, 106–115. [1.6]

Robinson, I., and Sheather, S. (1989). "Weighted Selection for the Multiset $X + X$ with Application to R-Estimates and Associated Confidence Limits," *J. Stat. Comp. Simul.* **31**, 19–35. [4.3]

Rocke, D. M., Downes, G. W., and Rocke, A. J. (1982). "Are Robust Estimators Really Necessary?," *Technometrics* **24**, 95–102. [4.7]

Ronchetti, E. (1979). *Robustheitseigenschaften ver Tests.* Dip. thesis. ETH, Zurich. [5.3]

Rousseeuw, P. J. (1981). "A New Infinitesimal Approach to Robust Estimation," *Z. Wahrsch. verw. Geb.* **56**, 127–132. [3.2, 3.4]

Rousseeuw, P. J. (1984). "Least Median of Squares Regression," *J. Am. Stat. Assoc.* **79**, 871–880. [7.6, 7.8]

Rousseeuw, P. J., Daniels, B., and Leroy, A. (1984a). "Applying Robust Regression to Insurance," *Insur. Math. and Econ.* **3**, 67–72. [7.1]

Rousseeuw, P. J., and Leroy, A. (1987). *Robust Regression and Outlier Detection.* New York: Wiley. [4.9, 7, 7.1, 7.8, C.5]

Rousseeuw, P. J., and Ronchetti, E. (1981). *The Influence Curve for Tests.* Res. Rep. 21. Fachgruppe fur Stat., ETH, Zurich. [5.3, 6.4]

Rousseeuw, P. J., and Yohai, V. (1984). "Robust regression by means of S-estimators," in *Robust and Nonlinear Time Series Analysis*, J. Franke, W. Härdle, and R. D. Martin, (Eds.). Lecture Notes in Statistics 26, Springer, New York, 256–272. [7.8]

Ruppert, D., and Carroll, R. J. (1980). "Trimmed Least Squares Estimation in the Linear Model," *J. Am. Stat. Assoc.* **75**, 828–838. [7.6, C.5]

Scott, D. W. (1979). "On Optimal and Data-Based Histograms," *Biometrika* **66**, 605–610. [4.3]

Serfling, R. J. (1980). *Approximation Theorems of Mathematical Statistics.* New York: Wiley. [1.4, 7.4, B.1]

Shapiro, S. S., and Wilk, M. B. (1965). "An Analysis of Variance Test for Normality (Complete Samples)," *Biometrika* **52**, 591–611. [3.6]

Sheather, S. J. (1986a). "A Finite Sample Estimate of the Variance of the Sample Median," *Stat. Prob. Let.* **4**, 337–342. [3.5]

Sheather, S. J. (1986b). "An Improved Data-Based Algorithm for Choosing the Window Width when Estimating the Density at a Point," *Comp. Stat. DA* **4**, 61–65. [4.3]

Sheather, S. J. (1987). "Assessing the Accuracy of the Sample Median: Estimated Standard Errors Versus Interpolated Confidence Intervals," in *Statistical Data Analysis Based on the L1 Norm and Related Methods* Y. Dodge, (Eds.). Amsterdam, North-Holland, 203–215. [3.7]

Sheather, S. J., and Hettmansperger, T. P. (1987). "Estimating the Standard Error of Robust Regression Estimates," *Proceedings of STATCOMP '87*, LaTrobe University, Melbourne, Australia. [7.6]

Sheather, S. J., and McKean, J. W. (1987). "A Comparison of Testing and Confidence Interval Methods for the Median," *Stat. Prob. Lett.* **6**, 31–36. [4.6]

Siddiqui, M. M. (1963). "Optimum Estimators of the Parameters of Negative Exponential Distributions from One or Two Order Statistics," *Ann. Math. Stat.* **34**, 117–121. [2.3]

Siegel, A. F. (1982). "Robust Regression Using Repeated Medians," *Biometrika* **69**, 242–244. [7.8]

Silverman, B. W. (1986). *Density Estimation for Statistics and Data Analysis.* London: Chapman and Hall. [4.3]

Simkin, C. G. F. (1978). "Hyperinflation and Nationalist China," in *Stability and Inflation.*, (Eds.). New York: Wiley. [7.1]

Stather, C. R. (1981). *Robust Estimation Based on Hellinger Distance Methods.* Ph.D. thesis, La Trobe Univ., Melbourne. [1.6]

Staudte, R. G. (1980). *Robust Estimation.* Queen's Papers in Pure and Applied Mathematics, no. 53. A. J. Coleman, and P. Ribenboim (Eds.). Kingston, Ont., Canada: Queen's Univ. [2.3, 3.2, 4.4]

Staudte, R. G., Guiguet, M., and Collyn D'Hooghe, M. (1984). "Additive Models for Dependent Cell Populations.," *J. Theor. Biol.* **109**, 127–146. [2.1]

Staudte, R. G., Woodward, J., Fincher, G., and Stone, B. (1983). "Water-Soluble β–D–Glucans from Barley (Hordeum vulgare) Endosperm, III. Distribution of Cellotriosyl and Cellotetraosyl Residues," *Carbohyd. Polym.* **3**, 299–312. [1.1]

Stigler, S. M. (1969). "Linear Functions of Order Statistics," *Ann. Math. Stat.* **40**, 770–788. [3.1, 4.3]

Stigler, S. M. (1973). "Simon Newcombe, Percy Daniell, and the History of Robust Estimation 1885–1920," *J. Am. Stat. Assoc.* **68**, 872–879. [1.2]

Stigler, S. M. (1977). "Do Robust Estimators Work with Real Data?," *Ann. Stat.* **5**, 1055–1098. [4.7, C.3]

Student (W. S. Gosset) (1908). "The Probable Error of a Mean," *Biometrika* **6**, 1–25. [5.1]

Tibshirani, R. (1988). "Variance Stabilization and the Bootstrap," *Biometrika* **75**, 433–444. [2.6, 3.5]

Tukey, J. (1958). "Bias and Confidence in Not Quite Large Samples," *Ann. Math. Stat.* **29**, 614. [3.7]

C 6222495 **JOHN WILEY & SONS, INC.**

I C333708 1 WILEY DRIVE

04/22/92 SOMERSET NJ 08875

SENT WITH COMPLIMENTS OF: KAREN BEDNARSKI 0071

QUAN	AUTHOR AND SHORT TITLE	ISBN	LOCATION
1	STAUCTE ESTIMATICN W DSK	C471855472	A165 25298

Tukey, J. W. (1960). "A Survey of Sampling from Contaminated Distributions," in *Contributions to Probability and Statistics* I. Olkin, et al., (Eds.). 448–485. [1.2, 2.3, 4.4]

Tukey, J. W. (1977). *Exploratory Data Analysis.* Reading, MA: Addison-Wesley. [3.7]

Tukey, J. W., and McLaughlin, D. H. (1963). "Less Vulnerable Confidence and Significance Procedures for Location Based on a Single Sample: Trimming/Winsorization 1," *Sankhya A* **25**, 331–352. [4.3]

von Mises, R. (1947). "On the Asymptotic Distribution of Differentiable Statistical Functions," *Ann. Math. Stat.* **18**, 309–348. [B.4]

Welch, B. L. (1937). "The Significance of the Difference between Two Means when the Population Variances are Unequal," *Biometrika* **29**, 350–362. [6.1, 6.2]

Welsch, R. E. (1980). "Regression Sensitivity Analysis and Bounded-Influence Estimation," in *Evaluation of Econometric Models* J. Kmenta and J. B. Ramsey, (Eds.). 153–167, New York: Academic Press. [7.6]

Welsh, A. H. (1984). *Some Problems in Adaptive Estimation.* Ph.D. thesis, Australian Nat. Univ., Canberra. [1.6]

Welsh, A. H. (1986). "On the Use of the Empirical Distribution and Characteristic Function to Estimate Parameters of Regular Variation," *Austral. J. Stat.* **28**, 173–181. [1.6]

Welsh, A. H. (1987). "The Trimmed Mean in the Linear Model, (plus Discussion and Rejoinder)," *Ann. Stat.* **15**, 20–45. [7.8, C.5]

Wilcoxon, F. (1945). "Individual Comparisons by Ranking Methods," *Biometrics* **1**, 80–83. [5.1, 5.2]

Wilson, E. B. (1952). *An Introduction to Scientific Research.* New York: McGraw-Hill [1.1]

Wu, C. F. J. (1986). "Jackknife, Bootstrap, and Other Resampling Methods in Regression Analysis, (plus Discussion and Rejoinder)," *Ann. Stat.* **14**, 1261–1350. [7.8]

Yohai, V. J. (1987). "High Breakdown Point and High Efficiency Robust Estimates for Regression," *Ann. Stat.* **15**, 642–656. [7.8]

Yuen, K. K. (1974). "The Two-Sample Trimmed t for Unequal Population Variances," *Biometrika* **61**, 165–170. [6.5]

Yuen, K. K., and Dixon, W. J. (1973). "The Approximate Behaviour and Performance of the Two-Sample Trimmed t," *Biometrika* **60**, 369–374. [6.5]

Author Index

Subject Index

Italicized page numbers in the index give the location of definitions for the corresponding entries.

GUTTMAN, WILKS, and HUNTER • Introductory Engineering Statistics, *Third Edition*
HAHN and SHAPIRO • Statistical Models in Engineering
HALD • Statistical Tables and Formulas
HALD • Statistical Theory with Engineering Applications
HAND • Discrimination and Classification
HEIBERGER • Computation for the Analysis of Designed Experiments
HOAGLIN, MOSTELLER and TUKEY • Exploring Data Tables, Trends and Shapes
HOAGLIN, MOSTELLER, and TUKEY • Understanding Robust and Exploratory Data Analysis
HOCHBERG and TAMHANE • Multiple Comparison Procedures
HOEL • Elementary Statistics, *Fourth Edition*
HOEL and JESSEN • Basic Statistics for Business and Economics, *Third Edition*
HOGG and KLUGMAN • Loss Distributions
HOLLANDER and WOLFE • Nonparametric Statistical Methods
HOSMER and LEMESHOW • Applied Logistic Regression
IMAN and CONOVER • Modern Business Statistics
JESSEN • Statistical Survey Techniques
JOHNSON • Multivariate Statistical Simulation
JOHNSON and KOTZ • Distributions in Statistics
Discrete Distributions
Continuous Univariate Distributions—1
Continuous Univariate Distributions—2
Continuous Multivariate Distributions
JUDGE, GRIFFITHS, HILL, LÜTKEPOHL and LEE • The Theory and Practice of Econometrics, *Second Edition*
JUDGE, HILL, GRIFFITHS, LÜTKEPOHL and LEE • Introduction to the Theory and Practice of Econometrics, *Second Edition*
KALBFLEISCH and PRENTICE • The Statistical Analysis of Failure Time Data
KASPRZYK, DUNCAN, KALTON, and SINGH • Panel Surveys
KAUFMAN and ROUSSEEUW • Finding Groups in Data: An Introduction to Cluster Analysis
KEENEY and RAIFFA • Decisions with Multiple Objectives
KISH • Statistical Design for Research
KISH • Survey Sampling
KUH, NEESE, and HOLLINGER • Structural Sensitivity in Econometric Models
LAWLESS • Statistical Models and Methods for Lifetime Data
LEAMER • Specification Searches: Ad Hoc Inference with Nonexperimental Data
LEBART, MORINEAU, and WARWICK • Multivariate Descriptive Statistical Analysis: Correspondence Analysis and Related Techniques for Large Matrices
LINHART and ZUCCHINI • Model Selection
LITTLE and RUBIN • Statistical Analysis with Missing Data
McNEIL • Interactive Data Analysis
MAGNUS and NEUDECKER • Matrix Differential Calculus with Applications in Statistics and Econometrics
MAINDONALD • Statistical Computation
MALLOWS • Design, Data, and Analysis by Some Friends of Cuthbert Daniel
MANN, SCHAFER and SINGPURWALLA • Methods for Statistical Analysis of Reliability and Life Data
MARTZ and WALLER • Bayesian Reliability Analysis
MASON, GUNST, and HESS • Statistical Design and Analysis of Experiments with Applications to Engineering and Science
MILLER • Beyond ANOVA, Basics of Applied Statistics
MILLER • Survival Analysis
MILLER, EFRON, BROWN, and MOSES • Biostatistics Casebook

Applied Probability and Statistics (Continued)

MONTGOMERY and PECK · Introduction to Linear Regression Analysis
NELSON · Accelerated Testing, Statistical Models, Test Plans, and Data Analyses
NELSON · Applied Life Data Analysis
OCHI · Applied Probability and Stochastic Processes in Engineering and Physical Sciences
OSBORNE · Finite Algorithms in Optimization and Data Analysis
OTNES and ENOCHSON · Digital Time Series Analysis
PANKRATZ · Forecasting with Univariate Box-Jenkins Models: Concepts and Cases
POLLOCK · The Algebra of Econometrics
RAO and MITRA · Generalized Inverse of Matrices and Its Applications
RÉNYI · A Diary on Information Theory
RIPLEY · Spatial Statistics
RIPLEY · Stochastic Simulation
ROSS · Introduction to Probability and Statistics for Engineers and Scientists
ROUSSEEUW and LEROY · Robust Regression and Outlier Detection
RUBIN · Multiple Imputation for Nonresponse in Surveys
RUBINSTEIN · Monte Carlo Optimization, Simulation, and Sensitivity of Queueing Networks
RYAN · Statistical Methods for Quality Improvement
SCHUSS · Theory and Applications of Stochastic Differential Equations
SEARLE · Linear Models
SEARLE · Linear Models for Unbalanced Data
SEARLE · Matrix Algebra Useful for Statistics
SKINNER, HOLT, and SMITH · Analysis of Complex Surveys
SPRINGER · The Algebra of Random Variables
STAUDTE and SHEATHER · Robust Estimation and Testing
STEUER · Multiple Criteria Optimization
STOYAN · Comparison Methods for Queues and Other Stochastic Models
STOYAN, KENDALL and MECKE · Stochastic Geometry and Its Applications
THOMPSON · Empirical Model Building
TIJMS · Stochastic Modeling and Analysis: A Computational Approach
TITTERINGTON, SMITH, and MAKOV · Statistical Analysis of Finite Mixture Distributions
UPTON and FINGLETON · Spatial Data Analysis by Example, Volume I: Point Pattern and Quantitative Data
UPTON and FINGLETON · Spatial Data Analysis by Example, Volume II: Categorical and Directional Data
VAN RIJCKEVORSEL and DE LEEUW · Component and Correspondence Analysis
WEISBERG · Applied Linear Regression, *Second Edition*
WHITTLE · Optimization Over Time: Dynamic Programming and Stochastic Control, Volume I and Volume II
WHITTLE · Systems in Stochastic Equilibrium
WILLIAMS · A Sampler on Sampling
WONNACOTT and WONNACOTT · Econometrics, *Second Edition*
WONNACOTT and WONNACOTT · Introductory Statistics, *Fourth Edition*
WONNACOTT and WONNACOTT · Introductory Statistics for Business and Economics, *Third Edition*
WOOLSON · Statistical Methods for the Analysis of Biomedical Data

Tracts on Probability and Statistics

AMBARTZUMIAN · Combinatorial Integral Geometry
BIBBY and TOUTENBURG · Prediction and Improved Estimation in Linear Models
BILLINGSLEY · Convergence of Probability Measures
KELLY · Reversibility and Stochastic Networks
TOUTENBURG · Prior Information in Linear Models